Advanced Praise

"Our Quality Management System is based on the process management principles defined by Mr. Gaal in this book. We are a small company, employing about forty people in the aerospace related contract manufacturing business. The processes and responsibilities are identified and the instructions on how to do them are stated in understandable language. My employees follow the instructions and I don't have to worry what's going on when I am not around. The quality operating procedures 006, 007, and 009 replace Mil-I-45208 inspection system, as a functional inspection procedure for whatever there is to know how to take care of controlling products as they go through the many operations. This book has the tell-me-how-to-do-it. And this documented quality system, by defining and integrating the product related processes, created the overall framework for us to keep focused on the most important part of our business, — satisfying our customers. Our customers are very satisfied with the quality of our products and the business is growing due in large measure to this documented quality system."

Zbigniew Stanek, President, Aero-Craft Corporation

"The method for resolution of customer complaints is so thoroughly handled by the author of this book that I don't think anybody would ever lose a customer for non-responsiveness. What is really good here is that solving problems is not a superficial matter, but an in-depth analysis to go to the source of the problem and correct it. Furthermore, you cannot make a mistake in doing it because the instructions, especially the flowcharts, are so clearly presented that even the most elementary mind would be able to follow it.

There is a natural evolution of operational sequencing going on in this book that I have found most interesting. The interrelated processes are so cohesively presented that one doesn't have to rely on theories to explain them for the reality of events speak for themselves. I can take this book and show it to any of my co-workers that this is how you should control the process you are working on. It is amazing how well every part of the many processes are integrated and referenced to maintain an orderly follow-through.

I truly enjoyed reading this book for it has a directed force behind it to do things right from the beginning to the end in maintaining organizational objectives. It would be a great business asset to any manufacturer to control the most vital part of its business, the very shop where a product is realized."

Eugene McMullan, National Sales Manager, Leon's Bakery

"The attention to detail, the logic of process and the conclusion of the final product leaves one with a sense of awe and fulfillment. One believes that he/she can develop and produce a system that will assure the quality of any manufactured product. The structure of the *Training Matrics,* encompassed in a compact 232-page manual allows the reader full latitude to adjust the process to whatever his needs present. With exacting detail and thoroughness, each process and action person is given a blueprint and flowchart that defies misinterpretation. Additionally, all items are totally cross-referenced at the point of introduction to allow the reader a different manner of learning and understanding. For me there are really four books in this offering with an Executive Summary. The four main books are *The Training Matrics, Quality Systems Manual, Quality Operating Procedures*, and *Forms*.

This work, in my opinion, provides the new or well-established manufacturer with a concrete, clear, superbly documented means of developing their own Quality Management System. It provides great latitude within its structure to account for individual needs, nuances, and style yet gives the rationale and direction necessary to ensure the development of an effective entity within the boundaries of the manufacturing plant. This has been a most enjoyable learning experience for me. I heartily recommend this masterpiece to all who truly seek a verified means of providing excellence in manufacturing."

R.C. Cormier, Adjunct Professor, University of Hartford

An electronic copy of the forms presented in *ISO 9001:2000 for Small Business: Implementing Process-Approach Quality Management* is available for downloading at the CRC Press website. Go to www.crcpress.com. Under "Electronic Products", click on the "Download and Update Page" link. Then click on the "download" link located next to the book title to access the forms.

ISO 9001:2000
for Small Business

Implementing Process-Approach Quality Management

Arpad Gaal

S^t_L

St. Lucie Press

Boca Raton   London   New York   Washington, D.C.

Library of Congress Cataloging-in-Publication Data

Gaal, Arpad.
    ISO 9001:2000 for small business: Implementing process-quality approach management/ by Arpad Gaal.
      p. cm.
     ISBN 1-57444-307-0 (alk. paper)
      1. Quality control—Standards. 2. Quality assurance—Standards. 3.
Manufactures—Quality control—Evaluation. 4. ISO 9000 Series Standards.

TS156 .G28 2001
658.5′62—dc21   2001019636

Visit the CRC Press Web site at www.crcpress.com

PREFACE

This book is written for quality practitioners in manufacturing shop environments to aid them in the understanding of how process-approach quality management is implemented and practiced to the rigorous requirements of ISO 9001/2000. It is a completely integrated system in which the organizational quality objectives are identified and interactively communicated in the fulfillment of internal and external customer requirements. This is not a theory book, but a fact-impelled reality, based on experience, derived from actual quality practice in manufacturing shops. Process-approach to quality management is the preferred technique by ISO 9001/2000. It embodies a total commitment from organizations to establish, implement, and maintain a documented quality system that controls process-driven interactions among all the internal and external parties in achieving conformity of a product to quality requirements.

To control process-driven interactions among these parties, I have established a system-wide integration of quality objectives and interactively structured them into all the core departmental work assignments, creating a quality throughput system. By doing this, I have interconnected the sequentially flowing work processes with the quality objective criteria. At the same time, I have also identified the beginning and end of these interacting work assignments to enable us to determine the process and quality boundaries for each. This, now, allows us to measure performance results in each work assignment as the flow-process is continuing toward product realization. By knowing the performance results, we can promptly determine whether the process or the quality requirements have been met and accordingly implement continual improvements. The entire quality system's procedures in this book have been designed to support this type of organizational interaction with all the contractual parties, maintaining integration and coherence in achieving internal and external customer satisfaction.

As a management representative in the sense of ISO commitments, I have taken the lessons learned from being an ISO Rep and coupled them with the lessons learned from my forty-years shop experience in manufacturing and produced this book. It contains the necessary instructions, procedures, and examples, all integrated and accurately communicated to the relevant work centers. My intention was to ensure that others would be able to take this book and convert and implement their quality systems along the model of this process-approach quality management system, without outside help. This process-approach quality system is specially tailored for small businesses in the manufacturing sector, incorporating all the quality system's objectives in line with ISO 9001/2000 requirements. To ensure operational effectiveness, I have implemented and debugged this system in real manufacturing environments. This is now the operating system with several Connecticut based companies. The works contain the quality manual, quality operating procedures, and appropriate forms. To ensure effective implementation, I have created some 235 pages of flowcharts as a training aid, which can be converted to become the quality operating procedures in small shops to replace text-type procedures.

This is a TQM system for small manufacturing firms, incorporating the P-D-C-A provisions as a preventive quality tool and also as a corrective action tool.

v

DEDICATION

To my beautiful wife, Maria, who complained a lot about my wasting time on this book, but still endured it and stayed faithful. To my loving son, Adam, who most of the time wanted to escape my presence when I talked about the contents of this book, but still willingly proofread it. To my loving daughter, Judy, always willing, but short on time, nevertheless, can't wait to read the book. I thank the Almighty God that the book is finished.

ABOUT THE AUTHOR

Among all the college courses one can take to earn a degree in his chosen field, Quality System Design and Management was not one of them. Arpad left college to learn it in industry. He was researching why errors and mistakes occurred in just about anything man did for others. Besides biblical connotations, he found that most of the time the causes were the lack of adequate process instructions in relation to the expected work results – that clear-cut instruction of what the quality requirements of the job were. He published his first book on the subject in Hungarian in 1998.

Arpad runs his own business today and teaches small companies about the *special-cause variations* that responsible employees introduce into the work processes upstream/downstream that impact continuing assignments in product realization processes with devastating results on cost, cycle time, and customer satisfaction. These *special-cause variations* are the mistakes in planning and communication that employees unknowingly transfer cross-functionally without detection. When they are detected, the damage in work processes and products has been inflicted.

The book at hand is based on Arpad's forty years experience in manufacturing in the United States and Europe. It is a quality management system in line with ISO 9001/2000, interactively designed to remove special-cause variations as customer requirements are aligned with internal processes. It is also an integrated *process-approach* system, tailored for self-implementation in two to six months in cross-functional organizations and is fully convertible to meet one's own specific requirements.

ACKNOWLEDGEMENT

 Just when I finished this book, it dawned on me how little I know without the imagination, common sense, and good judgment of others. As I wrote these words down, another thing came to mind. How can I express my appreciation to so many people, living and dead, who unselfishly poured out their goodwill to help me, without hurting someone's feeling? I couldn't. Because of this, it would be unwise for me to give thanks to a few and leave out the many. If I did, I would be spending the rest of my life trying to justify my reasons. But, let it be known to all those who have given me advice, support, recommendation, criticism or rejection that my appreciation for their contributions is such that words cannot describe. Only my prayers can provide the due acknowledgment to each and all without leaving anybody stranded.

Established: 1986
New plant built: 1998
Number of employees: 40/50
Employee turnover: less than 1%
Product line: turbo and aircraft parts machining
Overall customer satisfaction rating: 5.7 sigma

Aero-Craft Corporation is located in Newington, Connecticut and is one of the companies where the Process Approach Quality Management System has been implemented and registered to ISO9000 and AS9000. Aero-Craft is a world-class company in small business. (Photo courtesy of Aero-Craft Corporation.)

TABLE OF CONTENTS

TITLE   **PAGE**

BOOK SECTION ONE
CONTENTS TO TRAINING METRICS

CONTENTS	**Page**

BOOK SECTION TWO
QUALITY SYSTEM MANUAL

CONTENTS   **Page**

BOOK SECTION THREE
QUALITY OPERATING PROCEDURES (QOP)

CONTENTS   Page

BOOK SECTION FOUR
FORMS MANUAL

CONTENTS   **Page**

APPENDIX

ISO for
9001:2000 Small Business

Implementing Process-Approach
Quality Management

INTRODUCTION

The Reason

The most important part of any manufacturing business is the shop. That's where the product is made and that's where the problems are concentrated. Problems come in documents, processes, and methods with different impact on product quality or the way we achieve it. This book will show you how to remove problems, but not necessarily with conventional methods. The method we are using is known as Process-Approach Quality Management, which manages product quality problems through prevention in every area in the product's life cycle. In the past, we have been focusing on problems most often found at the end of the line. This process-approach system will go to the front of the line instead and work with everybody who may not have realized that problems do flow down from process to process which, eventually, impact both the product and customer satisfaction. From here on, therefore, we will be working with <u>process owners</u> through documented procedures in order to implement preventive work habits.

One reason why anybody would want to write a book is to fulfill a need. Many years ago, when I needed advice on how to integrate organizational work processes to fall in line with process-approach controls, I found no publication on the market to help me. I took the challenge and spent seven years bringing it all together and issued it in this book, not realizing at the beginning that there would be a time when others would have a need for it also. That need has finally arrived due to the year 2000 revisions of the ISO 9000 standards. The book contains my forty years experience in manufacturing, the lessons that are extremely important in practicing process-approach controls.

The quality system procedures, integrated and issued in this book, have been designed to control work processes in a true production environment. This is how a process-approach system should be implemented. Depending on the type of work assignment, work result verification by process owners should be a requirement if we want to stem the flow of errors from process to process. In a process approach quality management, controlling product quality should not be confined to the manufactured parts alone, for all kinds of document errors also account for a substantial part of the total non-conformance figure. These, too, should be controlled not after their release, but at the time of issuance.

There is nothing in this book that has not been put through operational use to prove out the validity of the system to practical applications in controlling the processes to which the procedures apply. Those companies that implemented this quality management system and have registered to ISO 9002/1994 can attest to that.

Therefore, the procedures presented here are qualified for implementation in shop environments engaged in contract manufacturing services. Registration is not the main objective here, compatibility to ISO 9000 is. Any alteration or conversion to add or remove requirements from the presented material can be easily accomplished without upsetting the basic procedural framework. Pages have been devoted under the 'Guidance' section to accommodate the prospective user in this effort.

Inevitably, the question, how this can be possible, is going to be asked by all prospective users. Just as the ISO 9000 standards don't define the type of products they want the guidelines to control, neither does this quality management system. The reasons are well founded. When the work processes are maintained and controlled according to approved procedures, the final product will meet the customer's product quality requirements, regardless what type of product was manufactured. The work assignments we are controlling are directly tied to the evolution of processes in the work cycle and not to how the detailed work methods are executed. Efficient work methods depend on the experience and continued training of the work force and not on how the processes control product quality. In our process-approach system, the responsible departments have been charged individually and collectively to establish and implement the process steps needed to complete work assignments, ensuring every step of the way the maintenance of product quality, as required. The process steps in manufacturing shops are defined in Job Travelers or Operation Sheets, or in combination of the two, and the product specifications are defined in the drawings and attachments thereto. The skilled operators will know what to do without burdening them with faulty work methods instructions, which become the nitpicker's paradise during audits. Therefore, we have intentionally excluded them from the process-approach procedures. We are not in the business of work methods training, but rather in the business of controlling the processes which control the product on its journey to customer satisfaction.

The following departments in manufacturing shops are the responsible bodies to plan, implement, and execute the customers' Purchase Order requirements from the beginning to the end in the product's life-cycle. They fall under the control of this process-approach system. These departments are commonly known as core departments and they interact in the chain of events to carry out customer requirements.

Contracts – charged with negotiating agreements with customers and flowing down purchase order requirements to other departments;

Engineering – charged with laying out manufacturing processes, including software programs, processing work instructions, production folders, etc.;

Purchasing – charged with ordering materials and lining up subcontractors;

Production Planning – charged with scheduling the manufacturing processes;

Manufacturing – charged with processing the products;

Inspection – charged with controlling product quality.

One of the objectives of writing these procedures was to limit the bulk of documents that burden the quality system under ISO commitment. While documentation is important, not every detail, especially work methods, should be in quality documents, for that alone could choke production efficiency. Forgotten in this process is the fact that the workers who are putting the quality in the product read basically only the processing dedicated procedures. It is not the Tier levels of different documents that will guarantee acceptable work results from process owners in the chain of the process cycle. Rather, it is the way those Tier levels are integrated to link the quality requirements to the processes within the organization's overall commitment to product quality. In this effort, I have combined the various Tier levels into two – the Quality System Manual and the Quality Operating Procedures – to make the system suitable for small business applications, where low overhead and multiple job assignments dominate the business environment.

A process-approach quality management system must incorporate continuous improvement tools in every process that has anything to do with the product's processing cycle. That is the dominant objective of this process-approach system. As we read the quality objective procedures in this book, we would hardly recognize that they are in fact achieving not only product quality but also those aspects of the Total Quality Management principles that accomplish continuous improvement. The whole process-approach system is integrated in a closed-loop process control, enacted via the application of twelve Quality Operating Procedures. To give you a taste of it, I have listed below a few controlling provisions. Others will be experienced during implementation and application.

1. The product quality requirements are identified, documented, and implemented in process instruction procedures up front before any processing takes place.
2. The process is controlling the product and prevents the problems from migrating from work center to work center in the processing cycle.
3. All the members of the organization are process owners, accountable for their work assignments and working together as a team to prevent problems to take hold of the process.
4. Standard quality objectives are maintained in all the processes to ensure consistent product quality. When these objectives become ineffective, they are modified through just-in-time corrective/preventive action to make them effective again. (Management Reviews are not just-in-time corrective action tools.)
5. Non-conformance is handled as a continuous improvement tool, as it happens, where it happens. Documented procedures ensure just-in-time implementation of corrective/preventive action and follow-up.
6. Customer complaints and product returns are lessons learned, corrective action tools and they are handled as a process to determine cause and corrective action.
7. Supervisory process surveillance is a time honored, on-the-spot corrective action tool and is maintained through documented procedures.
8. Customer requirements and product quality awareness are maintained through regular training, according to organized procedures in the Training Metrics.

As we step back for a moment and look at the individual components of this process-approach system, we will recognize that they were designed by applying the General System Theory to integrate and improve organizational teamwork and commitment to effectively manage quality systems in manufacturing shops.

This book is a completely revised edition of the original text I published in Hungarian under ISBN 963 04 9683 6 in 1998 (Title: *Vilagpiac Minosegkovetelmenye*). The company name "MINFOR" used in this book's procedures is my registered company in Hungary. Lessons learned from the first edition made me realize that I needed a completely new layout for this edition to enable me to incorporate training methods, flowcharting, process ownership, continuous improvement tools, the handling of customer complaints and product returns and many others. To make this process-approach system a truly useful tool in small business quality management for the manufacturing shops, I had to implement it, debug it and register it. What you see here is a real operating system that you can also implement without any outside intervention. Small business cannot follow the documentation system big business is pursuing. They cannot afford it, much less maintain it.

The Purpose

The process of enacting product quality requirements is like a seed planted. It will not grow to perfection unless all the conditions for it have been taken care of. This is also true in a quality management system. The seed is the product and the conditions for growth are the organization's commitment to quality, ever more working toward perfection to attain the objectives for quality to nourish customer satisfaction, internally and externally.

When a product becomes non-conforming, let's forget for a moment the exhaustive effort of getting it right and look at the central point that evades our attention most of the time: at what cost will it be right? Here is just one example that will tell the story.

1. Contracts – the administrator was sick and stayed home for several days. He didn't leave anybody in charge to follow up things. A customer faxed in a purchase order line item change, altering configuration and material. The Receptionist ripped it off and placed it on the administrator's desk. Nobody else paid attention to it;
2. Engineering – still working to the requirements of the old purchase order, issued the continuing job traveler, the operation sheet, and the drawing, as it did it before;
3. Purchasing – ordered the same material without knowing there was a change;
4. Production Planning – scheduled production the way the part was run before, being unaware of changes;
5. Production – set the job up the way it used to run it before;
6. Inspection – checked the parts the way it always did. Accepted and released them for shipment;
7. The customer returned the product as "Non-conforming to specification requirements, due to purchase order amendments."

Sounds familiar? Similar blunders take place every day, in a thousand different varieties, in the real world of running a business. Can we fix it? I always have faith that we can. But it will take some effort on the part of top management to realign organizational commitments and departmental responsibilities to manage flow-down requirements to prevent problems from taking over the processes.

This is the centerpiece of what this book is all about, – controlling departmental responsibilities one after the other as the work process flows from contracts to engineering, to purchasing, to planning, to production, and to inspection. ISO 9001/2000 calls this type of setup, "process-approach" in managing product quality. I have never believed in anything else. For as long as we are part of an organization that has no process control in place, we tend to pass on our mistakes to the next guy in the process cycle, only to compound them progressively to impact cost and product quality. It's bad enough to have a problem anywhere in the processing cycle, but when it is at the end of the line, it could wipe out any profit the organization counted on making on the project.

This book will integrate departmental responsibilities under a command and control structure through the process-approach mechanism so that we can prevent the transfer of mistakes from department to department. The details on how to realize this are laid out in the procedural context, called Quality Operating Procedures (QOP). I wouldn't dare to call this book complete for everybody and for everything insofar as quality process controls are concerned. However, what's included here is more than adequate to effectively manage core departmental quality requirements for product quality enforcement in small businesses, engaged in contractual manufacturing relationships.

Writing procedures, to enforce process approach quality requirements in the evolution of work assignments, must focus the work instruction definitions on preventive quality measures. Preventive quality practices should not be only the aftermath of fault discoveries, but rather the prerequisite in planning activities that eventually become the specified work instruction. I have put extraordinary effort into writing my procedures to include preventive measures wherever processes would impact product quality.

Customers require product quality excellence even though the purchase orders issued by them may not always clearly define what they really want. A process-approach quality management system must provide the enforcement vehicle by which product quality requirements are imposed in the defined processes, whether customer purchase orders clearly state the product quality requirements or not. To ensure that these requirements are accurately flown down to the relevant processing centers, I have entrusted Quality and Engineering with the responsibility to make that happen. The fulfillment of customer satisfaction in manufacturing begins, first of all, by knowing up front what the customer's product quality requirements are.

In contractual relationship, the product quality requirements should be stipulated in the contract (the PO). If they aren't there, the customer doesn't know what those requirements are, or will accept those work results that the supplier is willing to build into the product. In order to ensure that quality requirements will be built into the product, they must be identified at the earliest possible moment the contract is issued, or even earlier at the RFQ phase. Once identified, the requirements must be made part of the work procedures (the Job Traveler and Operation Sheet) at all relevant process steps and built into the product. Then verified that the work-results demonstrate compliance. In order to make this happen, I have added documented procedures, enforcing this critical requirement as soon as the customers' purchase orders are issued and logged in. The process is known as 'Purchase Order Review' and it is carried out by Quality Engineering through the application of Form A-001, the first one of the many controlled forms used for work results demonstration.

The Objective

There should never be anxiety on the part of management over meeting customers' product quality requirements. Ensuring customer satisfaction must start with the evolution of work procedures and not through backtracking after the problems have been discovered at the end of the line. For, if we don't understand up front our customers' product quality requirements, we will not be able to build them into the product the first time around.

In any defined processing cycle there are only two types of products going through the product realization chain, – conforming and non-conforming products. Both must be identified and controlled wherever and whenever they happen in the processing cycle. Quality Operating Procedures 006 and 009 in this process-approach system handle these controls very effectively. These procedures realistically represent the most important controlling documents in any quality management system, for they are directly tied to controlling product acceptance and rejection in the entire processing cycle.

Many times, when non-conformance during processing has been identified, work continuation goes into a tailspin and everything stops in order to determine what happened. In the majority cases, historically, document errors of all sorts cause the non-conformance. Correcting the errors in documents could take anywhere from days to weeks, especially if the customer is the government. These are, of course, not as frequent as the internally generated document errors which cause the equipment, already in a production mode, to be put on hold, pending the outcome of the investigation. In many cases, this is completely unnecessary for the problems and causes become identifiable right on the spot and immediately could be resolved between engineering and quality in order to let production continue, if there were a procedure in place allowing them to do that. I have carefully considered remedying situations like this and created the procedural vehicle to take care of it. Quality Operating Procedure 002, Section Two, paragraph 3.03.1 provides the necessary steps needed to accomplish this. QOP 002 is the procedure that deals with the control of documents and data for the whole quality system.

The process-approach quality management system presented in this book had also been designed to enforce continuous improvement as the problems occur in all the areas of the operating quality system. This has been locked into the documented mechanism of the process-approach system. Every time nonconformance is reported and submitted for MRB review, corrective action has to be determined and identified. The Management Representative has to oversee the plan for implementation and follow-up to ensure effectiveness. He cannot close out an MRB, form (A-006), until corrective action has been effectively implemented. Experience tells me that failure to follow up corrective actions promptly causes unrelenting repetition of the same problem and becomes the beehive, emptied during management meetings. This is why I have locked in the follow-up requirement as a just-in-time action item. Furthermore, doing corrective action, based on periodic Management Review, is not at all effective in timely implementation of changes. The customer can't wait.

Very few things are more irksome to administrative and inspection personnel alike, than the lack of guidance as to what to do with the heap of documents generated during a day's work. I have given special attention to this matter. Quality Operating Procedure 002, Section Three (Handling and Retention) lists all the applicable documents of the quality system and gives guidance to each department as to where to file the pertinent papers, when completed.

A very important element of this process-approach system is to take care of customer complaints and the measurement of customer satisfaction in a coordinated fashion and on a regular basis. Because these subjects have product non-conformance attributes, especially in the manufacturing area, I have defined the responsibility and the administration for them under Quality functions. To simplify this undertaking, I have designed a single form to take care of both assignments (Form A-038). The details for this are covered in Quality Operating Procedure 009, Section Three. The application of this form in non-manufacturing areas is not advisable. While the manufacturing companies most of the time use measurable results to determine customer satisfaction, this cannot always be said of the service industry.

Not all the small companies want text-type procedures, such as the ones included here, in order to manage their operations effectively. Those that have only a few employees may prefer to guide themselves with flowcharts. That was one of the reasons why I have earmarked (shaded) the primary processes in the flowcharts presented in this book. By retaining the primary processes along with the connected instructions and the header (page plate) on each page, the flowcharts can serve as documented procedures with the quality manual and the forms in controlling one's manufacturing operations. A few clicks with the mouse can remove very neatly any unneeded portions in the flowcharts. The flowcharts have also been fully integrated with the quality manual and forms just as the text-type procedures have been, creating a 'global' linkage of the quality system.

To ensure that the Quality Management System procedures presented in this book qualify for operational implementation in a small business, manufacturing environment, I have worked with a Connecticut based company to make that a reality. After implementation, this company registered itself to ISO 9002/1994. The Certificate Registration Number is 253, issued by HSB Registration Services, a fully accredited member of ANSI/RAB.

I would like to go on explaining other innovative features embodied in the presented procedures that represent the process-approach quality management system, but find it unnecessary because they will be recognized upon application. Also, the "Guidance" section behind this page will cover the operational elements of the system. For a quick glance as to how the procedures are integrated with requirements and responsibility in this process-approach system, please look up Quality Operating Procedure (QOP) 012, Sections One and Two at the end of Book Section Three. It will give you a close-up view of the operational communicating tool that also satisfies ISO 9001/2000 requirements. For those companies wanting to align their ISO 9001/1994 quality system documents to meet the requirements of ISO 9001/2000, I have provided the identification of changes and at the same time given guidance how to implement them. These are located under Appendix Five.

CONTENTS TO GUIDANCE

End of Contents to Guidance

GUIDANCE

General Information

This book has four sections. Each is devoted to specific requirements.

Book Section One – The Training Metrics (pg 34)
Book Section Two – The Quality System Manual (pg 267)
Book Section Three – The Quality Operating Procedures (pg 291)
Book Section Four – The Forms (pg 452)

Basic explanation about each section –

1.0   The Training Metrics – (Book Section One)

1.1   Background

The word Quality in small business mostly means inspecting parts. Rarely would anybody think that it could be much more than that. When we look at a small business, regardless what it makes, it is made up of functional departments like big business, but much less of it. It is still run by people, – the hierarchy. How often when we inspect parts and find something wrong, do we point to higher-ups in the organization for the cause of it? It is much harder to solve problems upstream in a hierarchy than downstream from it. Take this fact from me. I have gone through forty-two companies in forty years. They made plastic parts, fasteners, diodes, sheet metal parts, printed wiring boards, optics, special processes and a host of other things. And I have found the same setup everywhere. We are more focused on getting parts out the door than controlling in-processes to avoid problems. We would rather rework the parts or even junk them than cure the source of the problem that caused it. Higher-ups have authority with immunity from blame. Most often the cause of problem points to the leadership for not being able to recognize that managing product quality pulls in with it the whole organization, for the cause of problems can be as many as there are people within an organization. So let's put the controls into the system and let the system control everybody alike, regarding product quality.

This is where **process approach** in managing quality brings everybody together within an organization in order to limit the transfer of problems from the beginning to the end, from one assignment to the next assignment. This is where every person within an organization becomes a process owner and responsible for controlling his/her own work process. Training every person within an organization is the key in understanding how process approach in product quality management works.

During a training session, I was explaining this process approach concept to people with administrative and supervisory responsibilities. The explanation alone didn't work. The people had only limited concept in recognizing the chain of events that took place in a product's life-cycle.

They could not follow the interconnecting steps of how a product is moved from one process to the next, from contracts to planning, to engineering, to purchasing, to production, to inspection, and out to the customer. They tried to put my words into a road map. And it worked for the first few minutes. Then, as time went on, people lost track of what comes after what. Yet, nobody spoke up, raised his hand, or jumped out of his seat to tell me, "I just can't get it!" This training session went on for a few days before I realized what was missing: – the picture. I suspended the training for a week.

At home, I started planning and drawing flowcharts. I took the entire quality system and broke it down into its structural elements – the General System Theory process. I was the student and not the teacher. If I couldn't understand it, nobody else could, I said. The wastebasket was filling up in the meantime. Time was pressing me to move, for the week was almost over. Pretending that I must get a good night's sleep on account of work tomorrow, I urged everybody in the family to go to bed early. We did.

As I heard my wife going into a deep, rhythmical breathing, I crept away from her. She was sleeping. Until that time, I was praying. In between, I was imagining how the flowcharts should be sequenced. Praying?! Yes, praying. When you can't see the light at the end of the tunnel, that's when you look up. That's where you find the Almighty Hand blocking the light. And that's how your prayers remove the blockage.

In my makeshift office, I had already sheets upon sheets of flowcharts. I looked at them in the silence of the night. Everybody was gone. Now, I took all the sheets and jammed them into the wastebasket. Childish! I said. I decided not to try running through the whole quality system, giving a little bit here and a little bit there. I changed direction and went back to the beginning – the process-approach. It made a lot of sense. This way accomplishing training, accomplishing process controls, and accomplishing product quality, I could focus, step by step, on what was needed to make the interconnection, making each step build on the work results of the previous steps. The commonly used statistical tools guided me in the right direction. (See book page 504)

This is how the whole Training Metrics and the Quality Operating Procedures in this book were developed – step by step, right from the beginning. My students, the leaders of the company, remarked after seeing the flowcharts, "How come, we didn't start out that way, anyway?" I was embarrassed. "Ignorance!" I replied.

1.2   What will the flowcharts do for you

They will give you the step by step understanding how a quality system is structured. How the processes of the system interrelate with each other. What-to-do and what not-to-do. The explanations, the recommendations, the advice, the instructions and the warning notes, all in all, lead the user to make meaningful interpretations about a quality system that is built on process-approach. The training metrics is a tool that guides the reader in understanding the primary processes, illustrating those assignments that are needed to integrate core departmental functions in carrying out flow-down requirements. As such, they have been closely integrated with the quality manual, the text-type quality operating procedures, and the forms issued in this book.

Besides being a training material for those in need to know how a quality system in a small-shop environment should work, the flowcharts can also be used as a checklist in determining the proper flow process and requirements of a given assignment. Anybody can use the training metrics to establish a quality system. It gives guidance on how to structure the quality manual – the paragraph numbering, the revision controls, the interfacing references, the page plating requirement and a lot more. Similarly, it gives guidance how to put together and control the process instructions (Quality Operating Procedures) and the forms. It shows how to integrate a quality management system that is tailored to fit the working conditions in small business. What it gives you is the real world, the hands-on experience as it happens, as you are running a business on a hectic schedule. What you will find here is what is practiced every day in getting the work done, according to both the customer and internal requirements in maintaining product quality for customer satisfaction.

1.3 How to use the flowcharts

Flowcharts are point to point indicators of how a process flows in line with planned expectations. In this book, we are using the flowcharts to give direction as to how the planned processes and assignments are accomplished in a quality system. To do this, we needed geometrical figures to express and differentiate what are primary, alternate, and parallel processes, as well as instructions, advice, notes and explanations. All in all, the flowcharting techniques tie every thing together to create that lasting picture to focus our attention on how we should do things.

Here are the flowcharting figures we use in the Training Metrics:

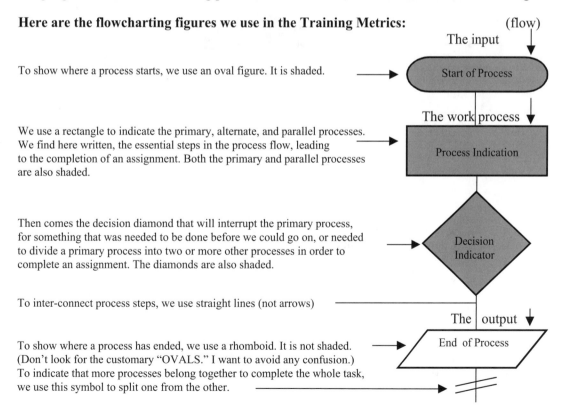

To show where a process starts, we use an oval figure. It is shaded.

We use a rectangle to indicate the primary, alternate, and parallel processes. We find here written, the essential steps in the process flow, leading to the completion of an assignment. Both the primary and parallel processes are also shaded.

Then comes the decision diamond that will interrupt the primary process, for something that was needed to be done before we could go on, or needed to divide a primary process into two or more other processes in order to complete an assignment. The diamonds are also shaded.

To inter-connect process steps, we use straight lines (not arrows)

To show where a process has ended, we use a rhomboid. It is not shaded. (Don't look for the customary "OVALS." I want to avoid any confusion.) To indicate that more processes belong together to complete the whole task, we use this symbol to split one from the other.

As we go from page to page, we use a circle to indicate that the process is still continuing.  Figure 1

Taken all together, we have shown the beginning, **the input;** then the step by step activity, **the process;** then the completion of the process, **the output.**

A parallel process example:

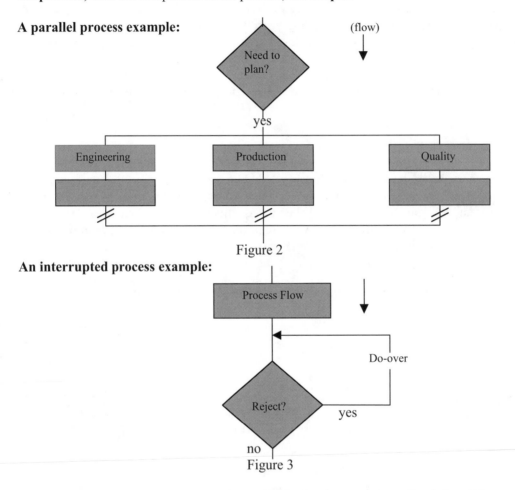

Figure 2

An interrupted process example:

Figure 3

In order to make flow diagrams understandable for the purposes of training, I have added instructions and other notations where they were needed. This would not have been possible without the use of additional flowcharting figures. In this application then, they become the just-right consulting tools without the presence of the consultant (See Figure 4 and 5). In reality, this is a question and answer session. We will find the use of this method interwoven even with the primary processes (See Figure 6). The examples below will illustrate this.

Most of the Training Metrics have <u>Basic Information</u> provided to understand the specifics that relate the questions posed by the decision diamonds. The answers to these are given inside the rhomboid figures. The flowcharting figures used with Basic Information are not shaded.

Here is how it starts:

Here is how it continues:

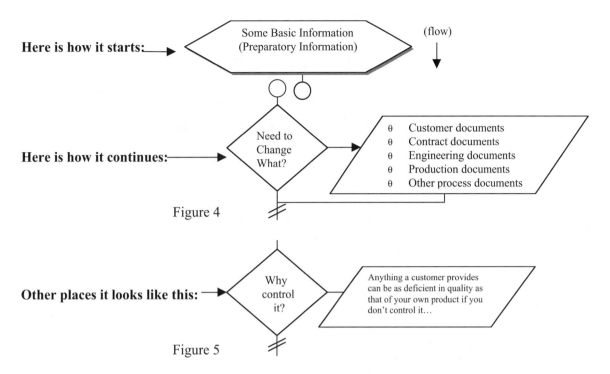

Figure 4

Other places it looks like this:

Figure 5

The question and answer interwoven with the primary process:
(When a decision diamond is part of the primary process, it is shaded)

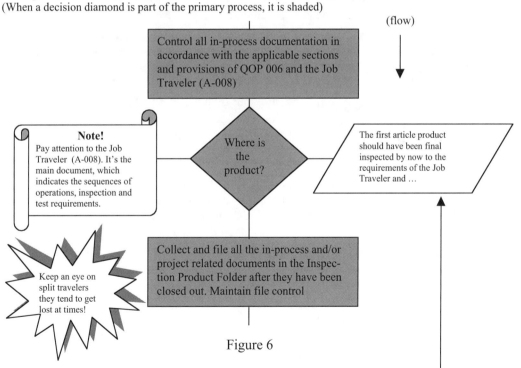

Figure 6

Note that the rhomboidal figure appears everywhere when a question posed by a decision diamond needs answer. This includes the primary process as well as the preparatory process.

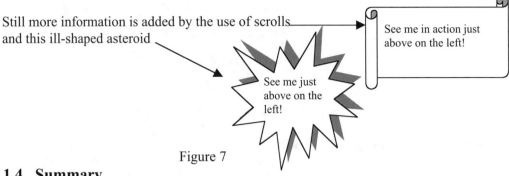

Still more information is added by the use of scrolls
and this ill-shaped asteroid

Figure 7

1.4 Summary

The Training Metrics takes the place of expensive consulting and it comes free with this book. There are twelve separately grouped Training Metrics issued in this book, corresponding to the twelve text-type Quality Operating Procedures (QOP) found in Book Section Three.

A few things are worth to remember when using the Training Metrics:

1. Each procedure in the Training Metrics contains full coverage for the corresponding sections found in the Quality Operating Procedures, and as needed, additional information as well.
2. Use the Training Metrics and the Quality Operating Procedures together, section for section, to understand the full scope of process requirements.
3. References are given everywhere for the application of forms. They are needed all the time, once a quality system has been implemented. They show the recorded work results – the evidence to prove compliance to requirements. Remember them.
4. The Training Metrics have been fully integrated with the requirements of the quality system manual in this book. They can serve in creating your own quality system; they can guide you to implement and maintain your system; they can be used as your own training metrics; and finally, they can be converted into your own process instruction procedures.

2.0 The Quality System Manual – (Book Section Two)

The manual in this book follows the requirements of ISO 9002/1994. It is intentionally constructed according to the clause numbering system of the Standard for a very good reason. ISO 9001/2000 ignores the descriptive layout of the quality system requirements as that is known in the manufacturing sector. ISO 9001/1994 covered this quite well. We are talking about core departmental responsibilities – Contracts, Purchasing, Engineering, Planning, Processing, and Inspection. These are the vital, functional departments that are interacting in the processes of a product's life-cycle. We can easily recognize and determine these functions as we look at the contents of the ISO 9001/1994. This is not the case in ISO 9001/2000. It tries to be everything to everybody and puts the quality requirements out of order without tying them to any specific industry. A small business owner would have to hire a consultant just to understand what the Standard is trying to say, never mind implementing it. Without the assistance of consultants, small business owners would be locked out of the benefits that ISO 9001/2000 should bring about through its TQM framework.

Therefore, in this book, ISO 9001/2000 implementation takes place through process-approach quality management, the way business owners run their businesses in manufacturing. Any business owner in the manufacturing sector can take this book and implement his quality system from it without consultants. Additionally, companies that are registered to ISO 9002/1994 can also convert their system, simply by following the explanations given under Appendix Five at the end of the book. The only caution I should give you, however, is that if you need to incorporate sector specific requirements, follow the technique I have used in incorporating AS 9000 in this book. If you don't need to, you will have an ISO compatible quality management system based on your manufacturing business processes, ensuring organizational focus on product quality, integrated performance measurement, and a locked-in continuous improvement.

The quality manual in this book identifies the product quality requirements from ISO 9002/1994, for that is the suitable framework for the maintenance of product quality in manufacturing. Within this framework, I have implemented the requirements of ISO 9001/2000. I have not "personalized" the manual because that doesn't seat well with the formal structure of the process-approach system in this book. Furthermore, the work instructions as to how to accomplish processes have been removed from the Quality Operating Procedures and replaced by process instructions. This development allows this quality system to be used as a model in the manufacturing sector.

Inasmuch as the manual follows the layout of 9002/1994, the main emphasis is placed on core departmental responsibilities and decidedly on the assignments within these departments. Why? The core departmental responsibilities in contracts, purchasing, engineering, planning, processing, and inspection are all grounded in product related activities all day. These are the departments that have to deal with the realities to move the contracted product from the purchase order phase to the shop floor and to the shipping dock, ensuring that product quality is maintained everywhere. To make all this happen in a constructive way and with as little problem as possible, the ground rules have been incorporated in every quality objective procedure. This is the process-approach system. This is where the Quality System Manual becomes a book of commitments. I have so worded this document that the requirements of the quality system are structurally linked to the Quality Operating Procedures which contain one or more Quality Objective Procedure.

No quality management system is easy to understand in terms of implementation and operation. For this reason, I have chosen only the core departmental assignments in managing quality requirements in small business, paving the way through the conversion methods to incorporate the non-product related areas of the business later, once the cost benefits of the process-approach system are realized.

2.1   How to use the Quality Manual in this book

The best way to use the Quality System Manual is to understand what each main clause (4.1, 4.2, 4.3, ... 4.7, 4.8, ... etc.) is stating. Then, go from there to the referenced Quality Operating Procedure. If there isn't any QOP given, the manual itself contains the process instructions in the sub-clauses. Example: paragraph 4.20 – Statistical Techniques.

It will not be long before you notice that the manual has incorporated a sector specific standard, AS 9000. This has been done for a couple of reasons. First, to cover the metal cutting industry, – shops in the aircraft parts machining sector would have a need for it. Second, it will serve as an example of how other sector specific standards can also be incorporated likewise. TL-9000, ISO 14001, and QS 9000 Standards can be similarly adapted. While I'm on this subject, I should note that the sector specific standards rely on ISO 9000 as the primary requirement in the maintenance of quality systems. The sector specific requirements are add-on requirements and they can be easily aligned with every requirement in the quality manual presented here. In my case, by the expansion of the sub-clauses, I was able to incorporate the requirements of AS9000. What this really means is that whether it is my manual or yours, the basic ISO 9000 requirements would have to be covered first for each clause, other than exclusions. Then would come the add-on requirements.

Just a few pages below (on pg 23), I will be spending some time with you explaining conversion techniques in all areas of the quality system in order to facilitate adding or taking away requirements.

2.2 Summary

We will not mention on this page how to put together a quality manual, for that is already given in the Training Metrics in QOP 002. Nevertheless, some things need to be highlighted.

1. Management must define its commitments and objectives to quality.
   (See Mission Statement in the QSM, clause 1.1)

2. The quality system manual must state in policy statements what are the responsibilities of the organization to product quality. (See clause 4.1.2.1)

3. The policy statements should be written in commending language (shall), short and clear. (Elaborate here only the essentials. Hold details to a minimum, for you have to give account as to how you carry them out.)

4. The manual must give references to other controlling documents.

5. The manual must be a controlled document. (See the headers (page plate) on each page in the manual.)

6. The manual must state the applicable Standard from which the requirements have been imposed. (Find this in the Mission Statement)

7. The quality system manual is a policy document. It must contain definition as to product quality responsibilities and objectives for the operating system. All statements expressed in the manual must be accounted for.

I could go on and on here and write easily thirty pages about manual construction. It would not help much without illustration. Instead of fragmenting the matter, I have presented the whole quality manual for you to look at. See it in Section Two in this book. Use it as your picture in ten thousand words. You can rewrite it or use it as it is, if it suits your particular product quality requirements. Don't forget to put your name on it.

3.0   The Quality Operating Procedures – (Book Section Three)

The Quality Operating Procedures (QOP) are process instructions. As such, they are the second Tier documents in this book, outlining and elaborating the quality objective requirements of the quality system manual. They have a structural layout of their own and are controlled no less than the manual is. In order to maintain the linkage in documents, the QOPs refer back to the manual to identify from where the requirements have been taken. They must also refer to those documents in which the documentation (record keeping) is carried out. This way, the linkage among documents is maintained forward and backward to ensure that the quality system is fully integrated. All this has been structured into the QOPs in this book to make them a practical document to be relied on as the source material in controlling work processes. QOP 012, Sections One and Two, gives you the put-thru index for quick navigation between procedures and requirements. Section One is the operational put-thru, while Section Two steps in to handle non-conforming cases. Book page numbers have been given throughout in order to facilitate finding the referenced materials.

On every page of every QOP there is a header (page plate) to indicate the document's controlling mechanism – issue and revision date, revision level, page number, document approval and references to the QSM and Forms.

To understand how to construct process instruction procedures, please refer to the "Training Metrics" in Book Section One under QOP 002 (pg 66,67,68). It's right after the manual's construction methods.

The level of detail in describing the process instructions in my procedures has been kept intentionally in line with the required skill level of the process owners. The instructions are not a beginner's procedure. If they were, they would lose practicality of application with skilled people. The work methods on how to carry out the identified processes have been removed ("work method" is also known in the industry as "work culture"). Processing related work methods instructions should be given in the Job Travelers and Operation Sheets, as required. Inspection related work methods instructions should be defined in the "Inspection Instruction Sheets" (IMS), as required. Work methods instructions are as varied as companies are. This process-approach management system or any other quality management system would flunk on arrival if it tried to standardize the work methods instructions and present it as a model.

QOP 006 (Control of Inspections, pg 127 or 357) and QOP 009 (Control of Non-conforming Products, pg 199 or 403) are the heart of the procedures. Through the application of these two process instructions, we regulate product quality enforcement of the business, for they control what is conforming and non-conforming. This should be the most familiar ground for all quality professionals. Under the military standards, product-quality was dominantly controlled through them. Therefore, these two sets of documents require most of our attention. They are long, all inclusive procedures and require the use of most of the forms.

The Quality Operating Procedures that support this quality management system have been designed with paper reduction in mind. Instead of creating individual operating procedures for every assignment identified in the system, I have combined those assignments that belong to the same work category under one operating procedure and called them quality objective procedures instead. Take for instance, Quality Operating Procedure 006. It has seven individual sections. Every assignment deals with an inspection requirement – first piece, first article, process control, final, receiving, customer source, and subcontractor source inspections. The other QOPs follow the same grouping setup as much as possible. Approximately 150 pages have been saved by this consolidation effort. When I think of the time that process owners wasted on reading inflated procedures, I place myself into their shoes, for many times I myself wore them.

The Quality Operating Procedures presented in this book are controlled and linked together by the application of a dedicated Master List. The Master List identifies each operating procedure and lists the sectional titles and provides revision status indication for each operating procedure. It has its own revision recording page to indicate changes impacting the individual procedures. This is how the 'Procedures Manual' is controlled. Apart from this, the individual operating procedures are also controlled, first by a page header, second by their individual revision control page. To carry out the steps in doing the changes and other relevant controls, I have issued Quality Operating Procedure 002 for that reason (pg 60 or 311). To do it even easier than following text instructions, I have provided flowcharting methods for guidance. It is located under "Training Metrics" in QOP 002 (pg 60). This QOP is the very longest of all procedures, for whatever you need to do with the quality system's documents, this is the place to find it. *(I would like to add a word of caution here, regarding the distribution of the Quality Operating Procedures. The procedures as a whole should be kept together in a binder, titled* <u>*Procedures Manual*</u>*. Otherwise, they would be scattered all over the place and controlling them would be impossible.)*

Writing process instructions is about the most demanding thing to do, for we are always facing the unknown – how long should the instruction part of it be. I have learned from my own experience that writing instructions to employees should be based in general on two basic requirements: (1) the general skill and experience level of the work force, (2) the complexity of the assignment. The first one is where we can easily be carried away and bore the work force to death with it. When an employee has to spend more time on understanding the process instruction than getting the work done, then that process instruction is counter-productive. This is why you will find that the QOPs in this book are just a little bit more than basic outlines of the steps needed to accomplish the identified process assignments. And that's the way it should be, for an experienced employee knows the steps of his/her process assignment better than what the written words can tell. In fact, all of us procedure writers should learn first how the employees are doing their work before we try to tell them how we would like them to do it. And just one more thing, the extent of the detail in process instructions can be the richest ground to every auditor to find that "You are not doing what you said you do!" Process instructions should not be training procedures. In my case, I have provided the Training Metrics for that purpose.

3.1 Summary

The Quality Operating Procedures are stand-alone documents, fulfilling the stated objectives of the Quality Manual. They are individually identified and controlled: 1. by a page header on every page (page plate), and 2. by its own revision control page. As a group, they are identified and controlled by Master List, which has its own header and revision page. When these procedures are issued to the floor, they should be held together in a binder and titled as <u>Procedures Manual</u>. Individual distribution of the Procedures Manual should be maintained, for we should always know who has them. As we make changes to them, we need to recall them. For this purpose, Form A-010 was created (pg 465).

Other information regarding the use of Quality Operating Procedures:

1. Use the Table of Contents in front of every procedure to quickly find the title of interest;
2. Use the Master List to identify QOP numbers, title and revision status (pg 36 or 293);
3. The interrelationship among documents is referenced everywhere in the procedures, as needed;
4. Paragraph sections that pertain to AS 9000 are indicated in "**bold**" throughout the procedures;
5. For each Quality Operating Procedure, there is a Training Metrics provided. It is identified with the same QOP number and title.

4.0   Forms – (Book Section Four)

There is no worthwhile quality system in any business without being able to prove that work results meet the customer's specification requirements. Record keeping shows the evidence of compliance and the use of the forms provides the means to accomplish it. For this reason, the forms, applied in conjunction with process instruction procedures, are tailored to meet the specific needs of individual quality objectives. They are identified and referenced, as required, in each quality objective procedure. I have kept them as a group, separated them from the individual procedures, only for the reasons of practical handling and control. Most of them are very familiar forms and are easily understood by quality professionals everywhere. What is not so easily understood, however, is how to control and retain them. I have not spared words or pages to ensure that this matter would leave no questions to the readers. Every form fulfills a need and is accordingly controlled. The application of the forms is described in each quality objective procedure as the work-specific tasks require them. Special attention has been given in QOP 002, Section Three (pg 85 or 328), to the filing and retention control of every form identified in this book. This is also an ISO requirement this time around.

Record keeping is the most hated chore in running a quality system, for it requires accuracy and completeness. Being fully aware of this from my own experience as inspector and documentation auditor, I have paid particular attention in designing the forms user-friendly, making data recordings to fall in line with process flow requirements. But, as with anything we do, we think we are doing great, then somebody else comes along who can do it much better. I encourage all users to evaluate each form and change it in line with the requirements of his/her own needs.

All the forms used in this book have been identified and organized in order that we can easily recognize what they stand for. To facilitate the reader in this area, each form has been numbered in sequential order – A-001, A-002, A-003, etc. and is given a title that comes as close to the process application as possible. References to the use of forms have been given everywhere as required, both in the quality manual and in the process instruction procedures. Furthermore, easy identification and control is facilitated by the assignment of the Master List, which is located in front of the collection. To keep track of changes, as we occasionally have to make them, a revision control sheet is provided and is located right behind the Master List. The method of how to carry out changes to documents, including forms, is detailed in QOP 002, Section Two, and also in the Training Metrics bearing the same QOP number and section.

It is sometimes mind-boggling for inspection personnel and especially new people to know at once what all the quality forms stand for and how, when, and where they are applied. In order to help to find quickly where the relevant information is on the application of a form, each form has been marked with a reference locator. You may look at the bottom left on each form and go from there to the referenced source material to find the details on its application.

Changes to the forms issued in this book are inevitable, for there are so many different ways products can be controlled. One of the reasons why the forms have been provided in this book was to give a complete guidance to the readers as to how they should be established, identified, integrated, and controlled in a process-approach quality management system designed for small businesses. The other was to see how they are integrated with the procedures.

5.0   A Few Words on Implementing this Process-Approach System

The toughest decision we have to make when it comes to implementation is where to start. Changing over from the old system to the new system carries with it the inherent uncertainties stemming from not having experience with the new system. Fortunately, the quality system presented in this book is not a new system. Although there are innovations in it, like the application of Form A-001 "Purchase Order Review Sheet," and A-009, "Amendment to Procedures," the basics are based on already proven practices, repackaged to bring organizational interaction together to create a process approach system. What is found in this system should not be new at all to those who operated to Mil-Q-9858 and Mil-I-45208. What will be somewhat unfamiliar, however, is the universally hated requirement of accurate and complete documentation of work results. This too, should not be taken as overbearing, for I have removed unnecessary paperwork and basically limited record keeping to product quality enforcement. So, before implementing anything this book offers, the user should first know his current system in order to be able to realign it with the procedural norms presented here. Then, start the implementation in the following order:
(Use the Training Metrics to help the implementation process)

1. Start at the very beginning, the Contract Department. Implement first all four sections of Quality Operating Procedure 001, given in Book Section Three – Control of RFQ; Purchase Order Review; Verbal Purchase Orders; and Amendment to Contract. The reason for this is to prevent the flow-down of errors right from the beginning.

2. Then, implement Section Five (Receiving Inspection) from Quality Operating Procedure 006. With this move, you will limit the flow-in of deficient products from the outside.

3. Then, implement Section Four (Final Inspection) from Quality Operating Procedure 006. This will prevent the outflow of deficient parts.

4. While implementing the above three steps, issue form A-033 (Operator Product Verification Record) to those process owners who are responsible to put quality in the product. This move will cut down the generation of internally flowing non-conforming parts. (See the application of Form A-033 in QOP 006, Section One, par 3.01, pg 135 or 364.)

5. Now, take care of all the rejects by implementing Quality Operating Procedure 009, Control of Non-conforming Products, and start the corrective action and follow-up process.

6. By this time there should be plenty of records generated. Implement now Section Three (Handling and Retention of records) from Quality Operating Procedure 002.

7. Now, implement the rest of the Quality Operating Procedures in line with work processing requirements. It will be a must to implement them as they are integrated within the provisions of the already implemented operating procedures.

I have given you here the very same methods that I am using even today in implementing the same system presented in this book. The usual six months to a year implementation time, I am managing within two to three months. The biggest drawback to faster implementation I have encountered was creating organizational unity to integrate the system. Old habits are tough to give up by certain employees. Handle it as raising kids. Show and tell over and over again.

6.0   Conversion Methods

The basics…

Think for a moment and ask yourself: "What type of product am I supplying to my customers?" The answer to this question will determine the type of Standard(s) you need to have. Sector specific requirements are add-on requirements to the basic ISO requirements.

6.1   The quality manual in this book fulfills the requirements of:
1. ISO 9002/1994 and 9001/2000;
2. AS 9000 (is a sector specific requirement for aircraft-parts makers);
3 The clause layout of the quality manual is identical to the layout of ISO 9002/1994. The requirements and updating features to ISO 9001/2000 are shown in Appendix One, pg 499, and implemented as explained in Appendix Five, pg 509, at the end of the book.
4 The sector specific requirements (AS9000 in our case) have been added at the end of the main clauses in the sub-clauses, following the way other sector specific standards have added their own requirements.

If we wipe out the sub-clause information from the manual that relates to AS 9000, we have created a basic all-purpose quality manual that is compatible with ISO 9001/2000. When a requirement of the new Standard doesn't apply to a company's product or service it provides, it can be excluded. This is what I did also in paragraph 4.4 and 4.19. When a requirement of the

Standard is omitted, however, the reason for its omission must be stated in the manual. The manual in this book has been provided for guidance also in establishing, implementing, integrating and maintaining a process-approach quality system in core departmental activities in manufacturing. *(Omission of requirements is strictly controlled by clause 1.2 of ISO 9001/00.)*

6.2   Three areas to be concerned with when making changes to the manual:

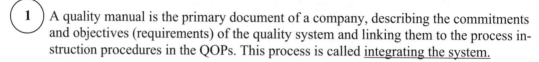

(**1**) A quality manual is the primary document of a company, describing the commitments and objectives (requirements) of the quality system and linking them to the process instruction procedures in the QOPs. This process is called <u>integrating the system.</u>

When we add requirements to or take away requirements from the manual (as with sector specific standards), we also have to add or take away the process instruction references and the instructions to it in the lower-tier procedures. This is how we flow down the addition or removal of requirements from the quality manual and the process instruction procedures. When we add requirements to the manual and don't define how we are going to comply with it, either in the manual or in the process instruction procedures, we are in violation. When we are removing a requirement and don't wipe out the process instructions and references to it, we are again in violation. This is the easiest way to confuse the system and also flunk the audit. So, when we make changes, we must cross check the system's documents in order to ensure that we have made an accurate and complete reconciliation (resolved the differences) in the flow-down process. Part of this reconciliation process may or may not impact the system's forms. When it does, we either revise the existing form or add a new form.

The entire process-approach system presented in this book is a very simple setup. Two different but inter-linked manuals embody the control of the entire product quality requirements in a small business – the quality manual and the operating procedures manual. The quality manual states the requirements. The process instructions in each quality objective procedure implement and maintain those requirements. Navigation between procedures is straightforward. From the Quality Manual, we look up the references and go to the Quality Operating Procedure(s). The process instruction procedures for each quality objective will identify its own form(s) needed to record work results. When making changes, we should follow these flow-down steps in the same manner. By doing so, we can add or remove any clause, paragraph, or sector specific requirement from the manual and the quality operating procedure with just a few clicks of the mouse.

(**2**) The quality manual must give references to lower-tier documents to indicate where the process instructions or other relevant documents can be found.

In the presented quality manual, these references to the Quality Operating Procedures (QOP) or to the forms are given under each applicable clause, as required.

(**3**) Furthermore, a quality manual must be a controlled document. It must indicate revision level, issuance date, and approval. Its pages must be numbered. And as you make changes to it, they must be recorded. To successfully incorporate the changes in all the distributed manuals, they should be tracked as to the individuals who have them.

Please turn to the manual now (found under Book Section Two, pg 267) and check out these three controlling conditions circled above. They will have to be remembered as long as you operate a quality system, for these are standard rules for controlling changes, linkage and integration in procedures, whether it's my manual or yours. For the details as to how to carry them out, please refer to QOP 002, Section Two (pg 71 or 317). The Training Metrics in Book Section One, under the same QOP identification, will flowchart the process steps for you and also give you additional advice.

7.0   Controlling Flow-down Requirements:

In a process-approach quality system, performance responsibility of one process owner interacts with the performance responsibility of another process owner. The whole processing cycle of a contracted product is interdependent from the beginning (contracts) to the end (shipping). The procedures in this book, established to regulate this type of quality management, follow the same interdependency from the beginning to the end. So, as we make changes in one area, we must first consider the impact they make elsewhere in the chain of events, before we can implement them. This is called the flow-down impact verification. As such, a bunch of things have to be done one after the other to ensure that we carry the changes through in a controlled manner.

Let's go through the process now to show you how AS 9000 was added to the procedures. Similarly, you can also add or remove any other sector specific requirement or the requirements of your own particular business by following the same method. A word of caution must be added here, regarding AS9000 requirements. Most of these requirements are plain, commonsense requirements that are actually needed if we want to run an effective, truly coordinated quality management system, – not just in the metal cutting business, but elsewhere also. So, when chopping these requirements away, please consider first what will happen without them. Truthfully, I have done the same thing and incorporated a number of them in the very procedures presented in this book.

It was already noted that sector specific requirements are add-on requirements to ISO 9001/1994 and that they are imposed by the expansion of sub-clauses. The imposed requirements begin universally under clause number **4 – Quality System Requirements.**

7.1   Let's demonstrate what we did. Example one:

For AS9000, the first requirement then is sub-clause 4.1.2.4 – "Suppliers having a quality assurance activity performed by an individual process owner (e.g., operator, buyer, planner) shall have procedures that define the specific tasks and responsibilities that are authorized and the corresponding requirements and training necessary to perform those tasks."

We can go ahead and write pages upon pages to tell in the quality manual how we would be enforcing this. That would be a foolish thing to do, for we can answer this in one sentence. Our manual under sub-clause 4.1.2.4 says:

4.1.2.4 **SAE AS9000** (notice the sub-clause in the Standard is identical to the sub-clause in the manual) "The requirements of AS9000 have been incorporated throughout the product's processing cycle by the application of Quality Operating Procedures."

This is exactly what we have done in the process instructions, the Quality Operating Procedures. We have defined the "specific tasks and responsibilities" to the process owners. Regardless, whether it's AS9100, TL9000, QS9000, or ISO 9000, controlling the work processes that are impacting product quality cannot be done without process instruction procedures. Otherwise, non-conformance will take over the process.

A very important objective we must remember when we deal with sector specific standards: The nature of the business we are engaged in. The reason why we want to impose additional sector specific requirements is because we are already supplying that product to the markets. A potato grader would not need AS9000, but an aircraft parts manufacturer would. In this manner then, the process instructions given to the employees should reflect the imposition of any additional requirement beyond those already given to fulfill the basic requirements of the ISO Standard. These additional requirements are called complementary requirements to the basic ISO requirements.

In paragraph 4.1.2.4 above, we have satisfied in the manual the sector specific requirement by stating how we are going to fulfill it. The rest belongs in the process instruction procedure. That is all a quality manual should say.

7.2 Let's go to a more difficult requirement. Example two:

For **AS9000** the requirement says:

4.5.3.1 Document Change Incorporation: "The supplier shall establish a process to ensure the timely review, distribution, implementation and maintenance of all authorized and released drawings, standards, specifications, planning, and changes. The supplier shall maintain a record of change effectivity and, when required, shall coordinate these effectivities with the customer."

Is this a strange, never heard or done requirement? Of course not. How would anybody be able to satisfy the customer's purchase order requirements, if he didn't have a process established to review, distribute, implement and maintain the drawings, standards, specifications, planning, and changes, throughout the entire production cycle? Nobody could do it without screwing it up. The fact is, however, that in a lot of cases, we do not have written procedures establishing how to do it. We are doing it by the traditional way: "I told you how to do it!" Hence, confusion, mistakes, delays, and arguments pervade our actions. In reality, this requirement should be a routinely established practice, not only for AS9000, but for every business whose business is to satisfy contract services. And that is exactly what we have done in this book, flown down the customer's requirements everywhere within the organization to ensure compliance to the "Document Change Incorporation." Here is what we are saying in our manual:

4.5.3.1 SAE AS9000 – amendments to customer documents are controlled through the provisions of QOP 001, Section Four (pg 52 or 307). The internal implementation of those amendments are carried out in accordance with the provisions of QOP 002, Section Two, paragraph 3.03, as required (pg 73 or 317).

As you can see, we don't elaborate, for that belongs in the lower-tier procedure. We have identified the requirement "Document Change Incorporation" and stated where the process instruction for it can be found. Plain and simple.

But now, let's go to QOP 002 operating procedure and also to the equivalent training metrics and find out what we have done with this "effectivity."

Without repeating here the process instructions given in QOP 001, Section Four and in QOP 002, Section Two, we will give you a short rundown as to what we have done. For the full details, however, please go to the identified procedures.

We have taken this "Document Change Incorporation" and made it a standard practice for any business that is engaged in contractual relationship with a customer, irrespective of AS9000 or any other sector specific standard. Why? Any business engaged in supplying products to customers is routinely involved in document changes as a result of "amendment to a contract" or amendments to its own internal documents as a result of the contract amendment. This whole change process to documents is an everyday bugger in the life of running a business. So why not make it then part of the process-approach, a core responsibility, in order to prevent the "I told you so" confusion associated with it. We have made this document change process a quality management tool and created for it the process instruction enforcement (controls) on how to carry it out. (This is what I meant, when I said above—plain, commonsense requirement.)

While the process instructions in QOP 002 (text) give you the formalities on what to do and how to control this very important objective, it is very difficult to train employees from it. There are too many steps involved here in implementing document changes, and it is not easy to keep the sequence going without skipping something. In order to facilitate the training of employees and to understand better how to walk through the various process steps, I provided the training metrics to work hand-in-hand with the text-type procedures. During training, you can follow the illustrative techniques of flowcharting, step by step. Use it also as your checklist to verify results.

7.3 Let's talk about an even bigger challenge. Example three:

Every company is required to comply with EPA and OSHA imposed requirements to a degree, depending on the type of business it runs – impacted by the number of employees, type of equipment and facility, product it makes, and its location. Most companies manage these affairs separately from their quality management system. There is nothing wrong doing it that way, except the cumulative costs associated with setting up separate procedures and administration to do it. In a small business this could be a back-breaker. Under a process approach system not only the product quality requirements but also the regulated EPA and OSHA requirements can be effectively documented through an additional operating procedure. This could be done, similarly, as we have executed the enclosure of AS9000 and followed it by the flow-down requirements in the process instructions and forms. The new standard (ISO 9001/2000) sees the future of quality management systems as an all-embracing policy document to combine all the organizational responsibilities under one unified and integrated management system. Makes a lot of sense.

As we look at the manual's title page, we read, " The provisions of SAE AS9000 are incorporated as a complementary requirement when imposed by contract." Similar wording can be lined up for each additional standard you wish to incorporate in your quality manual. Then, impose the specific requirements of the selected standard in the manual's sub-clauses under each applicable main clause as we did it for AS9000. Writing new operating procedures and creating new forms to satisfy the record keeping requirements would every bit depend on the standard chosen. As you go through combining different sector specific requirements under one management system, I strongly urge you to collect first from everybody all the existing papers that people already have started or implemented. Then, review them in order to understand what you already have and what else you would need to have in complying with the requirements of the selected standard. As the new procedures take effect, the old procedures would have to be locked up as people would rather fall back on the way it used to be than follow new requirements. Training, training, and more training is the only way to overcome this problem.

The techniques of incorporating EPA (ISO 14001) or OSHA requirements in the quality manual is not different from what I have done with AS9000. What would be different, however, is the process instruction on how to carry it out. These would have to be tailored as to the impact of EPA and OSHA requirements have on the type of business a company is engaged in. Because of this wide application, the requirements of EPA and OSHA have not been included in the quality manual and no operating procedures have been provided in this book for them.

While sector specific standards impose their requirements as complementary to the ISO Standard's, the EPA and OSHA requirements stand over and above all the requirements as federal, state, and local laws impose them under regulatory or statutory requirements. The ISO as well as the Sector Standards expect compliance with them not by detailed imposition but by reference to them. As such, it is up to the local company to implement and enforce them according to the requirements of the type of business it runs. So it stands to reason that a company may impose EPA or OSHA requirements separately from its quality management system or in combination thereof.

We have gone through three examples above to show you the methods by which you can make changes in your quality manual to fall in line with requirements needed to operate your business. Whether it's a sector specific requirement or a quality objective of your own, the quality manual in this book is structured to lend itself for provision and text changes without upsetting the fundamental construction requirements. One major task you will need to remember is that when you make changes in this manual to suit your needs, flow down the requirements in all lower-tier procedures. The rest of the controls are found in QOP 002. You can add completely new requirements and new quality operating procedures and forms so long as they are fully reconciled to indicate linkage in the succession of documents.

Keep in mind that if you are just setting up a process-approach system, you do not have to worry about controlling revision indications. Revision indication control takes effect only after you have released the quality system procedures for implementation or to the registrar. It is advised, however, that you make changes one at a time and completely flow the requirements down into the lower-tier procedures, so that you don't lose track of what belongs where. With this small advice, I have told you my own agonies in trying to do too many things at once. I had to re-do this whole book in order to flow-down correctly all the changes I made to it.

7.4 Summary

As changes are made to the quality manual, an immediate ripple effect takes place that may or may not impact the lower-tier procedures. That depends on what had been changed. Change to the clause requirements in the manual imposes change to the process instructions and conceivably to the forms also, whether the instructions are in the manual or issued separately. Rewriting the text to suit one's style of speech, without changing requirements and references, doesn't impact anything, except your well being. Removing requirements from the quality manual requires the removal of the applicable process instructions and possibly the impacted form(s). Whether you add or take away requirements from the manual, the flow-down requirements cannot be ignored, for the entire process-approach system would have enforcement problems. The operating procedures' and the forms' manuals are controlled separately by Master List and their Revision Sheets. When changes affect the included documents in these manuals, their Master Lists must also be updated. The quality manual has its own Revision Sheet to record changes. QOP 002, Section Two, explains how to carry out the changes. For training purposes, use the identical QOP from the Training Metrics, found in Book Section One.

8.0 Steps to follow when making changes to the quality manual or procedures

Note: Consider carefully the revision dates you are posting on documents and the effective implementation of changes into the work processes. They should coincide with each other. Conduct the necessary training ahead of time and not after the changes have been implemented. This would avoid confusion within your organization.

Phase One
1. determine what's needed to be changed. Review the change impact as it affects the manual, the process instructions, and the forms. (This is the flow-down impact analysis);
2. then, make copies of the impacted pages from the master documents and enter the changes on these pages. Flow down the change(s) from the manual to the process instructions and the forms, as applicable;
3. have the change(s) reviewed by someone else knowledgeable of the documented system;
4. review these corrections, if any, and revise the already made changes accordingly;
5. dry-run the intended change(s) through the operating system and adjust the text to suit the needs of the process. (This is the reality check. Don't skip it!);

Phase Two
6. now take the <u>masters</u> of the manual, the quality operating procedures, and the forms;

7. first, delete the original text in the manual and write in the new text from the draft. Second, do the same thing in the impacted quality operating procedure(s) and the form(s). Then, highlight the changes just made. If the change is a new requirement, add a new sub-clause number in the manual, in sequential order, under the main clause and write in the new text there. Follow the same method in the quality operating procedure. Then, define the process instruction in the QOP, as required. But, if there is no QOP referenced in the manual, there are two things you can do. Either create a new QOP or include the process instruction requirement in the manual, right under the sub-clause (see clause 4.18 on pg 289). Now, revise or create the appropriate form(s) to cover any recording requirements resulted from the change. In case of problems regarding form construction, turn to the Training Metrics, QOP 002, Section One, pg 69. In case of sequencing problems, please review the QSM and the QOPs to observe how AS9000 was added.

Phase Three

8. update the Revision Sheet(s) in the quality manual and the impacted quality operating procedure(s) to indicate the nature of the change(s);

9. then, update the header (page plate) on every page in the manual to indicate the new revision date and revision number. If page numbers increase, adjust the sequence accordingly;

10. then, update the header (page plate) on every page of the affected quality operating procedure to indicate the new revision date and revision number. If page numbers increase, adjust the sequence accordingly;

11. after that, update the header of the procedures' Master List and its revision sheet. Then, update the header of the forms' Master List and its revision sheet to indicate the new revision number and revision date. (These headers have different revision levels than the rest of the procedures for they control the quality operating procedures and the forms separately as a group.) Now, update in the body of the 'Revision History' sheet for both, by indicating the nature of change;

12. now, update in the body of the Master List for the Procedures Manual, under the applicable QOP number, to indicate the new revision number, revision date, and status. Do the same in the Master List of the forms also;

13. then, go to the applicable form(s) and update the new revision status of the individual form(s). Each form has the revision indication subscripted on the right bottom of the page(s);

14. then update on each form, as required, the where-to-find information on the application of the form(s). This is located on the left bottom of each form.

Note: I know that there is a lot to do here. But don't worry, I have given examples for these items as I debugged the system. Go to the "Revision History" of the QSM on page 271 and to the "Procedures Manual" on page 294 and see there how I implemented a number of changes.

Phase Four

15. conduct the necessary training for the employees to the extent required by the nature of the change(s) prior to the effective implementation of the change(s) into the work processes.

Phase Five

16. make enough copies of the revised quality manual, the quality operating procedures, and the forms to cover the replacement for all accounted documents issued earlier. The distribution of the quality manual and the quality operating procedures manual should be controlled by the application of Form A-010. The distribution of forms should be controlled as we issue them from centrally located file(s);

17. collect the obsolete quality manuals, the obsolete operating procedures, and the obsolete forms from all areas and replace them with the revised documents;

18. maintain record of distribution on Form A-010;

19. discard the obsolete documents, or put them in a secured place to prevent inter-mixing;

20. retain the draft pages (the copies you made from the masters on which you entered the changes) under separate file as long as they may be needed for background information;

21. replace in the field (including the registrar) all issued and <u>controlled</u> manuals. Request return of the obsolete manuals at the same time. Discard the obsolete manuals. <u>Uncontrolled</u> manuals, issued to certain customers for information only, don't require replacement.

22. keep all masters (hard and/or soft copies) in a secured place.

9.0   Changing the Training Metrics to become the Quality Operating Procedure for your quality system (AS9000 is excluded from the Flowcharts)

Background

Flowcharting to better understand quality related work processes and aiding process owners to ease the recognition of compliance requirements has become a universal tool in all types of businesses. Using flowcharts entirely as process instructions for the complete quality system procedures are not yet as common as they should be. I can only guess that work habits have a lot to do with it. Granted, the verbosity of the text-type procedures should never be duplicated in flowcharting, for making a hundred blocks per page in all directions would make it useless for process owners. The human eye is not created to scan a labyrinth and understand the details of it at once. Process owners don't have the time to do that. They have to mind production. Taking this into consideration, I opted, as much as possible, to the straight-line, vertical flow method in flowcharting. I mapped the whole process-approach system in this fashion to make it a practical tool for easy scanning by the workers. I have removed as many decision diamonds as I possibly could for the same reason, even at the expense of breaking some flowcharting rules.

9.1   Remove what's not needed from the flowcharts

Because I designed the flowcharts more for training purposes than for process instruction procedures, a few things may have to be removed from them to make them user-friendly. Not that they cannot be used as they are. The areas that deal with procedure construction and system installation and maintenance advice don't really belong to mapping work processes for product quality enforcement. You can easily see these as you compare the text-type procedures to the flowcharts, section by section. The question the user will have to answer is what's useful and practical. Do you need 'Basic Information'? Do you need the information on procedure construction methods? Or the advice given in the rhomboids, the scrolls, and the asteroids? If not, put the cursor on the edge line of the figure you want to remove and click the left button on the mouse once, then press the delete key, and it's gone.

The flowcharts presented in this book have been fully integrated with the quality manual, the text-type procedures and the forms. Each page has a header with the same controlling information as the text-type procedures. So, by removing the unneeded figures that don't relate to the primary processes (shaded) and to the connected instruction blocks, the compliance requirements will not be affected. The flowcharts follow the layout pattern of the text-type procedures. Section by section, title by title, requirement by requirement, they are parallel to the layout pattern with the text-type procedures. What is missing, however, from the flowchart procedures is the revision indication. This can be easily recognized when you look in the header on each page. The boxes for issue date, revision date and level are empty. If you want to use these flowcharts as a controlled process instruction procedure, you will need to add these to each page, as required. The revision history page to each QOP is already added, including the same Master List from the text-type procedures' manual. The revisions in both types of procedures are the same, even though the Training Metrics don't show them.

10.0   Miscellaneous

10.1   Keep a progressive outlook for your quality system

When you do conversion work to this quality manual, consider very carefully what you're removing. As you go through it and find that a requirement doesn't need to be in there because you aren't doing it, think again. Are you doing part of it? Will you be doing it tomorrow? As long as a requirement is in line with your business pursuit, but you don't at the moment have a need to enforce it, don't remove it. There are no rules on the books that say you can't be forward looking in your planning for the future. If you are in the metal fabrication business, the same advice applies to AS9000 also. Don't hesitate to keep the basic requirements in the manual because you are not doing it today. Prevent yourself from continually changing the manual to incorporate later the same provisions that you have just removed. Keep ahead of the continuous improvement aspirations made unmistakably clear in ISO 9001/2000. Any metal cutting shop can take the quality manual, the quality operating procedures, and the forms from this book and implement them as they are. Don't forget to put your name on them. The companies using this process-approach quality management system and registered to ISO 9002/1994 are Aircraft Parts manufacturers. They need one of the most complete product quality tools in the manufacturing business. With that in mind, you can rest assured that I have provided you with quality procedures covering the top of the industry from which you may want to come lower as your manufacturing operations determine it.

10.2   Make sure that your Management Representative is not just one of the guys

The role of the Management Representative is tantamount to the success or the failure in establishing, implementing and maintaining a quality management system. I tell you this because I am one. If this person lacks experience in practicing quality and the operational rigors of your business, effective and efficient results of all three prerequisites (how to establish, implement, and maintain) will only be marginal. I am putting myself on the line on this one, for there are a number of books on the market describing that this job can be filled by any manager in your business. Not so. Besides having both the quality experience and know-how of your type of operation, he must also possess overall understanding of the interactions of the quality objective requirements. He must have guts to deal with recalcitrant individuals, or else he/she will be the treadmill for exercise. Book knowledge alone will not do it.

10.3 Keep Management Reviews to address significant problems

Management Review is another area where operational details of the quality management system have been orchestrated to address problems that should have been solved as part of everyday responsibility. In this "process-approach" quality management system, corrective action and follow-up must be taken care of at once to bring about continuous improvement in the operating system. Effective and efficient production in line with product quality enforcement must be a proactive, consistent effort. Corrective action and follow-up of those problems that hinder this effort must be resolved promptly as they happen and where they happen within the departments that are responsible for maintaining the effectiveness of operations collectively. In the life of a small business, reviewing problems to improve the system cannot wait until top management sits down to hear them, for it would be highly damaging to the ongoing operations to delay satisfying the immediate needs of customers, both internal and external. Due to the sensitive nature of product turnaround requirements in small businesses, I have structured the quality system's procedures so as to build key requirements into the operating procedures that should ensure an unhindered production environment. Along with this reality in small business, I have also designed the Management Review criteria so it can be both a walk-me-thru process for improving system deficiencies on the spot and a periodic review process for addressing major issues. Through the provisions of Quality Operating Procedure 009 (Controlling Non-conforming Products), all system deficiencies can be cured even without waiting for Management Reviews. All it takes is to identify the deficiency on Form A-006, submit it to MRB (Material Review Board) to determine the cause and corrective action and touch base with top management to determine resources and implementation. Then, implement and follow up the corrective action requirement and close out the NCMR (Non-conforming Material Report). In order to have a record of the *Management Review*, the completion of Form A-034 (Management Review Status Record, pg 493) would be required. This type of agenda doesn't include a free lunch, but it surely works to thwart the festering problems lingering around, awaiting Management Review.

10.4 Book Summary

This book would not be complete without giving you additional information as to how Total Quality Management (TQM) and the all important Plan-Do-Check-Act cycle (developed by W. A. Shewhart) were incorporated as the principal drivers of this process-approach quality management system. For that information, please turn to Appendix Four, book page 503.

Companies registered under ISO 9002/1994 and seeking to convert their quality systems to meet the requirements of ISO 9001/2000 should turn to Appendix Five, book page 509. I have not only identified the changes, but also incorporated them into this process-approach quality management system. References and examples are provided for the reader as to where the changes belong, whether it's the quality manual or the operating procedures.

End of Guidance

BOOK SECTION ONE

DIAGRAMMATIC PROCEDURES

OVERVIEW TO THE DIAGRAMMATIC PROCEDURES

The training metrics are teaching tools, using flowcharting methods for many different learning objectives in the implementation and operation of the process-approach quality management system presented in this book. They are used in conjunction with the text quality operating procedures to enable quick comprehension as to how the specific processes flow from the beginning to the end in the realization of work assignments. To ensure objective guidance: instructions, basic subject information, advisement and individualized notes have been given where they were needed in line with the primary flow processes. In short, the training metrics are tutorial products to facilitate the implementation of core departmental assignments, based on process-approach quality management in manufacturing. But they are more. They can be easily turned into quality operating procedures using the flowcharting methods. They have been fully integrated with the text-type procedures and forms, and the quality system manual. For a quick review as to how the many processes are linked to the controlling procedures, please see QOP 012, Sections One (pg 250) and Two (pg 254). Note that page numbers have been added next to the references in order to facilitate easy navigation between documents.

TRAINING METRICS

CONTENTS   Page

MINFOR INCORPORATED	Iss dt. 1/1/99	Rev dt. 7/16/99	Pg.
QUALITY OPERATING PROCEDURE	Sign	Rev # 1	1
REQUIREMENT: QUALITY SYSTEM MANUAL SEC: 4.01 – 4.20			
SUBJECT: MASTER LIST			
QOP ALL DOCUMENT REQ'D: QOP 001 TROUGH QOP 012			

QOP	TITLE	REV	DATE	STATUS
001	Contract Review	0	1/1/99	New
	Sec 1 Control of RFQ			
	Sec 2 Purchase Order Review			
	Sec 3 Verbal Purchase Orders			
	Sec 4 Amendment to Contract			
002	Document and Data Control	0	1/1/99	New
	Sec 1 Approval and Issue			
	Sec 2 Document and Data Changes			
	Sec 3 Handling and Retention			
003	Control of Purchases (Internal)	0	1/1/99	New
	Sec 1 Approval and Issue of Purchase Orders			
	Sec 2 Evaluation of Subcontractors			
004	Control of Customer Supplied Product	0	1/1/99	New
005	Product Identification and Traceability	0	1/1/99	New
006	Inspection and Test Control	0	1/1/99	New
	Sec 1 First Piece Inspection			
	Sec 2 First Article Inspection			
	Sec 3 In-Process Inspection			
	Sec 4 Final Inspection			
	Sec 5 Receiving Inspection			
	Sec 6 Customer Source Inspection			
	Sec 7 Source Inspection at Subcontractors			
007	Control of Inspection, Measuring and Test Equipment	0	1/1/99	New
008	Inspection and Test Status	0	1/1/99	New
009	Control of Non-conforming Products	0	1/1/99	New
	Sec 1 Reporting Process			
	Sec 2 Disposition and Follow-up Process			
	Sec 3 Customer Complaints and Product Return			
010	Internal Quality Audits	0	1/1/99	New
011	Management Review	0	1/1/99	New
012	Performance Standard, Processing Control	0	7/16/99	New
	Sec 1 Processing Control (Put-Through)			
	Sec 2 Handling Non-conformance (Put-Through)			
	Sec 3 Developmental Process Control			

MINFOR INCORPORATED		Iss dt. 1/1/99	Rev dt. 7/16/99	Pg.
QUALITY OPERATING PROCEDURE	Sign		Rev # 1	2
REQUIREMENT: QUALITY SYSTEM MANUAL SEC: 4.1 – 4.20				
SUBJECT: MASTER LIST				
QOP All DOCUMENT REQ'D: QOP 001 through QOP 012				

REVISION HISTORY

Rev Date	Rev No	Description	Approval
7/16/99	1	Added QOP 012	AG

OVERVIEW TO QOP 001
(CONTRACT REVIEW)

Fulfilling customer satisfaction starts at the earliest phase when we establish a relationship with our customers. Most of us don't realize this fact until later when we find out that the customer's product specification many times is riddled with mistakes. The customer doesn't pay for downtime that you incur as a result of these mistakes. A process-approach quality management system starts right here, ensuring that you work with the customer up front, at the quoting phase of the job, so that you don't quote a job with all its mistakes already in it. Section One of this QOP is dealing with the quotation process, the RFQ.

Section Two of this QOP is dealing with controlling the Purchase Order Review, giving you a second opportunity to make sure that you take care of those mistakes that either you or the customer may have made between the time the RFQ was reviewed and the Purchase Order issued. Although many times Quality is not part of the job quoting process, this time it takes a serious review of the issued Purchase Order. It is at this time Quality determines what is really needed to make sure that product quality verification is planned for during the entire built-cycle of the product and that everybody becomes aware of it up front so that they can also plan for it. It is at this time Quality summarizes these requirements and issues them on Form A-001. This is the time when preventive quality starts rolling into the processing documents issued by Engineering.

Section Three controls how verbal agreements should be handled in order that we avoid confrontation with the customer later over what was said versus what should have been said. Again, preventive quality plays a major roll here to avoid problems later.

Section Four takes care of those issues that have everything to do with customer initiated document changes. It tells you how to control and implement them in order that customer satisfaction is maintained for both the organization and the buyer.

DIAGRAMMATIC PROCEDURE – QOP 001

CONTRACT REVIEW

CONTENTS   **Page**

DIAGRAMMATIC PROCEDURE – QOP 001

CONTRACT REVIEW

CONTENTS (continued)   **Page**

MINFOR INCORPORATED	Iss dt.	Rev dt.		Pg.
QUALITY OPERATING PROCEDURE	Sign	Rev #		3/21
REQUIREMENT: QUALITY SYSTEM MANUAL SEC: 4.3				
SUBJECT: Contract Review				
QOP 001 DOCUMENT REQ'D: This page				

REVISION HISTORY

Rev Date	Rev No	Description	Approval

TRAINING METRICS	Iss dt	Rev dt	Pg.
QUALITY OPERATING PROCEDURE	Sign	Rev #	4/21
REQUIREMENT: Evaluate the RFQ – Cost and Capability, QSM 4.3.2.			
SUBJECT: RFQ Review (The Quoting Process)			
QOP 001 |DOCUMENT REQ'D: RFQ, technical data package, forms A-004, A-005			

SECTION ONE

1.0 PURPOSE

To evaluate 'Request For Quote' (RFQ) in order to determine cost and capability.

2.0 APPLICATION

This procedure shall apply to accepted RFQs. Turned down RFQs shall not be controlled.

3.0 PROCEDURE

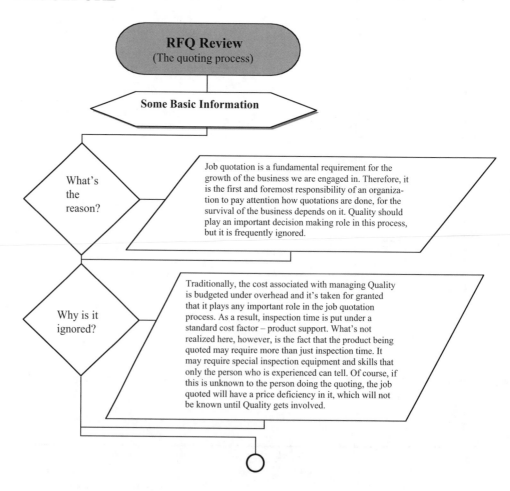

RFQ Review
(The quoting process)

Some Basic Information

What's the reason?

Job quotation is a fundamental requirement for the growth of the business we are engaged in. Therefore, it is the first and foremost responsibility of an organization to pay attention how quotations are done, for the survival of the business depends on it. Quality should play an important decision making role in this process, but it is frequently ignored.

Why is it ignored?

Traditionally, the cost associated with managing Quality is budgeted under overhead and it's taken for granted that it plays any important role in the job quotation process. As a result, inspection time is put under a standard cost factor – product support. What's not realized here, however, is the fact that the product being quoted may require more than just inspection time. It may require special inspection equipment and skills that only the person who is experienced can tell. Of course, if this is unknown to the person doing the quoting, the job quoted will have a price deficiency in it, which will not be known until Quality gets involved.

TRAINING METRICS	Iss dt	Rev dt	Pg.
QUALITY OPERATING PROCEDURE	Sign	Rev #	5/21
REQUIREMENT: Evaluate the RFQ – Cost and Capability, QSM 4.3.2.			
SUBJECT: RFQ Review (The Quoting Process)			
QOP 001 DOCUMENT REQ'D: RFQ, technical data package, forms A-004, A-005			

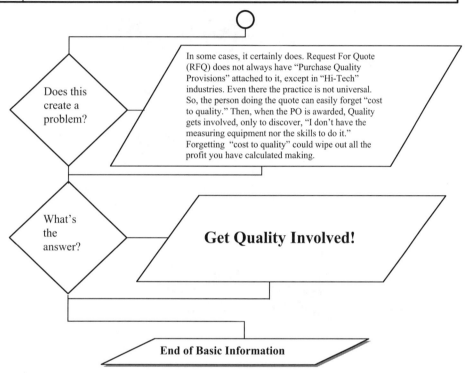

Does this create a problem?

In some cases, it certainly does. Request For Quote (RFQ) does not always have "Purchase Quality Provisions" attached to it, except in "Hi-Tech" industries. Even there the practice is not universal. So, the person doing the quote can easily forget "cost to quality." Then, when the PO is awarded, Quality gets involved, only to discover, "I don't have the measuring equipment nor the skills to do it." Forgetting "cost to quality" could wipe out all the profit you have calculated making.

What's the answer?

Get Quality Involved!

End of Basic Information

TRAINING METRICS	Iss dt	Rev dt	Pg.
QUALITY OPERATING PROCEDURE	Sign	Rev #	6/21
REQUIREMENT: Evaluate the RFQ – Cost and Capability, QSM 4.3.2.			
SUBJECT: RFQ Review (The Quoting Process)			
QOP 001 DOCUMENT REQ'D: RFQ, technical data package, forms A-004, A-005			

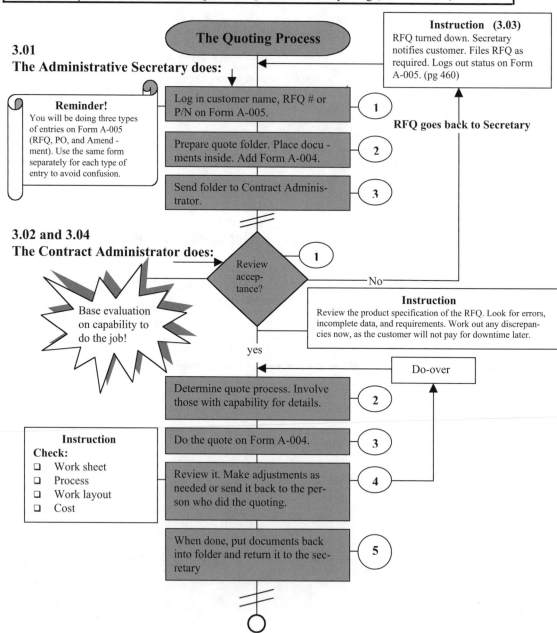

The Quoting Process

Instruction (3.03)
RFQ turned down. Secretary notifies customer. Files RFQ as required. Logs out status on Form A-005. (pg 460)

3.01
The Administrative Secretary does:

Reminder!
You will be doing three types of entries on Form A-005 (RFQ, PO, and Amend-ment). Use the same form separately for each type of entry to avoid confusion.

Log in customer name, RFQ # or P/N on Form A-005.   ①

Prepare quote folder. Place docu-ments inside. Add Form A-004.   ②

Send folder to Contract Adminis-trator.   ③

RFQ goes back to Secretary

3.02 and 3.04
The Contract Administrator does:

Review accep-tance?   ①

Base evaluation on capability to do the job!

No

Instruction
Review the product specification of the RFQ. Look for errors, incomplete data, and requirements. Work out any discrepan-cies now, as the customer will not pay for downtime later.

yes

Do-over

Determine quote process. Involve those with capability for details.   ②

Instruction
Check:
❑   Work sheet
❑   Process
❑   Work layout
❑   Cost

Do the quote on Form A-004.   ③

Review it. Make adjustments as needed or send it back to the per-son who did the quoting.   ④

When done, put documents back into folder and return it to the sec-retary   ⑤

TRAINING METRICS	Iss dt	Rev dt	Pg.
QUALITY OPERATING PROCEDURE	Sign	Rev #	7/21
REQUIREMENT: Evaluate the RFQ – Cost and Capability, QSM 4.3.2.			
SUBJECT: RFQ Review (The Quoting Process)			
QOP 001 DOCUMENT REQ'D: RFQ, technical data package, forms A-004, A-005			

The Administrative Secretary does:
3.05

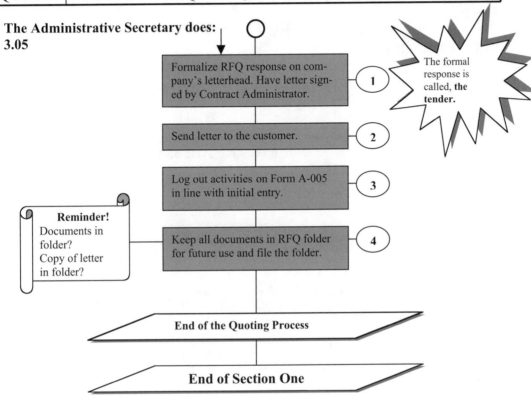

Formalize RFQ response on company's letterhead. Have letter signed by Contract Administrator.

1

The formal response is called, **the tender.**

Send letter to the customer.

2

Log out activities on Form A-005 in line with initial entry.

3

Reminder!
Documents in folder?
Copy of letter in folder?

Keep all documents in RFQ folder for future use and file the folder.

4

End of the Quoting Process

End of Section One

TRAINING METRICS	Iss dt	Rev dt	Pg.
QUALITY OPERATING PROCEDURE	Sign	Rev #	8/21
REQUIREMENT: Evaluation of Purchase Order Prior to Acceptance, QSM 4.3.2.			
SUBJECT: Purchase Order Review and Processing Preparation			
QOP 001 │ DOCUMENT REQ'D: PO, Tender, Forms A-001, A-003, A-005, A-011, A-031			

SECTION TWO

1.0　PURPOSE

To ensure that the Purchase Order content information agrees with that of the tender and to determine and document quality requirements imposed in the PO.

2.0　APPLICATION

The purchase order requirements shall apply to all activities affecting documents and processes during the entire processing cycle of the identified product.

3.0　PROCEDURE

3.01
The Administrative Secretary does:

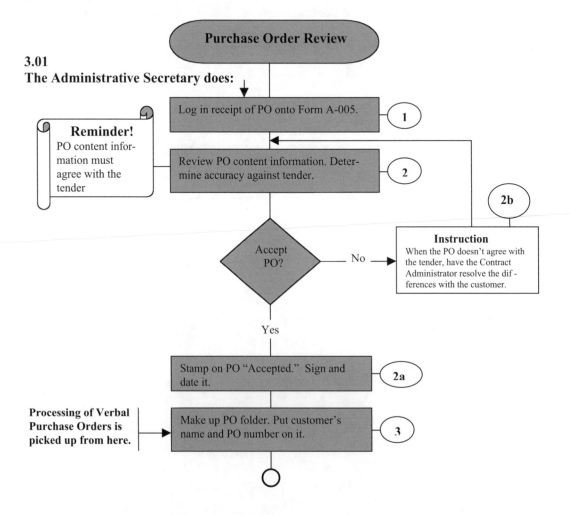

TRAINING METRICS	Iss dt	Rev dt	Pg.
QUALITY OPERATING PROCEDURE	Sign	Rev #	9/21
REQUIREMENT: Evaluation of Purchase Order Prior to Acceptance, QSM 4.3.2.			
SUBJECT: Purchase Order Review and Processing Preparation			
QOP 001 |DOCUMENT REQ'D: PO, Tender, Forms A-001, A-003, A-005, A-011, A-031			

Fill out Form A-011. Record on it the Customer sent document numbers and their revision levels. Attach this form to the inside of the front cover of the PO folder.   **4**

Make two copies of the PO and its attachment(s) now. Send one copy to Engineering and the other copy to Quality.   **5a and 5b**

Instruction
From now on keep updating Forms A-011 and A-031 when customer makes changes to the listed docu-ments. Then, follow up with the distribution and retrieval process as before.

Fill out Form A-031. Record on this form the part number and revision of the documents that have been distributed (new) and retrieved (obsolete).   **6**

Place the "Accepted" PO, attachments, and Form-031 in the folder and file it. Put the tender back in the RFQ file also.   **7**

Instruction
Repeat all the above steps for each customer PO. Multiple POs from the same customer require the same individual log-in, handling, and log-out.

Log out the PO activity on Form A-005 by recording the "Acceptance" status.   **8**

3.02
The Quality Engineer does:

Review the PO, sent by Contracts, to understand the quality requirements.   **1**

Instruction
Add product identification marking, workmanship, and shipping requirements onto A-001, as required.

Transcribe the requirements onto Form A-001. Add others as needed. If no requirements given in PO, impose those that are needed to control product qual-ity throughout the processing cycle.   **2**

Make two copies of A-001. Send one to Engineering, the other to Contracts (file copy). Don't reissue A-001 on repeated POs, unless amendments impact product and/or quality.   **3**

TRAINING METRICS	Iss dt	Rev dt	Pg.
QUALITY OPERATING PROCEDURE	Sign	Rev #	10/21
REQUIREMENT: Evaluation of Purchase Order Prior to Acceptance, QSM 4.3.2.			
SUBJECT: Purchase Order Review and Processing Preparation			
QOP 001 DOCUMENT REQ'D: PO, Tender, Forms A-001, A-003, A-005, A-011, A-031			

4 — Make up Inspection Product Folder. Place all documents inside. Maintain this folder on file throughout the entire processing cycle. Retain it as long as required by internal needs and the customer's PO requirement.

5 — Plan and carry out your work assignments in order to meet all the controlling requirements for an effective product quality realization for the contracted product.

3.03
The Process Engineer does:

1 — Review the PO and related technical documents sent by Contracts and also review Form A-001 which was sent by Quality.

Special Instruction
If this PO is a new job and you don't have the process capability developed for it, determine the need for process qualification and do it in line as defined under QOP 012, Sec Three. (pg 263)

Instruction
As part of the processing layout, you must include in the final procedures the requirements for workmanship, product handling, cleaning and preservation, safety notes, and storage, as applicable.

2 — Plan the processing layout per PO requirements. Include quality requirements per Form A-001.

3 — Prepare and approve all required documents for product realization. (Job Traveler, Operation Sheet, Software Program, etc.)

Instruction
The preparation of Job Travelers require extensive examination of the product's entire processing requirements. Define all the necessary preventive steps that relate to workmanship, product handling, cleaning and preservation.

4 — Maintain master documents in Engineering Folder(s).

Instruction
Keep track of PO amendments. Don't release any Production Folder if it requires amendment update. Refer to Section Four of this QOP for details as to how to do it. (See pg 52)

5 — Prepare Production Folder(s) as required. Ensure current revision levels in all the included documents.

Instruction
Cost control plays a critical roll as to how the production layout for the contracted product is planned. Ensure the control of cost in your throughput planning in line with established pricing – the RFQ.

6 — Follow up your planning regarding the readiness of special tooling, gages, and setup related equipment (fixtures). Do this prior to releasing a Production Folder.

7 — Release Production Folder(s) to Production Planning as required in line with production scheduling requirement.

TRAINING METRICS	Iss dt	Rev dt	Pg.
QUALITY OPERATING PROCEDURE	Sign	Rev #	11/21
REQUIREMENT: Evaluation of Purchase Order Prior to Acceptance, QSM 4.3.2.			
SUBJECT: Purchase Order Review and Processing Preparation			
QOP 001 DOCUMENT REQ'D: PO, Tender, Forms A-001, A-003, A-005, A-011, A-031			

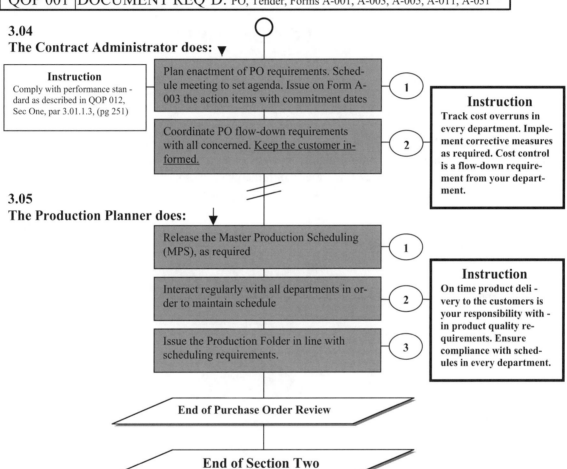

3.04
The Contract Administrator does: ▼

Instruction
Comply with performance stan - dard as described in QOP 012, Sec One, par 3.01.1.3, (pg 251)

1. Plan enactment of PO requirements. Schedule meeting to set agenda. Issue on Form A-003 the action items with commitment dates

2. Coordinate PO flow-down requirements with all concerned. Keep the customer informed.

Instruction
Track cost overruns in every department. Implement corrective measures as required. Cost control is a flow-down requirement from your department.

3.05
The Production Planner does: ↓

1. Release the Master Production Scheduling (MPS), as required

2. Interact regularly with all departments in order to maintain schedule

3. Issue the Production Folder in line with scheduling requirements.

Instruction
On time product deli - very to the customers is your responsibility with - in product quality requirements. Ensure compliance with schedules in every department.

End of Purchase Order Review

End of Section Two

TRAINING METRICS	Iss dt	Rev dt	Pg.
QUALITY OPERATING PROCEDURE	Sign	Rev #	12/21
REQUIREMENT: Evaluation of Purchase Order Prior to Acceptance, QSM 4.3.2.			
SUBJECT: Purchase Order Review **(Verbal Purchase Orders)**			
QOP 001 DOCUMENT REQ'D: PO, Forms A-001, A-003, A-005, A-011, A-031			

SECTION THREE

1.0 PURPOSE

To initiate processing and quality planning, based on customer's verbal commitment, pending issuance of formal purchase order.

2.0 APPLICATION

When a verbal commitment from a customer has been formalized and he/she has issued a Purchase Order, the requirements thereof shall apply to all activities affecting processes and documents during the entire production cycle of the contracted product.

3.0 PROCEDURE

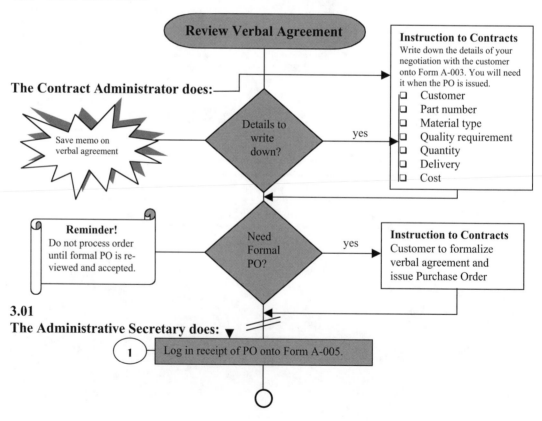

Review Verbal Agreement

The Contract Administrator does:

Save memo on verbal agreement

Details to write down?

yes

Instruction to Contracts
Write down the details of your negotiation with the customer onto Form A-003. You will need it when the PO is issued.
- Customer
- Part number
- Material type
- Quality requirement
- Quantity
- Delivery
- Cost

Reminder!
Do not process order until formal PO is reviewed and accepted.

Need Formal PO?

yes

Instruction to Contracts
Customer to formalize verbal agreement and issue Purchase Order

3.01
The Administrative Secretary does:

1   Log in receipt of PO onto Form A-005.

TRAINING METRICS	Iss dt	Rev dt	Pg.
QUALITY OPERATING PROCEDURE	Sign	Rev #	13/21
REQUIREMENT: Evaluation of Purchase Order Prior to Acceptance, QSM 4.3.2.			
SUBJECT: Purchase Order Review **(Verbal Purchase Orders)**			
QOP 001 DOCUMENT REQ'D: PO, Forms A-001, A-003, A-005, A-011, A-031			

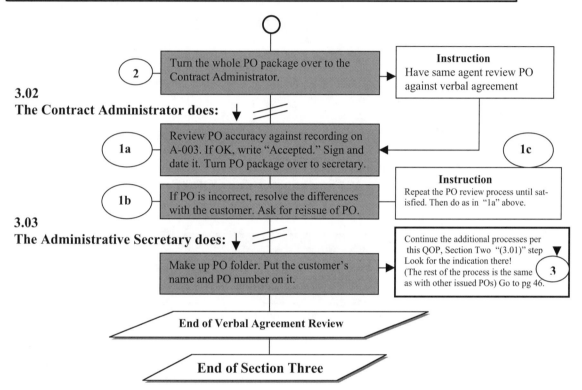

3.02
The Contract Administrator does:

2 — Turn the whole PO package over to the Contract Administrator.

Instruction
Have same agent review PO against verbal agreement

1a — Review PO accuracy against recording on A-003. If OK, write "Accepted." Sign and date it. Turn PO package over to secretary.

1c

1b — If PO is incorrect, resolve the differences with the customer. Ask for reissue of PO.

Instruction
Repeat the PO review process until satisfied. Then do as in "1a" above.

3.03
The Administrative Secretary does:

Make up PO folder. Put the customer's name and PO number on it.

Continue the additional processes per
 this QOP, Section Two "(3.01)" step
Look for the indication there!
(The rest of the process is the same
as with other issued POs) Go to pg 46.   **3**

End of Verbal Agreement Review

End of Section Three

TRAINING METRICS	Iss dt	Rev dt	Pg.
QUALITY OPERATING PROCEDURE	Sign	Rev #	14/21
REQUIREMENT: Amendment to Purchase Orders, QSM 4.3.3			
SUBJECT: Processing Amendments			
QOP 001 |DOCUMENT REQ'D: PO,RFQ, Forms A-001,A-003,A-004,A-005,A-006,A-009,A-011,A-031			

SECTION FOUR

1.0   PURPOSE

To establish documented procedures for implementing contract amendments and the resulting internal document changes.

2.0   APPLICATION

Contract amendments and the resulting changes shall apply to all the affected product(s) and document(s). Changes to internal documents shall be implemented by the use of form A-009, as required. (See QOP 002, Sec Two, par 3.03.1, pg 73)

3.0   PROCEDURE

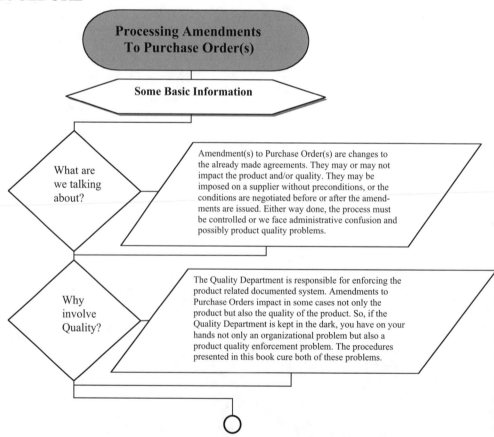

Processing Amendments
To Purchase Order(s)

Some Basic Information

What are we talking about?

Amendment(s) to Purchase Order(s) are changes to the already made agreements. They may or may not impact the product and/or quality. They may be imposed on a supplier without preconditions, or the conditions are negotiated before or after the amendments are issued. Either way done, the process must be controlled or we face administrative confusion and possibly product quality problems.

Why involve Quality?

The Quality Department is responsible for enforcing the product related documented system. Amendments to Purchase Orders impact in some cases not only the product but also the quality of the product. So, if the Quality Department is kept in the dark, you have on your hands not only an organizational problem but also a product quality enforcement problem. The procedures presented in this book cure both of these problems.

TRAINING METRICS	Iss dt	Rev dt	Pg.
QUALITY OPERATING PROCEDURE	Sign	Rev #	15/21
REQUIREMENT: Amendment to Purchase Orders, QSM 4.3.3			
SUBJECT: Processing Amendments			
QOP 001 DOCUMENT REQ'D: PO,RFQ, Forms A-001,A-003,A-004,A-005,A-006,A-009,A-011,A-031			

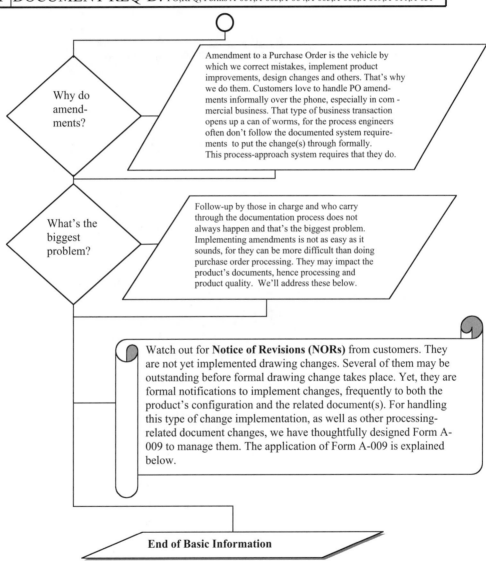

Why do amendments?

Amendment to a Purchase Order is the vehicle by which we correct mistakes, implement product improvements, design changes and others. That's why we do them. Customers love to handle PO amendments informally over the phone, especially in commercial business. That type of business transaction opens up a can of worms, for the process engineers often don't follow the documented system requirements to put the change(s) through formally. This process-approach system requires that they do.

What's the biggest problem?

Follow-up by those in charge and who carry through the documentation process does not always happen and that's the biggest problem. Implementing amendments is not as easy as it sounds, for they can be more difficult than doing purchase order processing. They may impact the product's documents, hence processing and product quality. We'll address these below.

Watch out for **Notice of Revisions (NORs)** from customers. They are not yet implemented drawing changes. Several of them may be outstanding before formal drawing change takes place. Yet, they are formal notifications to implement changes, frequently to both the product's configuration and the related document(s). For handling this type of change implementation, as well as other processing-related document changes, we have thoughtfully designed Form A-009 to manage them. The application of Form A-009 is explained below.

End of Basic Information

TRAINING METRICS	Iss dt	Rev dt	Pg.
QUALITY OPERATING PROCEDURE	Sign	Rev #	16/21
REQUIREMENT: Amendment to Purchase Orders, QSM 4.3.3			
SUBJECT: Processing Amendments			
QOP 001 DOCUMENT REQ'D: PO,RFQ, Forms A-001,A-003,A-004,A-005,A-006,A-009,A-011,A-031			

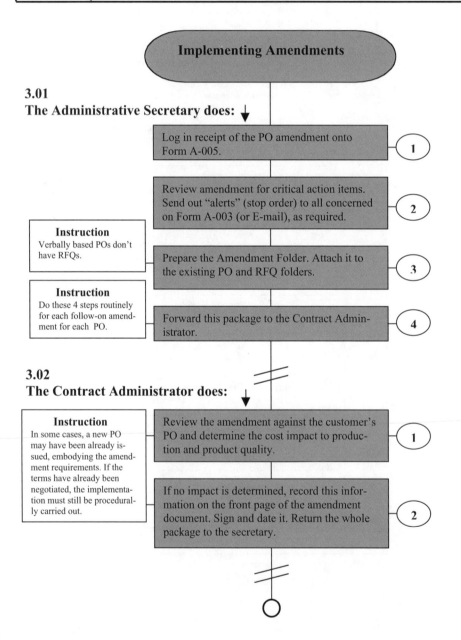

Implementing Amendments

3.01
The Administrative Secretary does:

> Log in receipt of the PO amendment onto Form A-005. — 1

> Review amendment for critical action items. Send out "alerts" (stop order) to all concerned on Form A-003 (or E-mail), as required. — 2

Instruction
Verbally based POs don't have RFQs.

> Prepare the Amendment Folder. Attach it to the existing PO and RFQ folders. — 3

Instruction
Do these 4 steps routinely for each follow-on amendment for each PO.

> Forward this package to the Contract Administrator. — 4

3.02
The Contract Administrator does:

Instruction
In some cases, a new PO may have been already issued, embodying the amendment requirements. If the terms have already been negotiated, the implementation must still be procedurally carried out.

> Review the amendment against the customer's PO and determine the cost impact to production and product quality. — 1

> If no impact is determined, record this information on the front page of the amendment document. Sign and date it. Return the whole package to the secretary. — 2

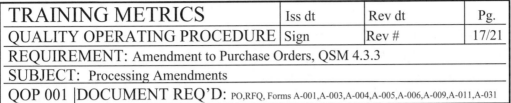

TRAINING METRICS	Iss dt	Rev dt	Pg.
QUALITY OPERATING PROCEDURE	Sign	Rev #	17/21
REQUIREMENT: Amendment to Purchase Orders, QSM 4.3.3			
SUBJECT: Processing Amendments			
QOP 001 |DOCUMENT REQ'D: PO,RFQ, Forms A-001,A-003,A-004,A-005,A-006,A-009,A-011,A-031			

3.03
The Administrative Secretary does:

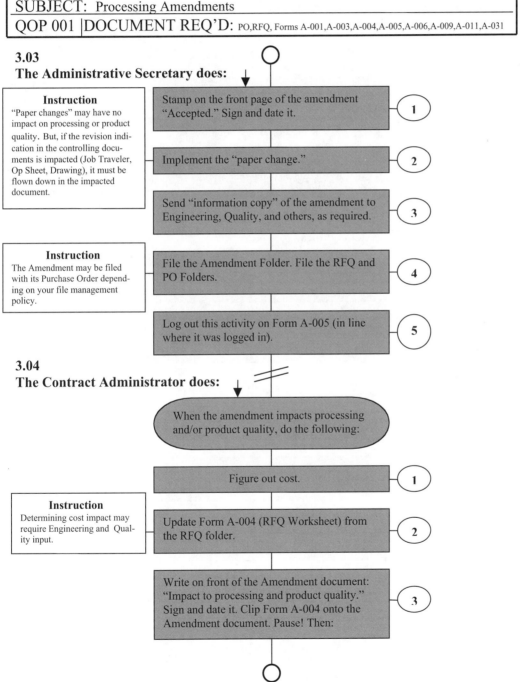

Instruction
"Paper changes" may have no impact on processing or product quality. But, if the revision indication in the controlling documents is impacted (Job Traveler, Op Sheet, Drawing), it must be flown down in the impacted document.

1. Stamp on the front page of the amendment "Accepted." Sign and date it.

2. Implement the "paper change."

3. Send "information copy" of the amendment to Engineering, Quality, and others, as required.

Instruction
The Amendment may be filed with its Purchase Order depending on your file management policy.

4. File the Amendment Folder. File the RFQ and PO Folders.

5. Log out this activity on Form A-005 (in line where it was logged in).

3.04
The Contract Administrator does:

When the amendment impacts processing and/or product quality, do the following:

1. Figure out cost.

Instruction
Determining cost impact may require Engineering and Quality input.

2. Update Form A-004 (RFQ Worksheet) from the RFQ folder.

3. Write on front of the Amendment document: "Impact to processing and product quality." Sign and date it. Clip Form A-004 onto the Amendment document. Pause! Then:

TRAINING METRICS	Iss dt	Rev dt	Pg.
QUALITY OPERATING PROCEDURE	Sign	Rev #	18/21
REQUIREMENT: Amendment to Purchase Orders, QSM 4.3.3			
SUBJECT: Processing Amendments			
QOP 001 |DOCUMENT REQ'D: PO,RFQ, Forms A-001,A-003,A-004,A-005,A-006,A-009,A-011,A-031			

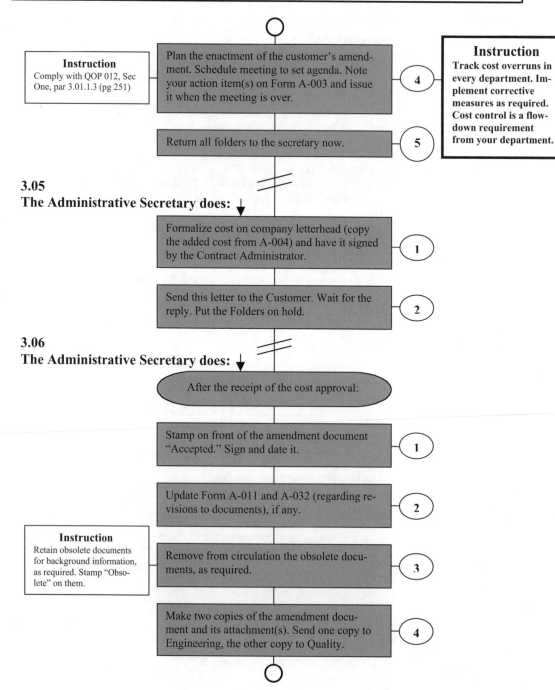

Instruction
Comply with QOP 012, Sec One, par 3.01.1.3 (pg 251)

Plan the enactment of the customer's amendment. Schedule meeting to set agenda. Note your action item(s) on Form A-003 and issue it when the meeting is over. ④

Instruction
Track cost overruns in every department. Implement corrective measures as required. Cost control is a flow-down requirement from your department.

Return all folders to the secretary now. ⑤

3.05
The Administrative Secretary does: ↓

Formalize cost on company letterhead (copy the added cost from A-004) and have it signed by the Contract Administrator. ①

Send this letter to the Customer. Wait for the reply. Put the Folders on hold. ②

3.06
The Administrative Secretary does: ↓

After the receipt of the cost approval:

Stamp on front of the amendment document "Accepted." Sign and date it. ①

Update Form A-011 and A-032 (regarding revisions to documents), if any. ②

Instruction
Retain obsolete documents for background information, as required. Stamp "Obsolete" on them.

Remove from circulation the obsolete documents, as required. ③

Make two copies of the amendment document and its attachment(s). Send one copy to Engineering, the other copy to Quality. ④

TRAINING METRICS	Iss dt	Rev dt	Pg.
QUALITY OPERATING PROCEDURE	Sign	Rev #	19/21
REQUIREMENT: Amendment to Purchase Orders, QSM 4.3.3			
SUBJECT: Processing Amendments			
QOP 001 |DOCUMENT REQ'D: PO,RFQ, Forms A-001,A-003,A-004,A-005,A-006,A-009,A-011,A-031			

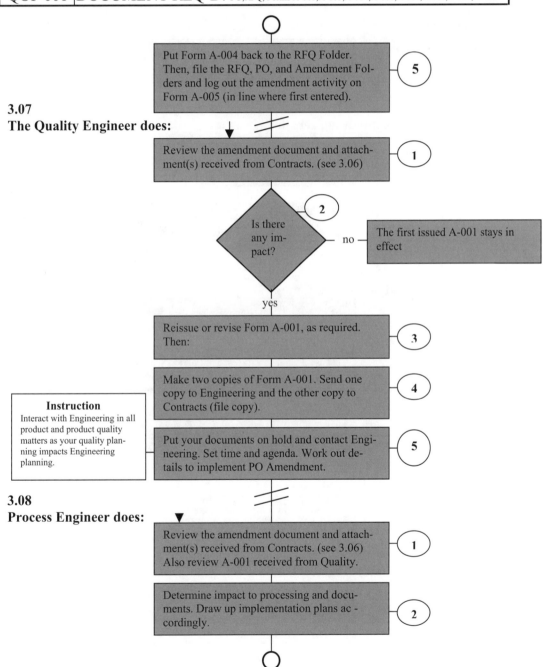

3.07
The Quality Engineer does:

3.08
Process Engineer does:

Instruction
Interact with Engineering in all product and product quality matters as your quality planning impacts Engineering planning.

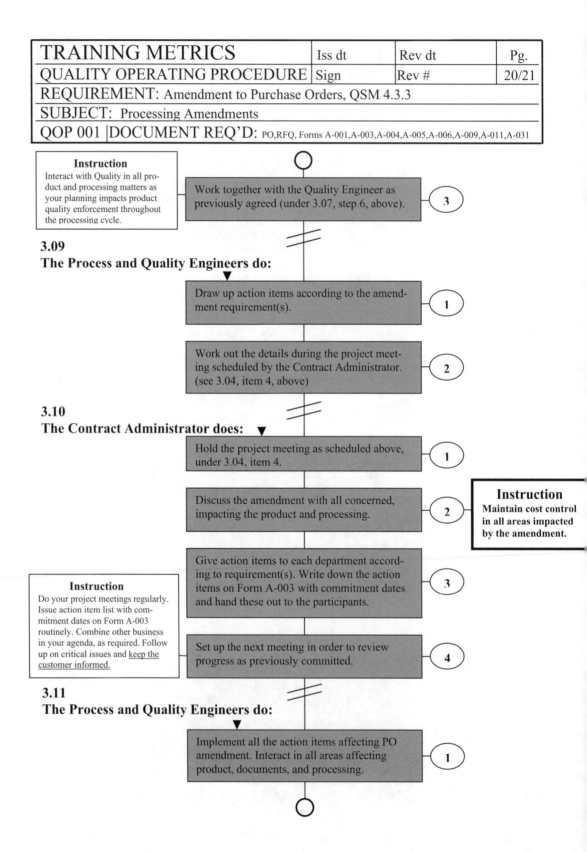

TRAINING METRICS	Iss dt	Rev dt	Pg.
QUALITY OPERATING PROCEDURE	Sign	Rev #	20/21
REQUIREMENT: Amendment to Purchase Orders, QSM 4.3.3			
SUBJECT: Processing Amendments			
QOP 001 DOCUMENT REQ'D: PO,RFQ, Forms A-001,A-003,A-004,A-005,A-006,A-009,A-011,A-031			

Instruction
Interact with Quality in all product and processing matters as your planning impacts product quality enforcement throughout the processing cycle.

Work together with the Quality Engineer as previously agreed (under 3.07, step 6, above). 3

3.09
The Process and Quality Engineers do:

Draw up action items according to the amendment requirement(s). 1

Work out the details during the project meeting scheduled by the Contract Administrator. (see 3.04, item 4, above) 2

3.10
The Contract Administrator does:

Hold the project meeting as scheduled above, under 3.04, item 4. 1

Discuss the amendment with all concerned, impacting the product and processing. 2

Instruction
Maintain cost control in all areas impacted by the amendment.

Give action items to each department according to requirement(s). Write down the action items on Form A-003 with commitment dates and hand these out to the participants. 3

Instruction
Do your project meetings regularly. Issue action item list with commitment dates on Form A-003 routinely. Combine other business in your agenda, as required. Follow up on critical issues and keep the customer informed.

Set up the next meeting in order to review progress as previously committed. 4

3.11
The Process and Quality Engineers do:

Implement all the action items affecting PO amendment. Interact in all areas affecting product, documents, and processing. 1

TRAINING METRICS	Iss dt	Rev dt	Pg.
QUALITY OPERATING PROCEDURE	Sign	Rev #	21/21
REQUIREMENT: Amendment to Purchase Orders, QSM 4.3.3			
SUBJECT: Processing Amendments			
QOP 001	DOCUMENT REQ'D: PO,RFQ, Forms A-001,A-003,A-004,A-005,A-006,A-009,A-011,A-031			

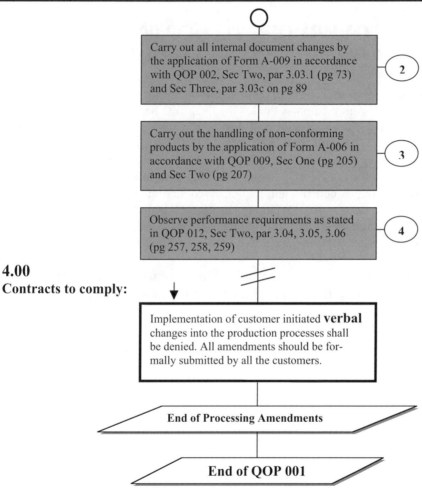

Carry out all internal document changes by the application of Form A-009 in accordance with QOP 002, Sec Two, par 3.03.1 (pg 73) and Sec Three, par 3.03c on pg 89 — ②

Carry out the handling of non-conforming products by the application of Form A-006 in accordance with QOP 009, Sec One (pg 205) and Sec Two (pg 207) — ③

Observe performance requirements as stated in QOP 012, Sec Two, par 3.04, 3.05, 3.06 (pg 257, 258, 259) — ④

**4.00
Contracts to comply:**

Implementation of customer initiated **verbal** changes into the production processes shall be denied. All amendments should be formally submitted by all the customers.

End of Processing Amendments

End of QOP 001

OVERVIEW TO QOP 002
(CONTROL OF DOCUMENT AND DATA)

The control of documents and data is prerequisite to be able to manage an effective quality system. Without it, you have no system. This QOP 002 has three sections in order to ensure that we manage the documents and data in line with a process-approach quality management system. We give you instructions on how to design and control the documents, comprising the quality system's procedures; give you instructions on how to carry out revisions, how to identify documents and how to retain them. In fact, this process-approach system is constructed, implemented, and maintained exactly the way it tells you how you should do yours. What is really good here is that whatever you need to know regarding document and data control is in one procedure. Other QOPs often refer you to come here and do your thing according to the stated requirements laid out for you.

There are so many steps involved here that I could easily write pages to tell you about them, but I won't do it here. Rather, I'll do it where it counts – inside the procedure. This QOP is one of the longest procedures, along with QOP 006 (inspection control, on pg 127), and QOP 009 (non-conformance control, on pg 199), to make sure that the documented system is effectively laid out, inter-linked, coordinated, and maintained for all the quality system's documents in the process-approach system.

DIAGRAMMATIC PROCEDURE   QOP 002

CONTROL OF DOCUMENT AND DATA

CONTENTS   **Page**

CONTROL OF DOCUMENT AND DATA

CONTENTS (continued)   **Page**

Section Three continued

TRAINING METRICS	Iss dt.	Rev dt.		Pg.
QUALITY OPERATING PROCEDURE	Sign	Rev #		3/33
REQUIREMENT: QUALITY SYSTEM MANUAL SEC: 4.5				
SUBJECT: CONTROL OF DOCUMENT AND DATA				
QOP 002 |DOCUMENT REQ'D: This QOP 002, Sec: 1, 2, and 3				

REVISION HISTORY

Rev Date	Rev No	Description	Approval

TRAINING METRICS	Iss dt.	Rev dt.	Pg.
QUALITY OPERATING PROCEDURE	Sign	Rev #	4/33
REQUIREMENT: Control of Documents and Data, QSM 4.5			
SUBJECT: Construction Methods, Approval and Issuance			
QOP 002 |DOCUMENT REQ'D: QOP 002 (text is in book Section Three) and this flowchart			

SECTION ONE

1.0   PURPOSE

To control the planning, approval and issuance of the quality system's documents.

2.0   APPLICATION

Planning, approval and issuance of the quality system's documents and data shall apply to all the identified documents within the Quality System Manual, the Quality Operating Procedures and the Forms.

3.0   PROCEDURE (Integrated Process Approach Quality Management)

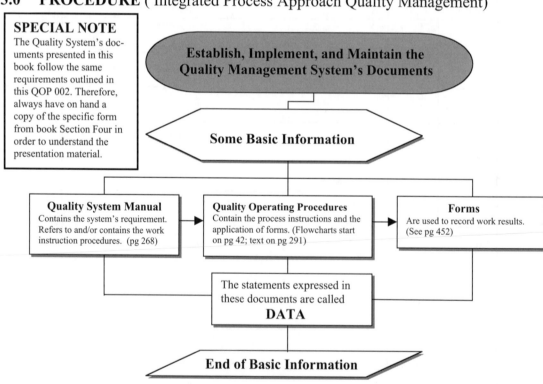

SPECIAL NOTE
The Quality System's documents presented in this book follow the same requirements outlined in this QOP 002. Therefore, always have on hand a copy of the specific form from book Section Four in order to understand the presentation material.

Establish, Implement, and Maintain the Quality Management System's Documents

Some Basic Information

Quality System Manual
Contains the system's requirement. Refers to and/or contains the work instruction procedures. (pg 268)

Quality Operating Procedures
Contain the process instructions and the application of forms. (Flowcharts start on pg 42; text on pg 291)

Forms
Are used to record work results. (See pg 452)

The statements expressed in these documents are called
DATA

End of Basic Information

TRAINING METRICS	Iss dt.	Rev dt.	Pg.
QUALITY OPERATING PROCEDURE	Sign	Rev #	5/33
REQUIREMENT: Control of Documents and Data, QSM 4.5			
SUBJECT: Construction Methods, Approval and Issuance			
QOP 002 |DOCUMENT REQ'D: QOP 002 (text is in book Section Three) and this flowchart			

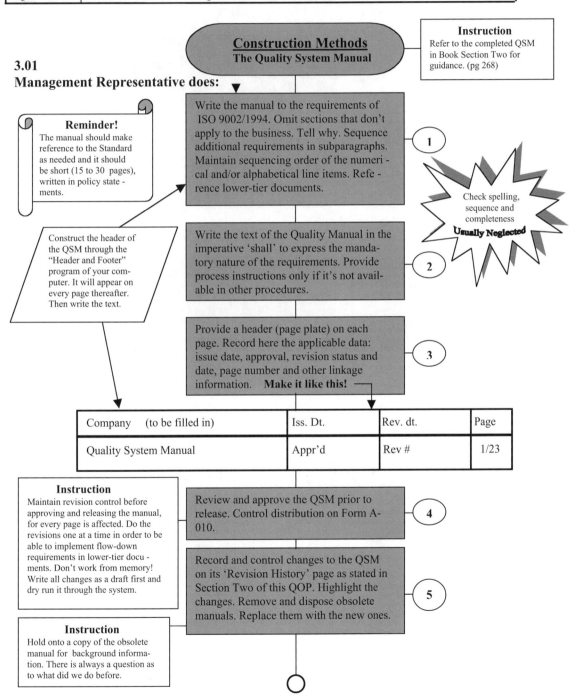

Construction Methods
The Quality System Manual

Instruction
Refer to the completed QSM in Book Section Two for guidance. (pg 268)

3.01
Management Representative does:

Reminder!
The manual should make reference to the Standard as needed and it should be short (15 to 30 pages), written in policy state-ments.

Write the manual to the requirements of ISO 9002/1994. Omit sections that don't apply to the business. Tell why. Sequence additional requirements in subparagraphs. Maintain sequencing order of the numeri-cal and/or alphabetical line items. Refe-rence lower-tier documents. **1**

Check spelling, sequence and completeness
Usually Neglected

Construct the header of the QSM through the "Header and Footer" program of your com-puter. It will appear on every page thereafter. Then write the text.

Write the text of the Quality Manual in the imperative 'shall' to express the manda-tory nature of the requirements. Provide process instructions only if it's not avail-able in other procedures. **2**

Provide a header (page plate) on each page. Record here the applicable data: issue date, approval, revision status and date, page number and other linkage information. **Make it like this!** **3**

Company (to be filled in)	Iss. Dt.	Rev. dt.	Page
Quality System Manual	Appr'd	Rev #	1/23

Instruction
Maintain revision control before approving and releasing the manual, for every page is affected. Do the revisions one at a time in order to be able to implement flow-down requirements in lower-tier docu-ments. Don't work from memory! Write all changes as a draft first and dry run it through the system.

Review and approve the QSM prior to release. Control distribution on Form A-010. **4**

Record and control changes to the QSM on its 'Revision History' page as stated in Section Two of this QOP. Highlight the changes. Remove and dispose obsolete manuals. Replace them with the new ones. **5**

Instruction
Hold onto a copy of the obsolete manual for background informa-tion. There is always a question as to what did we do before.

TRAINING METRICS	Iss dt.	Rev dt.	Pg.
QUALITY OPERATING PROCEDURE	Sign	Rev #	6/33
REQUIREMENT: Construction Methods, Control of Documents and Data, QSM 4.5			
SUBJECT: Approval and Issuance			
QOP 002 |DOCUMENT REQ'D: QOP 002 (text is in book Section Three) and this flowchart			

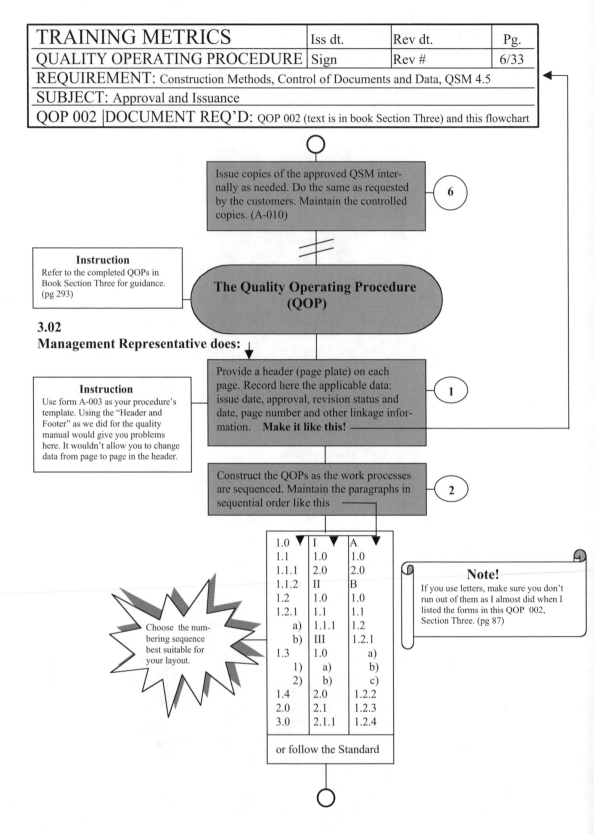

Issue copies of the approved QSM inter-nally as needed. Do the same as requested by the customers. Maintain the controlled copies. (A-010)   **6**

Instruction
Refer to the completed QOPs in Book Section Three for guidance. (pg 293)

The Quality Operating Procedure (QOP)

3.02
Management Representative does:

Provide a header (page plate) on each page. Record here the applicable data: issue date, approval, revision status and date, page number and other linkage infor-mation.   **Make it like this!**   **1**

Instruction
Use form A-003 as your procedure's template. Using the "Header and Footer" as we did for the quality manual would give you problems here. It wouldn't allow you to change data from page to page in the header.

Construct the QOPs as the work processes are sequenced. Maintain the paragraphs in sequential order like this   **2**

1.0	I	A
1.1	1.0	1.0
1.1.1	2.0	2.0
1.1.2	II	B
1.2	1.0	1.0
1.2.1	1.1	1.1
a)	1.1.1	1.2
b)	III	1.2.1
1.3	1.0	a)
1)	a)	b)
2)	b)	c)
1.4	2.0	1.2.2
2.0	2.1	1.2.3
3.0	2.1.1	1.2.4

or follow the Standard

Choose the num-bering sequence best suitable for your layout.

Note!
If you use letters, make sure you don't run out of them as I almost did when I listed the forms in this QOP 002, Section Three. (pg 87)

TRAINING METRICS	Iss dt.	Rev dt.	Pg.
QUALITY OPERATING PROCEDURE	Sign	Rev #	7/33
REQUIREMENT: Construction Methods, Control of Documents and Data, QSM 4.5			
SUBJECT: Approval and Issuance			
QOP 002 |DOCUMENT REQ'D: QOP 002 (text is in book Section Three) and this flowchart			

Instruction
Look up Quality Operating Procedure 006 in book Section Three and observe there the layout and how seven similar tasks (inspections) have been combined and controlled. (save time and paper)

Combine similar tasks under one procedure and give them a title and a separate section within the procedure. ③

Maintain linkage between the QSM and each QOP by referencing back and forth. ④

Don't reference rev. levels of documents when you write the manual and/or procedures!

Think of Updating!

Review and approve the procedures and associated form(s). Combine procedures into 'Procedures Manual,' then issue them. Control distribution on Form A-010. Procedures are proprietary documents, customers may not receive them. ⑤

Record and control change to a QOP on its individual 'Revision History' page. Update each page of the impacted QOP. Update with each change the 'Master List' that controls collectively all the procedures. Highlight the changes. ⑥

Instruction
Use a uniform system for highlighting changes. Italicize or make them **bold**.

Note
Revision control of the QOPs is defined in Section Two of this QOP. (pg 79)

Remove and dispose obsolete sections from the procedures and replace them with reviewed and approved sections. ⑦

Instruction
Hold onto a copy of the obsolete procedures for background information. There is always a question as to what did we do before.

Additional Basic Information
(on writing procedures)

Am I ready to write the process instructions? — No, first →

Instruction
Identify the work process. Understand what's being done. Write down the requirements.

TRAINING METRICS	Iss dt.	Rev dt.	Pg.
QUALITY OPERATING PROCEDURE	Sign	Rev #	8/33
REQUIREMENT: Construction Methods, Control of Documents and Data, QSM 4.5			
SUBJECT: Approval and Issuance			
QOP 002 │DOCUMENT REQ'D: QOP 002 (text is in book Section Three) and this flowchart			

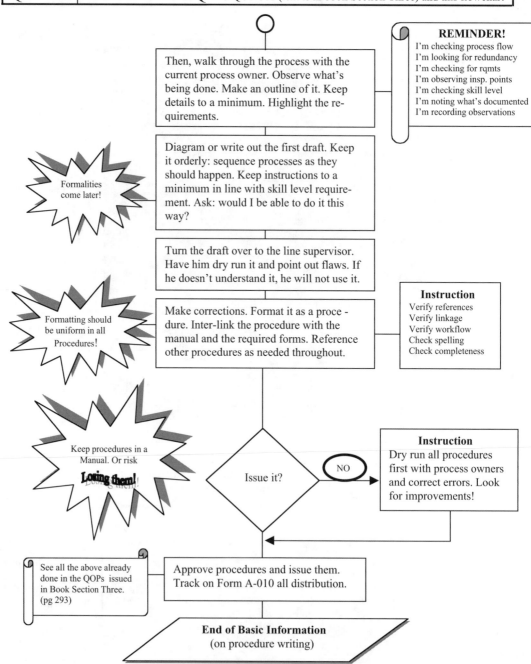

REMINDER!
I'm checking process flow
I'm looking for redundancy
I'm checking for rqmts
I'm observing insp. points
I'm checking skill level
I'm noting what's documented
I'm recording observations

Then, walk through the process with the current process owner. Observe what's being done. Make an outline of it. Keep details to a minimum. Highlight the requirements.

Formalities come later!

Diagram or write out the first draft. Keep it orderly: sequence processes as they should happen. Keep instructions to a minimum in line with skill level requirement. Ask: would I be able to do it this way?

Turn the draft over to the line supervisor. Have him dry run it and point out flaws. If he doesn't understand it, he will not use it.

Formatting should be uniform in all Procedures!

Make corrections. Format it as a proce - dure. Inter-link the procedure with the manual and the required forms. Reference other procedures as needed throughout.

Instruction
Verify references
Verify linkage
Verify workflow
Check spelling
Check completeness

Keep procedures in a Manual. Or risk **Losing them!**

Issue it?

NO

Instruction
Dry run all procedures first with process owners and correct errors. Look for improvements!

See all the above already done in the QOPs issued in Book Section Three. (pg 293)

Approve procedures and issue them. Track on Form A-010 all distribution.

End of Basic Information
(on procedure writing)

TRAINING METRICS	Iss dt.	Rev dt.	Pg.
QUALITY OPERATING PROCEDURE	Sign	Rev #	9/33
REQUIREMENT: Construction Methods, Control of Documents and Data, QSM 4.5			
SUBJECT: Approval and Issuance			
QOP 002 |DOCUMENT REQ'D: QOP 002 (text is in book Section Three) and this flowchart			

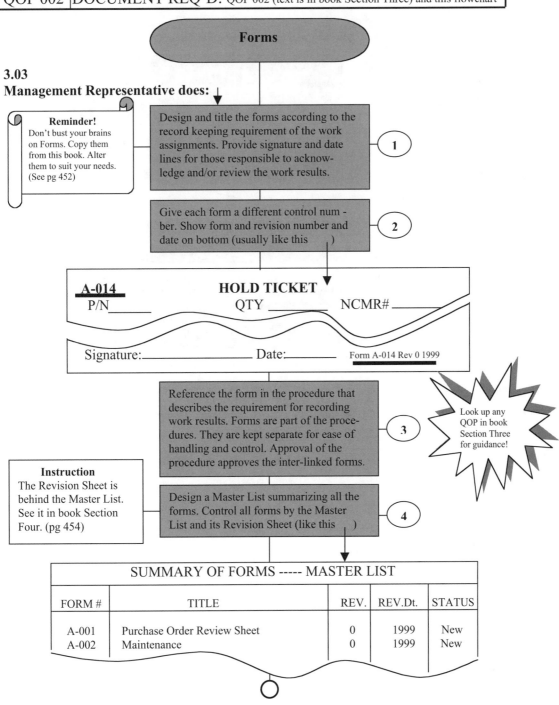

Forms

3.03
Management Representative does:

Reminder!
Don't bust your brains
on Forms. Copy them
from this book. Alter
them to suit your needs.
(See pg 452)

1 — Design and title the forms according to the
record keeping requirement of the work
assignments. Provide signature and date
lines for those responsible to acknowledge and/or review the work results.

2 — Give each form a different control num -
ber. Show form and revision number and
date on bottom (usually like this   )

A-014   **HOLD TICKET**
P/N_____   QTY _____   NCMR#_____

Signature:_____ Date:_____   Form A-014 Rev 0 1999

3 — Reference the form in the procedure that
describes the requirement for recording
work results. Forms are part of the procedures. They are kept separate for ease of
handling and control. Approval of the
procedure approves the inter-linked forms.

Look up any
QOP in book
Section Three
for guidance!

Instruction
The Revision Sheet is
behind the Master List.
See it in book Section
Four. (pg 454)

4 — Design a Master List summarizing all the
forms. Control all forms by the Master
List and its Revision Sheet (like this   )

SUMMARY OF FORMS ----- MASTER LIST				
FORM #	TITLE	REV.	REV.Dt.	STATUS
A-001	Purchase Order Review Sheet	0	1999	New
A-002	Maintenance	0	1999	New

TRAINING METRICS	Iss dt.	Rev dt.	Pg.
QUALITY OPERATING PROCEDURE	Sign	Rev #	10/33
REQUIREMENT: Construction Methods, Control of Documents and Data, QSM 4.5			
SUBJECT: Approval and Issuance			
QOP 002 |DOCUMENT REQ'D: QOP 002 (text is in book Section Three) and this flowchart			

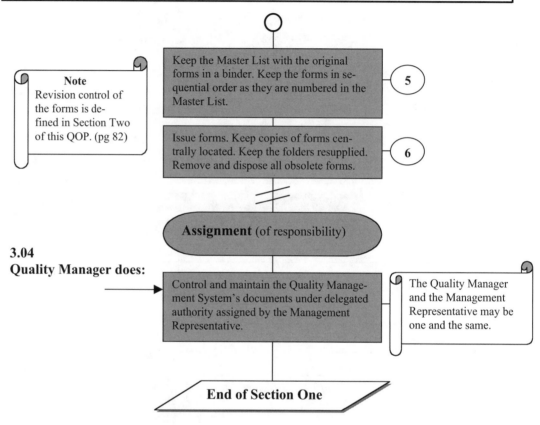

Note
Revision control of the forms is defined in Section Two of this QOP. (pg 82)

Keep the Master List with the original forms in a binder. Keep the forms in sequential order as they are numbered in the Master List.   **5**

Issue forms. Keep copies of forms centrally located. Keep the folders resupplied. Remove and dispose all obsolete forms.   **6**

Assignment (of responsibility)

3.04
Quality Manager does:

Control and maintain the Quality Management System's documents under delegated authority assigned by the Management Representative.

The Quality Manager and the Management Representative may be one and the same.

End of Section One

TRAINING METRICS	Iss dt.	Rev dt.	Pg
QUALITY OPERATING PROCEDURE	Sign	Rev #	11/33
REQUIREMENT: Control of Documents and Data, QSM 4.5.3			
SUBJECT: Document and Data Changes			
QOP 002 |DOCUMENT REQ'D: QOP 002 (text is in book Section Three) Forms A-009, A-010			

SECTION TWO

1.0 PURPOSE

To maintain control over document and data changes.

2.0 APPLICATION

The control over document and data changes shall apply to all the identified documents of the quality management system.

3.0 PROCEDURE

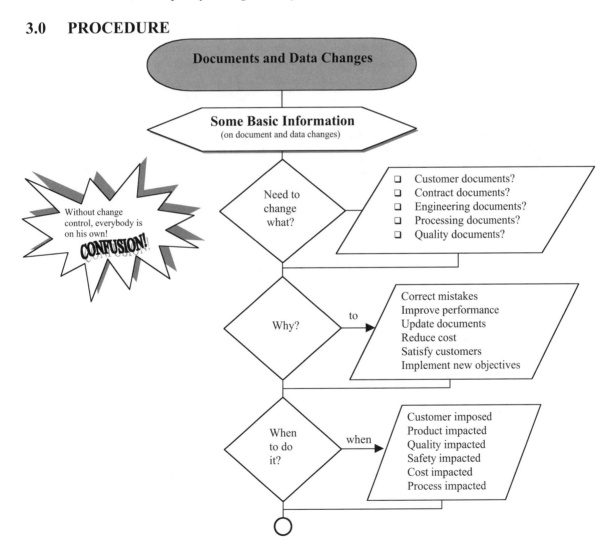

TRAINING METRICS	Iss dt.	Rev dt.	Pg.
QUALITY OPERATING PROCEDURE	Sign	Rev #	12/33
REQUIREMENT: Control of Documents and Data, QSM 4.5.3			
SUBJECT: Document and Data Changes			
QOP 002	DOCUMENT REQ'D: Form A-009, A-010		

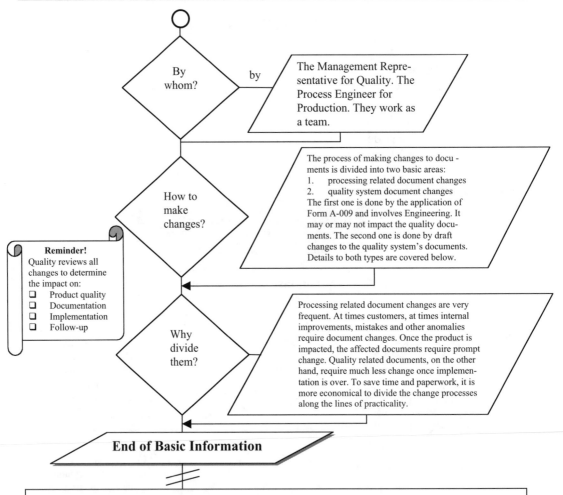

By whom?   by → The Management Representative for Quality. The Process Engineer for Production. They work as a team.

How to make changes? → The process of making changes to docu- ments is divided into two basic areas:
1.   processing related document changes
2.   quality system document changes
The first one is done by the application of Form A-009 and involves Engineering. It may or may not impact the quality documents. The second one is done by draft changes to the quality system's documents. Details to both types are covered below.

Reminder!
Quality reviews all changes to determine the impact on:
❑   Product quality
❑   Documentation
❑   Implementation
❑   Follow-up

Why divide them? → Processing related document changes are very frequent. At times customers, at times internal improvements, mistakes and other anomalies require document changes. Once the product is impacted, the affected documents require prompt change. Quality related documents, on the other hand, require much less change once implementation is over. To save time and paperwork, it is more economical to divide the change processes along the lines of practicality.

End of Basic Information

General Requirement

Errors, mistakes, and misleading instructions in documents are adversely affecting performance, processing efficiency, and product quality. Therefore, this process-approach system provides the tools to remove them when and where they occur. Process owners in all departments are required to report them as soon as they discover them to their immediate supervisors. The supervisors in turn are required to determine what immediate impact they have on the ongoing process, foremost on product quality, and take the necessary steps to eliminate them. QOP 012 (the Performance Standard), Section Two, paragraph 3.04 (pg 257) gives the required guidance as to how to go about curing these document deficiencies.

TRAINING METRICS	Iss dt.	Rev dt.	Pg.
QUALITY OPERATING PROCEDURE	Sign	Rev #	13/33
REQUIREMENT: Control of Documents and Data, QSM 4.5.3			
SUBJECT: Document and Data Changes			
QOP 002 |DOCUMENT REQ'D: Form A-009, A-010			

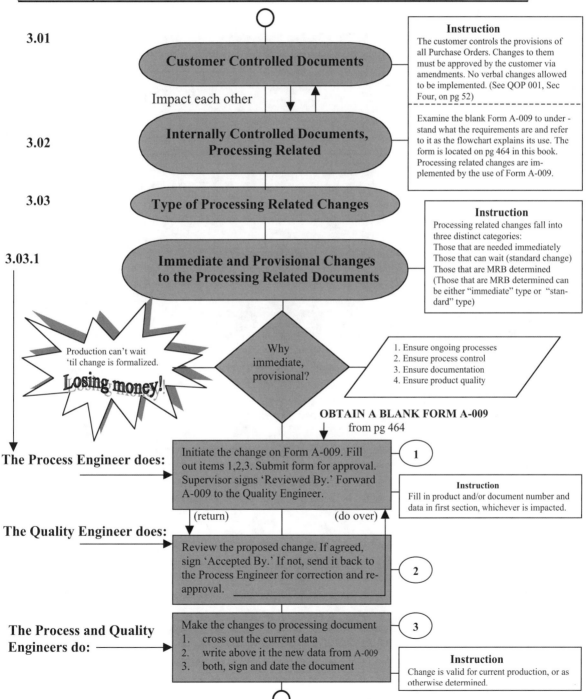

3.01

Customer Controlled Documents

Impact each other

Instruction
The customer controls the provisions of all Purchase Orders. Changes to them must be approved by the customer via amendments. No verbal changes allowed to be implemented. (See QOP 001, Sec Four, on pg 52)

3.02

Internally Controlled Documents, Processing Related

Examine the blank Form A-009 to under-stand what the requirements are and refer to it as the flowchart explains its use. The form is located on pg 464 in this book. Processing related changes are im-plemented by the use of Form A-009.

3.03

Type of Processing Related Changes

3.03.1

Immediate and Provisional Changes to the Processing Related Documents

Instruction
Processing related changes fall into three distinct categories:
Those that are needed immediately
Those that can wait (standard change)
Those that are MRB determined
(Those that are MRB determined can be either "immediate" type or "stan-dard" type)

Production can't wait 'til change is formalized.

Losing money!

Why immediate, provisional?

1. Ensure ongoing processes
2. Ensure process control
3. Ensure documentation
4. Ensure product quality

OBTAIN A BLANK FORM A-009
from pg 464

The Process Engineer does:

Initiate the change on Form A-009. Fill out items 1,2,3. Submit form for approval. Supervisor signs 'Reviewed By.' Forward A-009 to the Quality Engineer.

(1)

Instruction
Fill in product and/or document number and data in first section, whichever is impacted.

(return)   (do over)

The Quality Engineer does:

Review the proposed change. If agreed, sign 'Accepted By.' If not, send it back to the Process Engineer for correction and re-approval.

(2)

The Process and Quality Engineers do:

Make the changes to processing document
1. cross out the current data
2. write above it the new data from A-009
3. both, sign and date the document

(3)

Instruction
Change is valid for current production, or as otherwise determined.

TRAINING METRICS	Iss dt.	Rev dt.	Pg.
QUALITY OPERATING PROCEDURE	Sign	Rev #	14/33
REQUIREMENT: Control of Documents and Data, QSM 4.5.3			
SUBJECT: Document and Data Changes			
QOP 002 |DOCUMENT REQ'D: Form A-009, A-010			

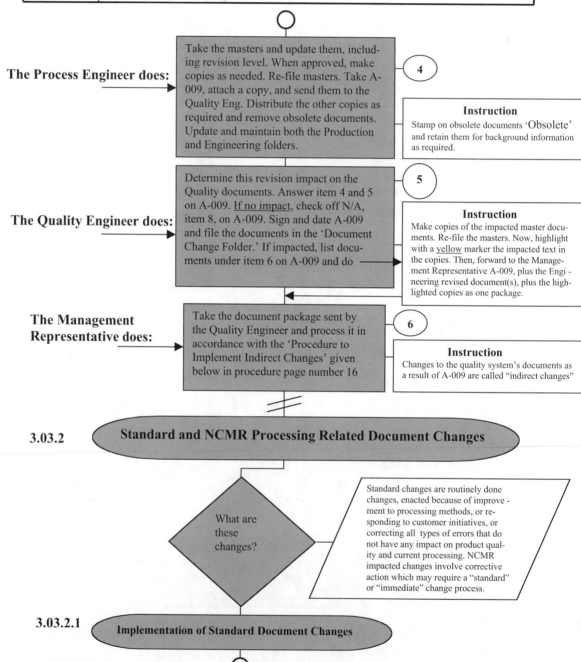

The Process Engineer does:

Take the masters and update them, including revision level. When approved, make copies as needed. Re-file masters. Take A-009, attach a copy, and send them to the Quality Eng. Distribute the other copies as required and remove obsolete documents. Update and maintain both the Production and Engineering folders.

4

Instruction
Stamp on obsolete documents 'Obsolete' and retain them for background information as required.

The Quality Engineer does:

Determine this revision impact on the Quality documents. Answer item 4 and 5 on A-009. If no impact, check off N/A, item 8, on A-009. Sign and date A-009 and file the documents in the 'Document Change Folder.' If impacted, list documents under item 6 on A-009 and do

5

Instruction
Make copies of the impacted master documents. Re-file the masters. Now, highlight with a yellow marker the impacted text in the copies. Then, forward to the Management Representative A-009, plus the Engineering revised document(s), plus the highlighted copies as one package.

The Management Representative does:

Take the document package sent by the Quality Engineer and process it in accordance with the 'Procedure to Implement Indirect Changes' given below in procedure page number 16

6

Instruction
Changes to the quality system's documents as a result of A-009 are called "indirect changes"

3.03.2   **Standard and NCMR Processing Related Document Changes**

What are these changes?

Standard changes are routinely done changes, enacted because of improve - ment to processing methods, or re- sponding to customer initiatives, or correcting all types of errors that do not have any impact on product qual- ity and current processing. NCMR impacted changes involve corrective action which may require a "standard" or "immediate" change process.

3.03.2.1   **Implementation of Standard Document Changes**

TRAINING METRICS	Iss dt.	Rev dt.	Pg.
QUALITY OPERATING PROCEDURE	Sign	Rev #	15/33
REQUIREMENT: Control of Documents and Data, QSM 4.5.3			
SUBJECT: Document and Data Changes			
QOP 002 |DOCUMENT REQ'D: Form A-009, A-010			

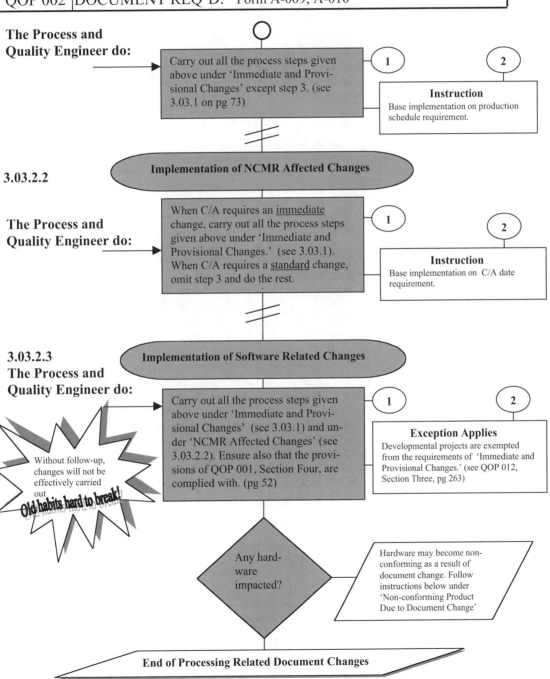

**The Process and
Quality Engineer do:**

Carry out all the process steps given above under 'Immediate and Provisional Changes' except step 3. (see 3.03.1 on pg 73)

1

2

Instruction
Base implementation on production schedule requirement.

Implementation of NCMR Affected Changes

3.03.2.2

**The Process and
Quality Engineer do:**

When C/A requires an <u>immediate</u> change, carry out all the process steps given above under 'Immediate and Provisional Changes.' (see 3.03.1). When C/A requires a <u>standard</u> change, omit step 3 and do the rest.

1

2

Instruction
Base implementation on C/A date requirement.

Implementation of Software Related Changes

**3.03.2.3
The Process and
Quality Engineer do:**

Carry out all the process steps given above under 'Immediate and Provisional Changes' (see 3.03.1) and under 'NCMR Affected Changes' (see 3.03.2.2). Ensure also that the provisions of QOP 001, Section Four, are complied with. (pg 52)

Without follow-up, changes will not be effectively carried out

Old habits hard to break!

1

2

Exception Applies
Developmental projects are exempted from the requirements of 'Immediate and Provisional Changes.' (see QOP 012, Section Three, pg 263)

Any hardware impacted?

Hardware may become non-conforming as a result of document change. Follow instructions below under 'Non-conforming Product Due to Document Change'

End of Processing Related Document Changes

TRAINING METRICS	Iss dt.	Rev dt.	Pg.
QUALITY OPERATING PROCEDURE	Sign	Rev #	16/33
REQUIREMENT: Control of Documents and Data, QSM 4.5.3			
SUBJECT: Document and Data Changes			
QOP 002 |DOCUMENT REQ'D: Form A-009, A-010			

4.0

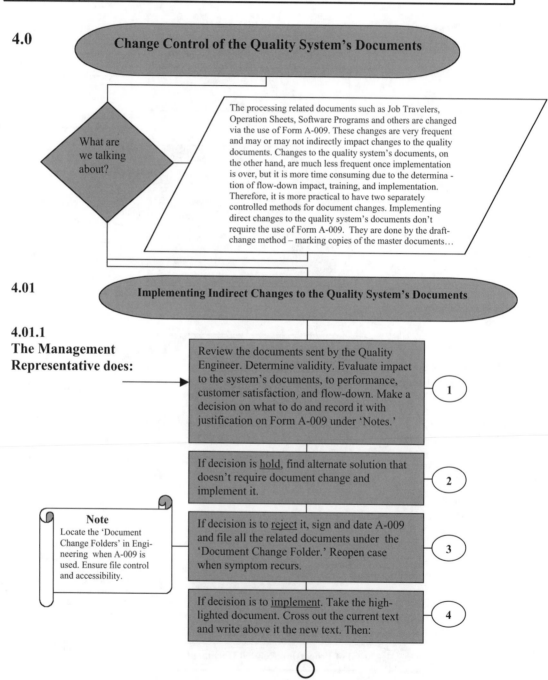

Change Control of the Quality System's Documents

What are we talking about?

The processing related documents such as Job Travelers, Operation Sheets, Software Programs and others are changed via the use of Form A-009. These changes are very frequent and may or may not indirectly impact changes to the quality documents. Changes to the quality system's documents, on the other hand, are much less frequent once implementation is over, but it is more time consuming due to the determina - tion of flow-down impact, training, and implementation. Therefore, it is more practical to have two separately controlled methods for document changes. Implementing direct changes to the quality system's documents don't require the use of Form A-009. They are done by the draft-change method – marking copies of the master documents...

4.01

Implementing Indirect Changes to the Quality System's Documents

4.01.1
The Management Representative does:

1 — Review the documents sent by the Quality Engineer. Determine validity. Evaluate impact to the system's documents, to performance, customer satisfaction, and flow-down. Make a decision on what to do and record it with justification on Form A-009 under 'Notes.'

2 — If decision is <u>hold,</u> find alternate solution that doesn't require document change and implement it.

Note
Locate the 'Document Change Folders' in Engineering when A-009 is used. Ensure file control and accessibility.

3 — If decision is to <u>reject</u> it, sign and date A-009 and file all the related documents under the 'Document Change Folder.' Reopen case when symptom recurs.

4 — If decision is to <u>implement</u>. Take the high-lighted document. Cross out the current text and write above it the new text. Then:

TRAINING METRICS	Iss dt.	Rev dt.	Pg.
QUALITY OPERATING PROCEDURE	Sign	Rev #	17/33
REQUIREMENT: Control of Documents and Data, QSM 4.5.3			
SUBJECT: Document and Data Changes			
QOP 002 |DOCUMENT REQ'D: Form A-009, A-010			

Instruction

Italicize or make "**bold**" all revision changes.

If product quality and/or processing is not impacted, the Management Representative can make corrections of typing and/or syntax errors to the text of the quality system's procedures. He/she must control the uniform distribution and retraction of obsolete documents.

The quality system's forms are part of the operating procedures by reference. They have been separated from them for reasons of handling and maintenance. Ensure that any change due to flow-down requirements is also implemented in the forms.

Determine flow-down requirement. When flow-down documents have been impacted, make a copy from their masters. Mark the copies by crossing out current text and writing above the new text. The marked documents will become the draft-change copies. Hold the masters as well as the marked copies at hand. — **5**

Evaluate the impact this change has on performance, customer satisfaction, and implementation. Set up training accordingly before formalizing and implementing the change(s). — **6**

Now, take all the impacted master documents and update them to the draft changes already made. Highlight the new data (italicize or make them **bold**). Complete A-009, item 7. Sign and date it. (Keep A-009 with the draft-change documents as you'll need them shortly for updating revision indication.) — **7**

Now, carry out the <u>revision indication</u> changes as stated below in par. 5.0. After finishing that task, come back here and continue with step 9a and b below. — **8**

Instruction

Replace the obsolete manuals with the revised manuals. Dispose obsolete manuals, as required. Those kept must be stamped <u>Obsolete</u>. Recall and replace manuals in the field, as required (controlled vs. uncontrolled)

Replace the obsolete quality operating procedures, the 'Master List,' and its 'Revision History' page in the Procedures Manual for each owner. Dispose the obsolete material, as required. Those kept must be stamped <u>Obsolete</u>.

Make the in-shop distribution of the revised documents according to Form A-010 'Document Issuance Control.' Retrieve at the same time all the obsolete documents. — **9a**

Make the distribution of the revised forms wherever shop files are. Collect obsolete forms at the same time and dispose them as required. Keep masters with other master forms. Keep the 'Master List' and its 'Revision History' page in front of the pile. — **9b**

End of Indirect Change Implementation

TRAINING METRICS	Iss dt.	Rev dt.	Pg.
QUALITY OPERATING PROCEDURE	Sign	Rev #	18/33
REQUIREMENT: Control of Documents and Data, QSM 4.5.3			
SUBJECT: Document and Data Changes			
QOP 002	DOCUMENT REQ'D: Form A-009, A-010		

4.02

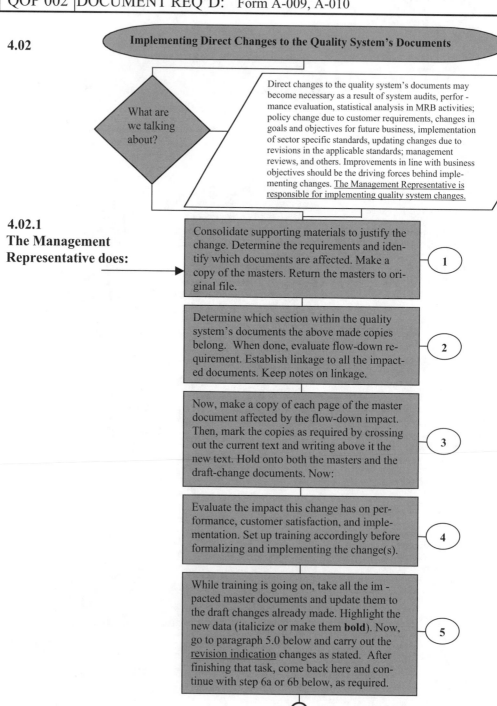

Implementing Direct Changes to the Quality System's Documents

What are we talking about?

Direct changes to the quality system's documents may become necessary as a result of system audits, perfor-mance evaluation, statistical analysis in MRB activities; policy change due to customer requirements, changes in goals and objectives for future business, implementation of sector specific standards, updating changes due to revisions in the applicable standards; management reviews, and others. Improvements in line with business objectives should be the driving forces behind imple-menting changes. <u>The Management Representative is responsible for implementing quality system changes.</u>

4.02.1
The Management Representative does:

1. Consolidate supporting materials to justify the change. Determine the requirements and iden-tify which documents are affected. Make a copy of the masters. Return the masters to ori-ginal file.

2. Determine which section within the quality system's documents the above made copies belong. When done, evaluate flow-down re-quirement. Establish linkage to all the impact-ed documents. Keep notes on linkage.

3. Now, make a copy of each page of the master document affected by the flow-down impact. Then, mark the copies as required by crossing out the current text and writing above it the new text. Hold onto both the masters and the draft-change documents. Now:

4. Evaluate the impact this change has on per-formance, customer satisfaction, and imple-mentation. Set up training accordingly before formalizing and implementing the change(s).

5. While training is going on, take all the im-pacted master documents and update them to the draft changes already made. Highlight the new data (italicize or make them **bold**). Now, go to paragraph 5.0 below and carry out the <u>revision indication</u> changes as stated. After finishing that task, come back here and con-tinue with step 6a or 6b below, as required.

TRAINING METRICS	Iss dt.	Rev dt.	Pg.
QUALITY OPERATING PROCEDURE	Sign	Rev #	19/33
REQUIREMENT: Control of Documents and Data, QSM 4.5.3			
SUBJECT: Document and Data Changes			
QOP 002 |DOCUMENT REQ'D: Form A-009, A-010			

Instruction
Replace the obsolete manuals with the revised manuals. Dispose obsolete manuals, as required. Those kept must be stamped Obsolete. Recall and replace manuals in the field, as required (controlled vs. uncontrolled)

Replace the obsolete quality operating procedures, the 'Master List,' and its 'Revision History' page in the Procedures Manual for each owner. Dispose the obsolete material, as required. Those kept must be stamped Obsolete.

Make the in-shop distribution of the revised documents according to Form A-010 'Document Issuance Control.' Retrieve at the same time all the obsolete documents.

6a

Instruction
Form A-010 has been designed to reassign documents in line with personnel changes. Update this form as required. (See pg 465)

Make the distribution of the revised forms wherever shop files are. Collect obsolete forms at the same time and dispose them as required. Keep masters with other master forms. Keep the 'Master List' and its 'Revision History' page in front of the pile.

6b

End of Implementing Direct Changes

Revision Indication Control of the Quality System's Documents

5.0

What are we talking about?

The quality system's documents are grouped in three separately controlled entities – the Quality Manual, the Quality Operating Procedures, and the Forms. Each has a built-in mechanism to control revision indication changes. The quality manual has its own 'Revision History' page and is located at the beginning pages. The quality operating procedures have two different 'Revision History' pages. One controls the revision indication of each procedure, the other is attached to a 'Master List' to control the Procedures Manual. The forms are controlled through their own 'Master List'"and its 'Revision History' page.

Note
Always update revision indication with direct or indirect change to the quality documents!

Revision Indication Change to the Quality System Manual

5.01
The Management Representative does:

1

Make a copy of the master of the 'Revision History' page of the manual. Cross out the current revision date and number. Write above it the new date and number. Now, go to the recording section and write in the required information under each column. This marked copy will be the draft-change document.

Definition
MASTER:
Originally approved and released document which may be issued as a hard copy, or stored as a soft copy, or both.
DRAFT-CHANGE:
Hand revised changes made on a copy of the master(s).

TRAINING METRICS	Iss dt.	Rev dt.	Pg.
QUALITY OPERATING PROCEDURE	Sign	Rev #	20/33
REQUIREMENT: Control of Documents and Data, QSM 4.5.3			
SUBJECT: Document and Data Changes			
QOP 002	DOCUMENT REQ'D: Form A-009, A-010		

2 Now, take the master manual and change the revision number and date in the header on every page according to the draft-change document. Update also the 'Revision History' page. Adjust page numbers as required. Do the same on the 'Contents' page also.

3 When finished, make as many copies of the manual as needed in line with the 'Document Issuance Control' form, A-010. Organize the pages in sequential order and bind them into individual manuals. Do the distribution in line with paragraph 4.02.1, step 6a, above. For the indirectly impacted changes, do the distribution in line with paragraph 4.01.1, step 9a, above. (See the instructions)

4 At this time, put the draft-change documents in order and file them in the 'Document Change Folder.' Take the master manual and put it back to its original file. Maintain file control.

End of Revision Indication Change of the Quality System Manual

5.02   **Revision Indication Change to the Quality Operating Procedures**

The Management Representative does:

1 Make a copy of the 'Revision History' page master of the affected QOP. Go to the header section and cross out the current revision date and number. Write above the effective date and number.

2 If other data is affected, do the same thing. Now, go to the recording section and write in the required information under each column's description. This page will be the draft-change copy.

TRAINING METRICS	Iss dt.	Rev dt.	Pg.
QUALITY OPERATING PROCEDURE	Sign	Rev #	21/33
REQUIREMENT: Control of Documents and Data, QSM 4.5.3			
SUBJECT: Document and Data Changes			
QOP 002 |DOCUMENT REQ'D: Form A-009, A-010			

3

Take the master procedure now and change the revision date and number in the header on every page as it is in the draft copy, done in step '1' above. Do other change(s) in the header, as required. Update also the 'Revision History' page, including the recording section. Adjust the page and 'Contents' numbers, as required.

4

The next phase is marking the 'Master List' and its 'Revision History' page. Make a copy from the masters. Start with the 'Revision History' page. Cross out the current date and number in the header and write above the effective date and number. Go now to the recording section and write in the required information under each column's description.

Note!

Consider the location of the 'Document Change Folders' carefully.

Suggestion: The indirectly im - pacted change records, which were implemented via the appli - cation of Form A-009, should be put in the Engineering area. Engi - neering needs to have easy access to them for background informa - tion. On the other hand, the di - rectly carried out change records, which were done on the draft copies, should be located in the Quality Department, for they are basically Quality related.

5

Now take the 'Master List' copy and cross out the current revision date and number and write above the same date and number as in its already marked 'Revision History' copy. Then, identify the procedure's number and title in the listing section and cross out there the current ' revision, date, and status' indi - cations and write above the effective revision and date as it appears in the already marked copy of the 'Revision History' of the affected operating procedure (done under step 1). Indicate revised under the 'status' column. These two pages, the 'Revision History' un - der step 4 and the 'Master List' under this step will be the draft copies.

6

Now, take the master of the 'Master List' and its 'Revision History' page and update them according to the revised data in the drafts. Then, do the next step below.

TRAINING METRICS	Iss dt.	Rev dt.	Pg.
QUALITY OPERATING PROCEDURE	Sign	Rev #	22/33
REQUIREMENT: Control of Documents and Data, QSM 4.5.3			
SUBJECT: Document and Data Changes			
QOP 002 |DOCUMENT REQ'D: Form A-009, A-010			

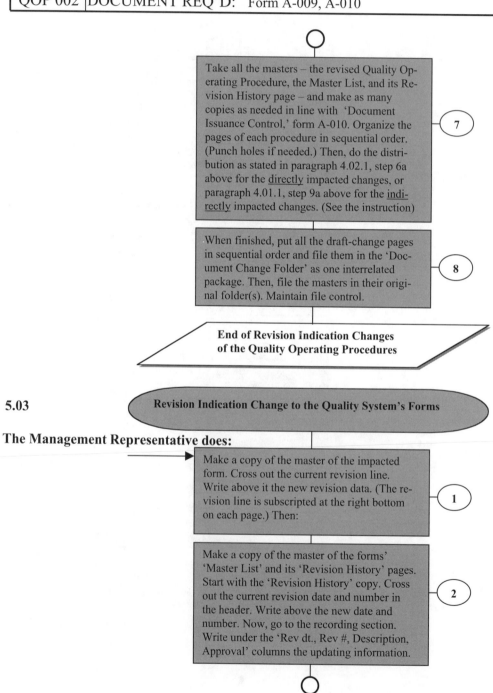

Take all the masters – the revised Quality Operating Procedure, the Master List, and its Revision History page – and make as many copies as needed in line with 'Document Issuance Control,' form A-010. Organize the pages of each procedure in sequential order. (Punch holes if needed.) Then, do the distribution as stated in paragraph 4.02.1, step 6a above for the <u>directly</u> impacted changes, or paragraph 4.01.1, step 9a above for the <u>indirectly</u> impacted changes. (See the instruction)   **7**

When finished, put all the draft-change pages in sequential order and file them in the 'Document Change Folder' as one interrelated package. Then, file the masters in their original folder(s). Maintain file control.   **8**

**End of Revision Indication Changes
of the Quality Operating Procedures**

5.03   **Revision Indication Change to the Quality System's Forms**

The Management Representative does:

Make a copy of the master of the impacted form. Cross out the current revision line. Write above it the new revision data. (The revision line is subscripted at the right bottom on each page.) Then: **1**

Make a copy of the master of the forms' 'Master List' and its 'Revision History' pages. Start with the 'Revision History' copy. Cross out the current revision date and number in the header. Write above the new date and number. Now, go to the recording section. Write under the 'Rev dt., Rev #, Description, Approval' columns the updating information.   **2**

TRAINING METRICS	Iss dt.	Rev dt.	Pg.
QUALITY OPERATING PROCEDURE	Sign	Rev #	23/33
REQUIREMENT: Control of Documents and Data, QSM 4.5.3			
SUBJECT: Document and Data Changes			
QOP 002 |DOCUMENT REQ'D: Form A-009, A-010			

Now go to the 'Master List' copy. Look up the form's number in the listing section. Cross out its current 'Rev., Dt., and Status' indications. Write above the new update from the already marked form (the sub-scripted change). Write Revised under the 'Status' column. These three pages will be the drafts.

(3)

Instruction
The three pages are:
1. the marked form
2. the marked 'Revision History'
3. the marked 'Master List'

Now, make the changes in the masters according to the drafts. When finished, make enough copies to cover the floor's and the central file's requirements. Do the distribution according to paragraph 4.02.1, step 6b, above, for the directly impacted changes (pg 78), or paragraph 4.01.1, step 9b, above, for the indirectly impacted changes (pg 76).

(4)

File all draft-change copies as one interrelated package in the 'Document Change Folder.' Then, file the masters in their original folders. Keep the 'Master List' and its 'Revision History' page in front of the forms' pile. Main - tain file control.

(5)

**End of Revision Indication Change
to the Quality System's Forms**

6.0

Handling Non-conforming Product's Resulting from Document Changes

6.01
The Quality Engineer does:

Implementation

Identify and tag the product using form A-012 and segregate it, as practical.

(1)

Instruction
When document changes are imposed as a result of C/A through the disposition process of NCMR, the Non-conforming product should be handled on the same NCMR.

Process completely Form A-006 (NCMR) in accordance with QOP 009, Sec One (pg 205) and Two (pg 207).

(2)

TRAINING METRICS	Iss dt.	Rev dt.	Pg.
QUALITY OPERATING PROCEDURE	Sign	Rev #	24/33
REQUIREMENT: Control of Documents and Data, QSM 4.5.3			
SUBJECT: Document and Data Changes			
QOP 002	DOCUMENT REQ'D: Form A-009, A-010		

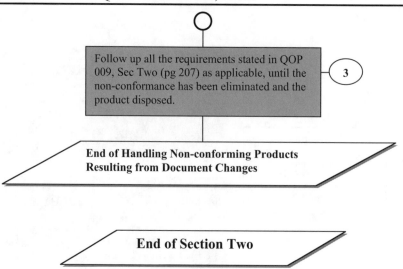

Follow up all the requirements stated in QOP 009, Sec Two (pg 207) as applicable, until the non-conformance has been eliminated and the product disposed.

3

End of Handling Non-conforming Products Resulting from Document Changes

End of Section Two

TRAINING METRICS	Iss dt.	Rev dt.	Pg.
QUALITY OPERATING PROCEDURE	Sign	Rev #	25/33
REQUIREMENT: Control of Documents and Data, QSM 4.16			
SUBJECT: Handling and Retention of Quality Records			
QOP 002 |DOCUMENT REQ'D: All records of the Quality System			

SECTION THREE

1.0   PURPOSE

To summarize the retention, maintenance, and responsibility in managing the Quality System's records.

2.0   APPLICATION

The hereunder written procedure shall apply to all the identified and documented records within the Quality System.

3.0   **PROCEDURE**

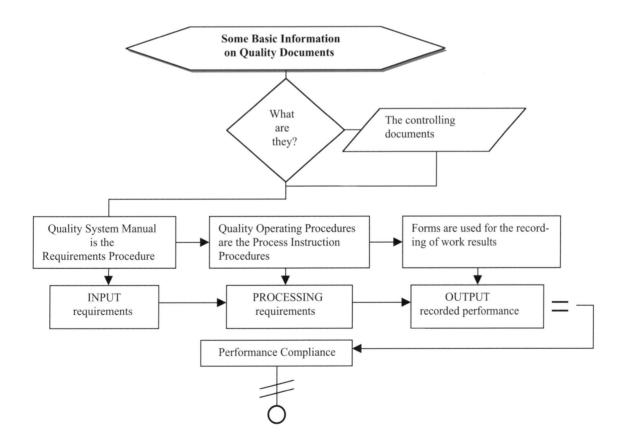

TRAINING METRICS	Iss dt.	Rev dt.	Pg.
QUALITY OPERATING PROCEDURE	Sign	Rev #	26/33
REQUIREMENT: Control of Documents and Data, QSM 4.16			
SUBJECT: Handling and Retention of Quality Records			
QOP 002 DOCUMENT REQ'D: All records of the Quality System			

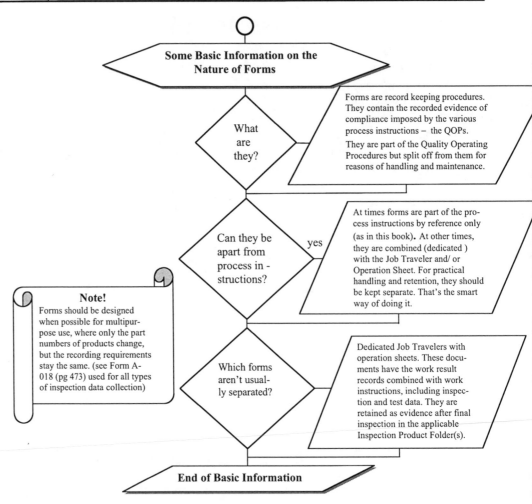

Some Basic Information on the Nature of Forms

What are they?

Forms are record keeping procedures. They contain the recorded evidence of compliance imposed by the various process instructions – the QOPs.

They are part of the Quality Operating Procedures but split off from them for reasons of handling and maintenance.

Can they be apart from process in - structions?   yes

At times forms are part of the process instructions by reference only (as in this book). At other times, they are combined (dedicated) with the Job Traveler and/ or Operation Sheet. For practical handling and retention, they should be kept separate. That's the smart way of doing it.

Note!
Forms should be designed when possible for multipur- pose use, where only the part numbers of products change, but the recording requirements stay the same. (see Form A- 018 (pg 473) used for all types of inspection data collection)

Which forms aren't usual- ly separated?

Dedicated Job Travelers with operation sheets. These docu- ments have the work result records combined with work instructions, including inspec- tion and test data. They are retained as evidence after final inspection in the applicable Inspection Product Folder(s).

End of Basic Information

TRAINING METRICS	Iss dt.	Rev dt.	Pg.
QUALITY OPERATING PROCEDURE	Sign	Rev #	27/33
REQUIREMENT: Control of Documents and Data, QSM 4.16			
SUBJECT: Handling and Retention of Quality Records			
QOP 002 |DOCUMENT REQ'D: All records of the Quality System			

(3. 0 continued)

General

The quality system's effectiveness is demonstrated through the control of its records which serve as evidence to prove product conformance or non-conformance to the specified requirements within the established and implemented system. (see Summary of Forms and Master List, pg 453)

Since record keeping is evidenced through the application of the various forms, the various departments within the organization shall be responsible for managing them in accordance with procedural requirements as described in the applicable QOPs.

To prove the effective operation of the quality system, subject records shall be periodically audited in accordance with documented procedures (see QOP 010, pg 224). Obsolete documents shall be retrieved from all affected areas. Their retention shall be determined according to need and those that are retained shall be stamped "Obsolete."

The quality system's records demonstrate the documentation of work results, recorded by those process owners who bear the responsibility for their accuracy and completeness each step of the way in the product's realization processes. The record retention responsibility, therefore, shall be assigned to those process owners who carry out the last step in the documentation, review, approval, or final handling of the designated record(s). According to work assignment responsibilities, as defined in the quality system's procedures, the retention of records shall be done as defined below.

Implementation

**3.01 The Management Representative
         does:**

Note!
All the forms are located in Book Section Four. They start on pg 455.

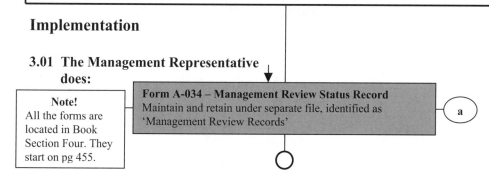

TRAINING METRICS	Iss dt.	Rev dt.	Pg.
QUALITY OPERATING PROCEDURE	Sign	Rev #	28/33
REQUIREMENT: Control of Documents and Data, QSM 4.16			
SUBJECT: Handling and Retention of Quality Records			
QOP 002 |DOCUMENT REQ'D: All records of the Quality System			

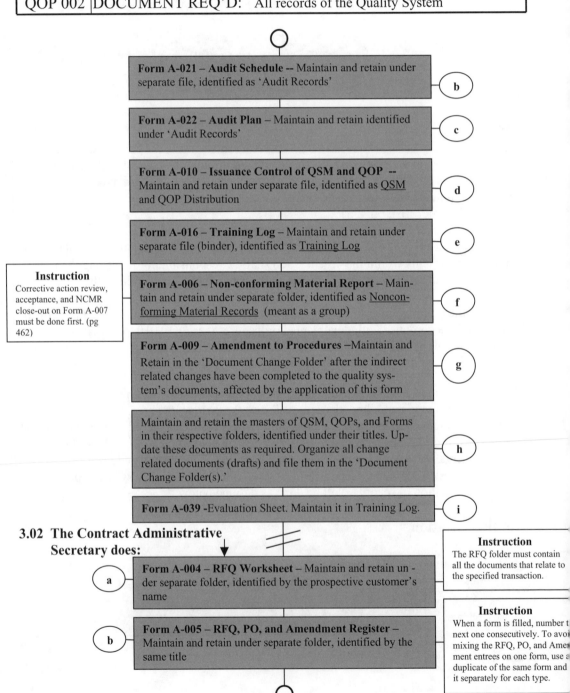

Form A-021 – Audit Schedule -- Maintain and retain under separate file, identified as 'Audit Records'   (b)

Form A-022 – Audit Plan – Maintain and retain identified under 'Audit Records'   (c)

Form A-010 – Issuance Control of QSM and QOP -- Maintain and retain under separate file, identified as QSM and QOP Distribution   (d)

Form A-016 – Training Log – Maintain and retain under separate file (binder), identified as Training Log   (e)

Instruction
Corrective action review, acceptance, and NCMR close-out on Form A-007 must be done first. (pg 462)

Form A-006 – Non-conforming Material Report – Maintain and retain under separate folder, identified as Nonconforming Material Records (meant as a group)   (f)

Form A-009 – Amendment to Procedures –Maintain and Retain in the 'Document Change Folder' after the indirect related changes have been completed to the quality system's documents, affected by the application of this form   (g)

Maintain and retain the masters of QSM, QOPs, and Forms in their respective folders, identified under their titles. Update these documents as required. Organize all change related documents (drafts) and file them in the 'Document Change Folder(s).'   (h)

Form A-039 -Evaluation Sheet. Maintain it in Training Log.   (i)

3.02 The Contract Administrative Secretary does:

(a) **Form A-004 – RFQ Worksheet** – Maintain and retain under separate folder, identified by the prospective customer's name

Instruction
The RFQ folder must contain all the documents that relate to the specified transaction.

(b) **Form A-005 – RFQ, PO, and Amendment Register** – Maintain and retain under separate folder, identified by the same title

Instruction
When a form is filled, number the next one consecutively. To avoid mixing the RFQ, PO, and Amendment entrees on one form, use a duplicate of the same form and it separately for each type.

TRAINING METRICS	Iss dt.	Rev dt.	Pg.
QUALITY OPERATING PROCEDURE	Sign	Rev #	29/33
REQUIREMENT: Control of Documents and Data, QSM 4.16			
SUBJECT: Handling and Retention of Quality Records			
QOP 002 DOCUMENT REQ'D: All records of the Quality System			

c — **Form A-011 – Control of Customer Documents** – Maintain and retain attached to the backside of the front cover of the respective customer's 'Purchase Order Folder'

d — **Form A-031 – Issue and Traceability of Customer Dwg.** Maintain and retain in the respective customer's 'Purchase Order Folder'

> **Instruction**
> When PO amendments issued by the customer, Quality may reissue this form. Place it in the 'PO Amendment Folder.' The PO and Amendment Folders may be combined, as required.

e — **Form A-001–Purchase Order Review Sheet** – Maintain and retain copy in the respective customer's 'Purchase Order Folder'

f — **Form A-015 – Record of Received Materials** – Maintain and retain in the respective customer's 'Purchase Order Folder' (You receive the original from Quality after release of the verified product.)

> **Instruction**
> Use the data recorded on this form to facilitate Production Planning. Looking for what's available for production, use the data recorded on this form as source information. It reflects current status of incoming products.

g — **Form A-028 – Packing Slip (Shipper)** – Maintain and retain a copy in the respective customer's 'Purchase Order Folder.' Original is shipped with the product to the customer

3.03 The Process Engineer does:

a — **From A-008 – Job Traveler** – Maintain and retain the master in the respective customer's 'Engineering Folder'

> **Instruction**
> When a Job Traveler is amended, update the master. The obsolete Job Traveler must be stamped 'Obsolete' and retained in the 'Document Change Folder.' Implement the changes via form A-009.

b — **Form A-001 – Purchase Order Review Sheet – and Form A-015 – Record of Received Materials --** Maintain and retain copies in the respective customer's Engineering Folder

c — **Form A-009 – Amendment to Procedures** – Maintain and retain this form in the 'Document Change Folder'

> **Instruction**
> Forms may be amended by Quality due to PO amendment or internal requirements. The data recorded on these forms must be implemented, as required, into the Job Traveler to ensure product quality enforcement in all the required processing steps.

3.04 The Purchasing Agent does:

a — **Form A-017 – Purchase Order (internal)** – Maintain and retain copy under the respective subcontractor's 'Purchase Order Folder.'

> Broom sticks, dust pans, soap and toilet paper are not usually product related. Forego QA review!

> **Instruction**
> Keep PO amendment(s) attached to the primary PO. Keep copy of product related correspondence done with the subcontractor in PO folder. Copy of released PO must show evidence of Quality approval prior to PO release to the subcontractor. This applies only to product related purchase orders.

TRAINING METRICS	Iss dt.	Rev dt.	Pg.
QUALITY OPERATING PROCEDURE	Sign	Rev #	30/33
REQUIREMENT: Control of Documents and Data, QSM 4.16			
SUBJECT: Handling and Retention of Quality Records			
QOP 002 |DOCUMENT REQ'D: All records of the Quality System			

3.05 The Quality Engineer does:

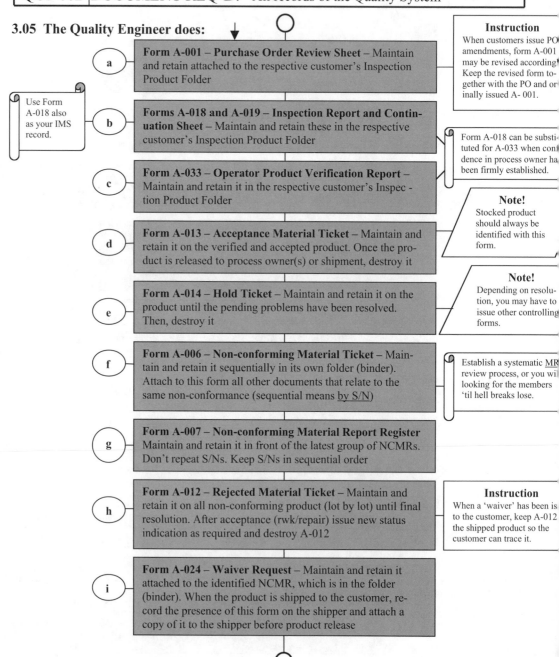

a

Form A-001 – Purchase Order Review Sheet – Maintain and retain attached to the respective customer's Inspection Product Folder

Instruction
When customers issue PO amendments, form A-001 may be revised according Keep the revised form together with the PO and or inally issued A-001.

Use Form A-018 also as your IMS record.

b

Forms A-018 and A-019 – Inspection Report and Continuation Sheet – Maintain and retain these in the respective customer's Inspection Product Folder

Form A-018 can be substituted for A-033 when confidence in process owner ha been firmly established.

c

Form A-033 – Operator Product Verification Report – Maintain and retain it in the respective customer's Inspection Product Folder

d

Form A-013 – Acceptance Material Ticket – Maintain and retain it on the verified and accepted product. Once the product is released to process owner(s) or shipment, destroy it

Note!
Stocked product should always be identified with this form.

e

Form A-014 – Hold Ticket – Maintain and retain it on the product until the pending problems have been resolved. Then, destroy it

Note!
Depending on resolution, you may have to issue other controlling forms.

f

Form A-006 – Non-conforming Material Ticket – Maintain and retain it sequentially in its own folder (binder). Attach to this form all other documents that relate to the same non-conformance (sequential means by S/N)

Establish a systematic MR review process, or you wi looking for the members 'til hell breaks lose.

g

Form A-007 – Non-conforming Material Report Register Maintain and retain it in front of the latest group of NCMRs. Don't repeat S/Ns. Keep S/Ns in sequential order

h

Form A-012 – Rejected Material Ticket – Maintain and retain it on all non-conforming product (lot by lot) until final resolution. After acceptance (rwk/repair) issue new status indication as required and destroy A-012

Instruction
When a 'waiver' has been is to the customer, keep A-012 the shipped product so the customer can trace it.

i

Form A-024 – Waiver Request – Maintain and retain it attached to the identified NCMR, which is in the folder (binder). When the product is shipped to the customer, record the presence of this form on the shipper and attach a copy of it to the shipper before product release

TRAINING METRICS	Iss dt.	Rev dt.	Pg.
QUALITY OPERATING PROCEDURE	Sign	Rev #	31/33
REQUIREMENT: Control of Documents and Data, QSM 4.16			
SUBJECT: Handling and Retention of Quality Records			
QOP 002 DOCUMENT REQ'D: All records of the Quality System			

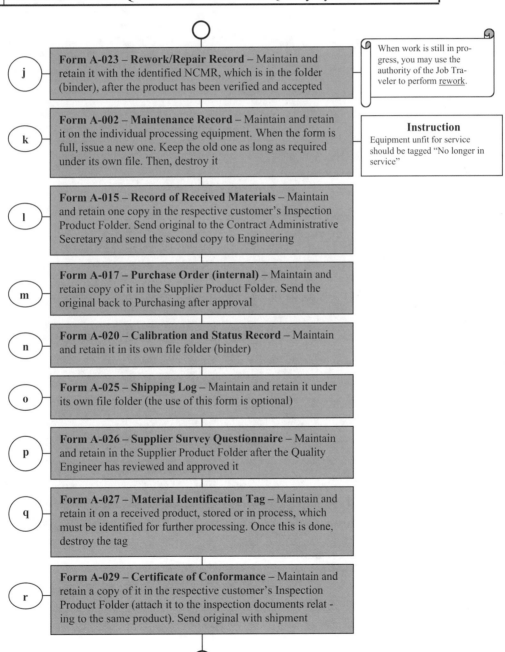

j — **Form A-023 – Rework/Repair Record** – Maintain and retain it with the identified NCMR, which is in the folder (binder), after the product has been verified and accepted

When work is still in progress, you may use the authority of the Job Traveler to perform <u>rework</u>.

k — **Form A-002 – Maintenance Record** – Maintain and retain it on the individual processing equipment. When the form is full, issue a new one. Keep the old one as long as required under its own file. Then, destroy it

Instruction
Equipment unfit for service should be tagged "No longer in service"

l — **Form A-015 – Record of Received Materials** – Maintain and retain one copy in the respective customer's Inspection Product Folder. Send original to the Contract Administrative Secretary and send the second copy to Engineering

m — **Form A-017 – Purchase Order (internal)** – Maintain and retain copy of it in the Supplier Product Folder. Send the original back to Purchasing after approval

n — **Form A-020 – Calibration and Status Record** – Maintain and retain it in its own file folder (binder)

o — **Form A-025 – Shipping Log** – Maintain and retain it under its own file folder (the use of this form is optional)

p — **Form A-026 – Supplier Survey Questionnaire** – Maintain and retain in the Supplier Product Folder after the Quality Engineer has reviewed and approved it

q — **Form A-027 – Material Identification Tag** – Maintain and retain it on a received product, stored or in process, which must be identified for further processing. Once this is done, destroy the tag

r — **Form A-029 – Certificate of Conformance** – Maintain and retain a copy of it in the respective customer's Inspection Product Folder (attach it to the inspection documents relating to the same product). Send original with shipment

TRAINING METRICS	Iss dt.	Rev dt.	Pg.
QUALITY OPERATING PROCEDURE	Sign	Rev #	32/33
REQUIREMENT: Control of Documents and Data, QSM 4.16			
SUBJECT: Handling and Retention of Quality Records			
QOP 002 |DOCUMENT REQ'D: All records of the Quality System			

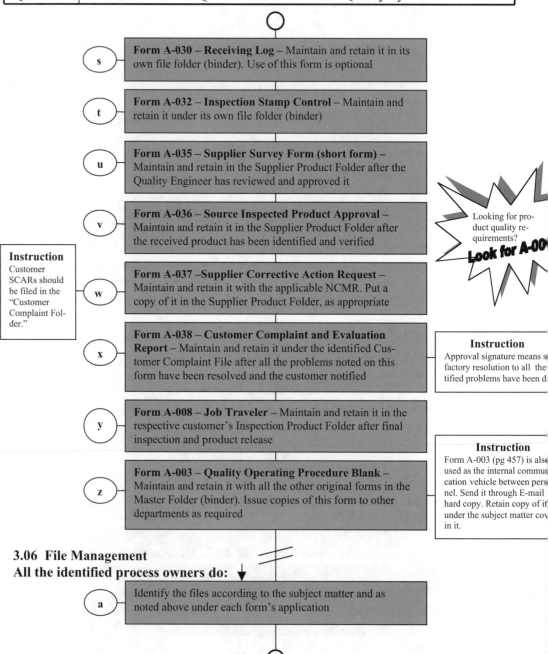

s — **Form A-030 – Receiving Log** – Maintain and retain it in its own file folder (binder). Use of this form is optional

t — **Form A-032 – Inspection Stamp Control** – Maintain and retain it under its own file folder (binder)

u — **Form A-035 – Supplier Survey Form (short form)** – Maintain and retain in the Supplier Product Folder after the Quality Engineer has reviewed and approved it

v — **Form A-036 – Source Inspected Product Approval** – Maintain and retain it in the Supplier Product Folder after the received product has been identified and verified

Looking for product quality requirements?
Look for A-00

Instruction
Customer SCARs should be filed in the "Customer Complaint Folder."

w — **Form A-037 –Supplier Corrective Action Request** – Maintain and retain it with the applicable NCMR. Put a copy of it in the Supplier Product Folder, as appropriate

x — **Form A-038 – Customer Complaint and Evaluation Report** – Maintain and retain it under the identified Customer Complaint File after all the problems noted on this form have been resolved and the customer notified

Instruction
Approval signature means s
factory resolution to all the
tified problems have been d

y — **Form A-008 – Job Traveler** – Maintain and retain it in the respective customer's Inspection Product Folder after final inspection and product release

Instruction
Form A-003 (pg 457) is also
used as the internal commu
cation vehicle between pers
nel. Send it through E-mail
hard copy. Retain copy of it
under the subject matter cov
in it.

z — **Form A-003 – Quality Operating Procedure Blank** – Maintain and retain it with all the other original forms in the Master Folder (binder). Issue copies of this form to other departments as required

3.06 File Management
All the identified process owners do:

a — Identify the files according to the subject matter and as noted above under each form's application

TRAINING METRICS	Iss dt.	Rev dt.	Pg.
QUALITY OPERATING PROCEDURE	Sign	Rev #	33/33
REQUIREMENT: Control of Documents and Data, QSM 4.16			
SUBJECT: File Management			
QOP 002 |DOCUMENT REQ'D: All records of the Quality System			

b — Arrange the documents in order of application. Group the documents in sequential order as they apply to a particular product, process, or assignment

Make sure when people remove sections from a file, they sign for the documents, or leave a note in the folder!

c — Provide easy access to the documents and allow authorized personnel to review them

d — Separate the active documents from inactive documents within the same folder. When the folders become unmanageable, remove those obsolete documents that are not needed for immediate background information and place them in storage. Store the documents in containers (boxes) and identify them in sequential order so that they can be traced back from where they have been removed

e — Provide safe location and protection from damage and un - authorized handling.

End of Handling and Retention of Quality Records

End of QOP 002 Flowcharting

OVERVIEW TO QOP 003
(CONTROL OF PURCHASES)

How often we find that the Purchasing Department was out of the quality loop because of its independent stature. Then the purchased item came in and was rejected. The inspection report indicated that the product was made by somebody who had absolutely no knowledge as to what kind of product quality the customer required. Now, you have to rework it at your expense, for you are out of time to meet the scheduled delivery requirement. Within a process-approach quality management system, the Purchasing Department becomes part of the team; after all we work for the same company and the same customer. Don't we? So let's bring this sacred cow into the fold in order that we all know our customers' requirements for product quality. QOP 003 is the procedure that requires the Purchasing Agent and the Subcontractor to understand the product quality requirements expected to be realized in purchased products. This procedure is packed with preventive quality requirements in order that we don't buy junk that would end up causing problems and wipe out any profit we calculated to make on the project.

DIAGRAMMATIC PROCEDURE – QOP 003

CONTROL OF PURCHASES (Internal)

CONTENTS **Page**

TRAINING METRICS	Iss dt.	Rev dt.	Pg.
QUALITY OPERATING PROCEDURE	Sign	Rev #	2/15
REQUIREMENT: QUALITY SYSTEM MANUAL SEC: 4.6			
SUBJECT: CONTROL OF PURCHASES (internal)			
QOP 003 DOCUMENT REQ'D: This page			

REVISION HISTORY

Rev Date	Rev No	Description	Approval

TRAINING METRICS	Iss dt.	Rev dt.	Pg.
QUALITY OPERATING PROCEDURE	Sign	Rev #	3/15
REQUIREMENT: Control of Purchases (Internal PO), QSM 4.6			
SUBJECT: Approval and Issue of Purchase Orders			
QOP 003 DOCUMENT REQ'D: QOP 003 (text), Form A-017,			

SECTION ONE

1.0 PURPOSE

To define the quality requirements for reviewing and approving the issuance of product related internal Purchase Orders. (Job quoting may be a preliminary requirement.)

2.0 APPLICATION

The provisions herein apply to the issuance of internal Purchase Orders for product related materials, processes (special) and other subcontracted services, which products will be delivered in compliance with the quality requirements of customers' purchase orders.

3.0 PROCEDURE

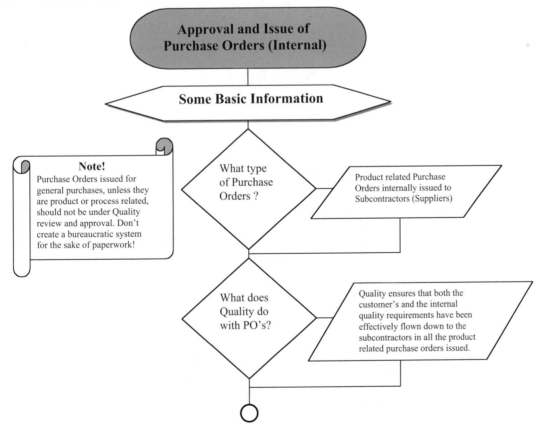

TRAINING METRICS	Iss dt.	Rev dt.	Pg.
QUALITY OPERATING PROCEDURE	Sign	Rev #	4/15
REQUIREMENT: Control of Purchases (Internal PO), QSM 4.6			
SUBJECT: Approval and Issue of Purchase Orders			
QOP 003 DOCUMENT REQ'D: QOP 003 (text), Form A-017,			

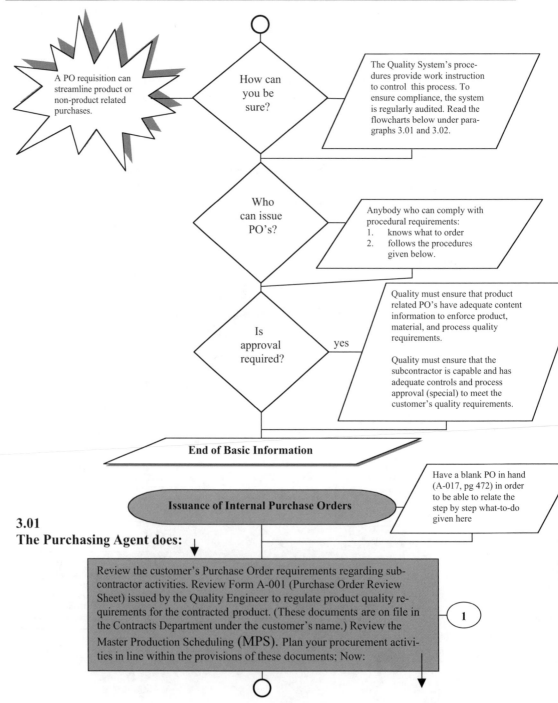

A PO requisition can streamline product or non-product related purchases.

How can you be sure?

The Quality System's procedures provide work instruction to control this process. To ensure compliance, the system is regularly audited. Read the flowcharts below under paragraphs 3.01 and 3.02.

Who can issue PO's?

Anybody who can comply with procedural requirements:
1. knows what to order
2. follows the procedures given below.

Is approval required?   yes

Quality must ensure that product related PO's have adequate content information to enforce product, material, and process quality requirements.

Quality must ensure that the subcontractor is capable and has adequate controls and process approval (special) to meet the customer's quality requirements.

End of Basic Information

Issuance of Internal Purchase Orders

Have a blank PO in hand (A-017, pg 472) in order to be able to relate the step by step what-to-do given here

3.01
The Purchasing Agent does:

Review the customer's Purchase Order requirements regarding subcontractor activities. Review Form A-001 (Purchase Order Review Sheet) issued by the Quality Engineer to regulate product quality requirements for the contracted product. (These documents are on file in the Contracts Department under the customer's name.) Review the Master Production Scheduling (MPS). Plan your procurement activities in line within the provisions of these documents; Now:

1

TRAINING METRICS	Iss dt.	Rev dt.	Pg.
QUALITY OPERATING PROCEDURE	Sign	Rev #	5/15

REQUIREMENT: Control of Purchases (Internal PO), QSM 4.6

SUBJECT: Approval and Issue of Purchase Orders

QOP 003 |DOCUMENT REQ'D: Form A-017

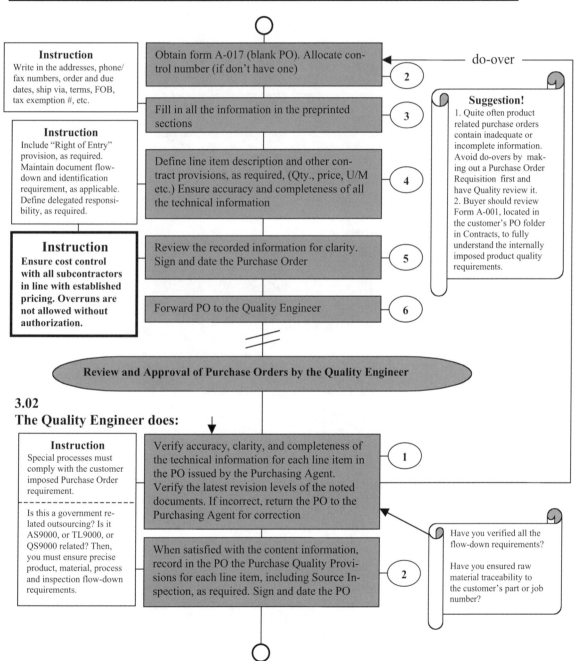

Instruction
Write in the addresses, phone/fax numbers, order and due dates, ship via, terms, FOB, tax exemption #, etc.

Instruction
Include "Right of Entry" provision, as required. Maintain document flow-down and identification requirement, as applicable. Define delegated responsibility, as required.

Instruction
Ensure cost control with all subcontractors in line with established pricing. Overruns are not allowed without authorization.

Obtain form A-017 (blank PO). Allocate control number (if don't have one) **2**

Fill in all the information in the preprinted sections **3**

Define line item description and other contract provisions, as required, (Qty., price, U/M etc.) Ensure accuracy and completeness of all the technical information **4**

Review the recorded information for clarity. Sign and date the Purchase Order **5**

Forward PO to the Quality Engineer **6**

do-over

Suggestion!
1. Quite often product related purchase orders contain inadequate or incomplete information. Avoid do-overs by making out a Purchase Order Requisition first and have Quality review it.
2. Buyer should review Form A-001, located in the customer's PO folder in Contracts, to fully understand the internally imposed product quality requirements.

Review and Approval of Purchase Orders by the Quality Engineer

3.02
The Quality Engineer does:

Instruction
Special processes must comply with the customer imposed Purchase Order requirement.

- - - - - - - - - - - - - -

Is this a government related outsourcing? Is it AS9000, or TL9000, or QS9000 related? Then, you must ensure precise product, material, process and inspection flow-down requirements.

Verify accuracy, clarity, and completeness of the technical information for each line item in the PO issued by the Purchasing Agent. Verify the latest revision levels of the noted documents. If incorrect, return the PO to the Purchasing Agent for correction **1**

When satisfied with the content information, record in the PO the Purchase Quality Provisions for each line item, including Source Inspection, as required. Sign and date the PO **2**

Have you verified all the flow-down requirements?

Have you ensured raw material traceability to the customer's part or job number?

TRAINING METRICS	Iss dt.	Rev dt.	Pg.
QUALITY OPERATING PROCEDURE	Sign	Rev #	6/15
REQUIREMENT: Control of Purchases (Internal PO), QSM 4.6			
SUBJECT: Approval and Issue of Purchase Orders			
QOP 003 |DOCUMENT REQ'D: Form A-017			

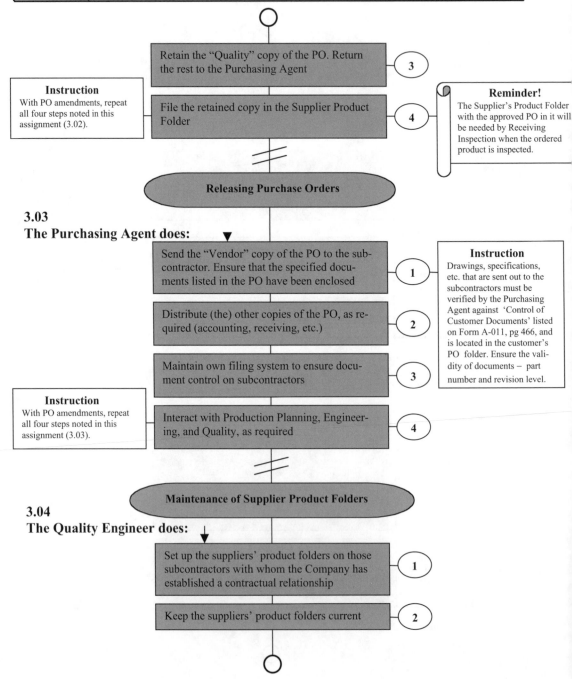

Instruction
With PO amendments, repeat all four steps noted in this assignment (3.02).

Retain the "Quality" copy of the PO. Return the rest to the Purchasing Agent — 3

File the retained copy in the Supplier Product Folder — 4

Reminder!
The Supplier's Product Folder with the approved PO in it will be needed by Receiving Inspection when the ordered product is inspected.

Releasing Purchase Orders

3.03
The Purchasing Agent does:

Send the "Vendor" copy of the PO to the sub-contractor. Ensure that the specified documents listed in the PO have been enclosed — 1

Distribute (the) other copies of the PO, as required (accounting, receiving, etc.) — 2

Maintain own filing system to ensure document control on subcontractors — 3

Instruction
With PO amendments, repeat all four steps noted in this assignment (3.03).

Interact with Production Planning, Engineering, and Quality, as required — 4

Instruction
Drawings, specifications, etc. that are sent out to the subcontractors must be verified by the Purchasing Agent against 'Control of Customer Documents' listed on Form A-011, pg 466, and is located in the customer's PO folder. Ensure the validity of documents – part number and revision level.

Maintenance of Supplier Product Folders

3.04
The Quality Engineer does:

Set up the suppliers' product folders on those subcontractors with whom the Company has established a contractual relationship — 1

Keep the suppliers' product folders current — 2

TRAINING METRICS	Iss dt.	Rev dt.	Pg.
QUALITY OPERATING PROCEDURE	Sign	Rev #	7/15
REQUIREMENT: Control of Purchases (Internal PO), QSM 4.6			
SUBJECT: Approval and Issue of Purchase Orders			
QOP 003 |DOCUMENT REQ'D: Form A-017			

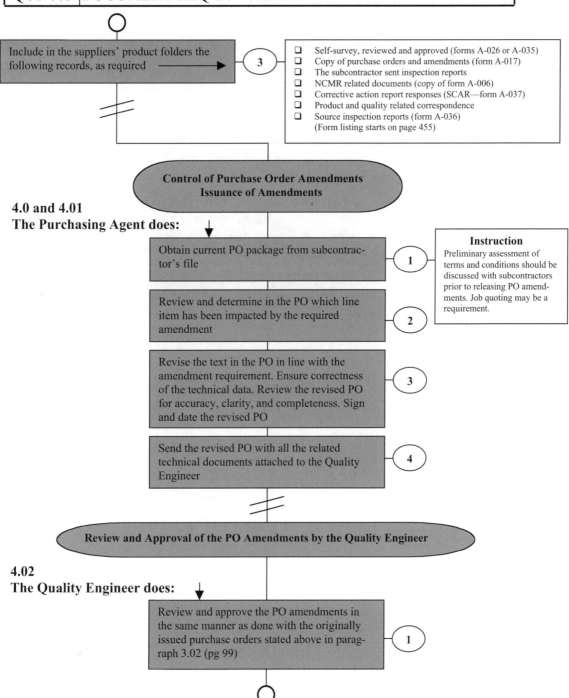

Include in the suppliers' product folders the following records, as required

3

❑ Self-survey, reviewed and approved (forms A-026 or A-035)
❑ Copy of purchase orders and amendments (form A-017)
❑ The subcontractor sent inspection reports
❑ NCMR related documents (copy of form A-006)
❑ Corrective action report responses (SCAR—form A-037)
❑ Product and quality related correspondence
❑ Source inspection reports (form A-036)
(Form listing starts on page 455)

**Control of Purchase Order Amendments
Issuance of Amendments**

**4.0 and 4.01
The Purchasing Agent does:**

Obtain current PO package from subcontractor's file

1

Instruction
Preliminary assessment of terms and conditions should be discussed with subcontractors prior to releasing PO amendments. Job quoting may be a requirement.

Review and determine in the PO which line item has been impacted by the required amendment

2

Revise the text in the PO in line with the amendment requirement. Ensure correctness of the technical data. Review the revised PO for accuracy, clarity, and completeness. Sign and date the revised PO

3

Send the revised PO with all the related technical documents attached to the Quality Engineer

4

Review and Approval of the PO Amendments by the Quality Engineer

**4.02
The Quality Engineer does:**

Review and approve the PO amendments in the same manner as done with the originally issued purchase orders stated above in paragraph 3.02 (pg 99)

1

TRAINING METRICS	Iss dt.	Rev dt.	Pg.
QUALITY OPERATING PROCEDURE	Sign	Rev #	8/15
REQUIREMENT: Control of Purchases (Internal PO), QSM 4.6			
SUBJECT: Approval and Issue of Purchase Orders			
QOP 003	DOCUMENT REQ'D: Form A-017		

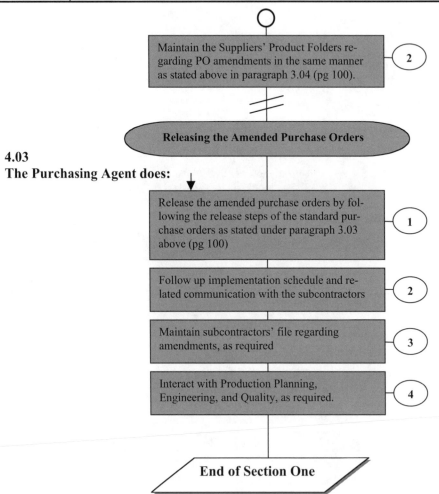

Maintain the Suppliers' Product Folders regarding PO amendments in the same manner as stated above in paragraph 3.04 (pg 100). — 2

Releasing the Amended Purchase Orders

4.03
The Purchasing Agent does:

Release the amended purchase orders by following the release steps of the standard purchase orders as stated under paragraph 3.03 above (pg 100) — 1

Follow up implementation schedule and related communication with the subcontractors — 2

Maintain subcontractors' file regarding amendments, as required — 3

Interact with Production Planning, Engineering, and Quality, as required. — 4

End of Section One

TRAINING METRICS	Iss dt.	Rev dt.	Pg.
QUALITY OPERATING PROCEDURE	Sign	Rev #	9/15
REQUIREMENT: Control of Purchases (Internal PO), QSM 4.6.2			
SUBJECT: Evaluation of Subcontractors			
QOP 003 |DOCUMENT REQ'D: Form A-026, A-035			

SECTION TWO

1.0 PURPOSE

To define the quality requirements regarding evaluation of subcontractors.

2.0 APPLICATION

The results of evaluation shall apply to the selection and approval of subcontractors. The evaluation process shall be based on subcontractors' capability to deliver acceptable products, and on evidence of documented procedures of quality requirements.

3.0 PROCEDURE

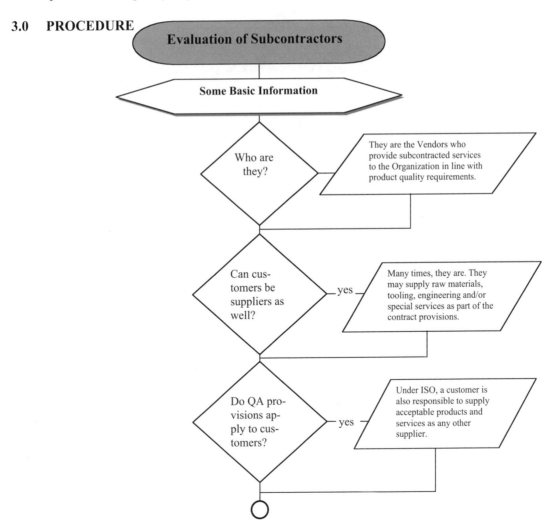

TRAINING METRICS	Iss dt.	Rev dt.	Pg.
QUALITY OPERATING PROCEDURE	Sign	Rev #	10/15
REQUIREMENT: Control of Purchases (Internal PO), QSM 4.6.2			
SUBJECT: Evaluation of Subcontractors			
QOP 003 DOCUMENT REQ'D: Form A-026, A-035			

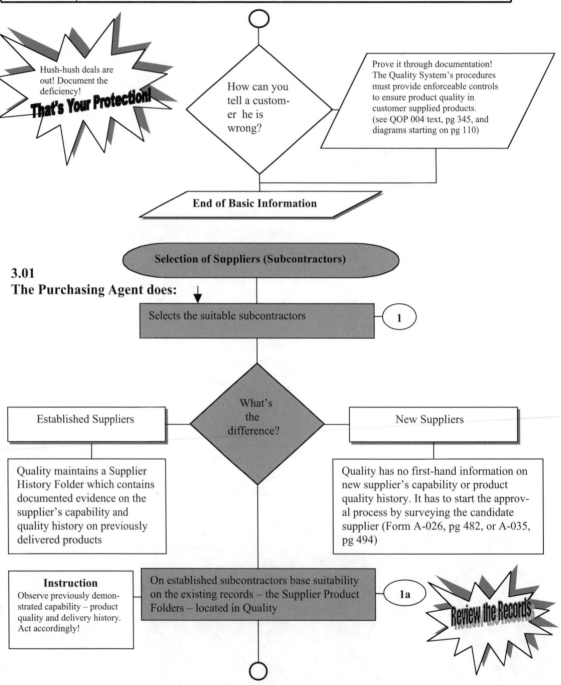

Hush-hush deals are out! Document the deficiency!
That's Your Protection!

How can you tell a customer he is wrong?

Prove it through documentation! The Quality System's procedures must provide enforceable controls to ensure product quality in customer supplied products.
(see QOP 004 text, pg 345, and diagrams starting on pg 110)

End of Basic Information

Selection of Suppliers (Subcontractors)

3.01
The Purchasing Agent does:

Selects the suitable subcontractors   1

What's the difference?

Established Suppliers

New Suppliers

Quality maintains a Supplier History Folder which contains documented evidence on the supplier's capability and quality history on previously delivered products

Quality has no first-hand information on new supplier's capability or product quality history. It has to start the approval process by surveying the candidate supplier (Form A-026, pg 482, or A-035, pg 494)

Instruction
Observe previously demonstrated capability – product quality and delivery history. Act accordingly!

On established subcontractors base suitability on the existing records – the Supplier Product Folders – located in Quality   1a

Review the Records

TRAINING METRICS	Iss dt.	Rev dt.	Pg.
QUALITY OPERATING PROCEDURE	Sign	Rev #	11/15
REQUIREMENT: Control of Purchases (Internal PO), QSM 4.6.2			
SUBJECT: Evaluation of Subcontractors			
QOP 003 DOCUMENT REQ'D: Form A-026, A-035			

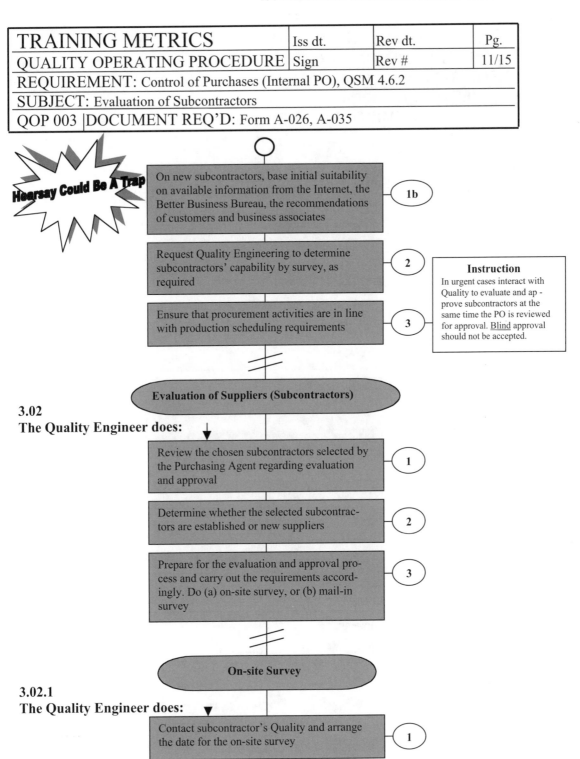

Hearsay Could Be A Trap

On new subcontractors, base initial suitability on available information from the Internet, the Better Business Bureau, the recommendations of customers and business associates **1b**

Request Quality Engineering to determine subcontractors' capability by survey, as required **2**

Ensure that procurement activities are in line with production scheduling requirements **3**

Instruction
In urgent cases interact with Quality to evaluate and ap-prove subcontractors at the same time the PO is reviewed for approval. <u>Blind</u> approval should not be accepted.

Evaluation of Suppliers (Subcontractors)

3.02
The Quality Engineer does:

Review the chosen subcontractors selected by the Purchasing Agent regarding evaluation and approval **1**

Determine whether the selected subcontrac-tors are established or new suppliers **2**

Prepare for the evaluation and approval pro-cess and carry out the requirements accord-ingly. Do (a) on-site survey, or (b) mail-in survey **3**

On-site Survey

3.02.1
The Quality Engineer does:

Contact subcontractor's Quality and arrange the date for the on-site survey **1**

TRAINING METRICS	Iss dt.	Rev dt.	Pg.
QUALITY OPERATING PROCEDURE	Sign	Rev #	12/15
REQUIREMENT: Control of Purchases (Internal PO), QSM 4.6.2			
SUBJECT: Evaluation of Subcontractors			
QOP 003 DOCUMENT REQ'D: Form A-026, A-035			

Details for project specific information are found in the Purchase Order Re - view Sheet, form A-001, in the Job Traveler, and in the applicable drawings.

2 — Review the project specifics that would be required for capability determination in doing the survey at the subcontractor

3 — Prepare the necessary documents to take in order to facilitate the survey (on form A-026 or A-035, whichever is more suitable). Go to the subcontractor as scheduled

Not everything is black and white during surveys!

Prepare Yourself!

The survey could indicate numerous problems in procedural areas. Ask yourself if they would affect your product quality requirement. Can C/A resolve the issues? Can the purchase be made contingent on supplier's C/A?

4 — Conduct the quality survey by following the questions listed on the selected survey form. Check off each item according to finding. Take notes, as required

5 — Review the survey results with the subcontractor's Quality Representative. Point out areas of nonconformance, if any, and inform him/her that approval is contingent on C/A. Or inform him/her that there has been no violations found. Prepare to conclude the survey

6 — Sign and date the survey form and give a copy to the subcontractor's representative

Don't sound more upbeat than the circumstances are! Don't promise what you can't deliver!

Instruction
Issuance and approval of Purchase Order(s) to subcontractors with outstanding C/A requirements shouldn't be done until the identified noncompliance has been corrected and implemented. Issue form A-037 (SCAR, pg 496) in order to verify compliance.

7 — Back home, submit the completed survey to supervision for final approval (auditing) and issue trip report on survey results to all concerned

8 — Make up the Supplier Product Folder as required and file in it the relevant survey related documents, including the trip report.

Mail-in Survey

3.02.2
The Quality Engineer does:

1 — Mail/Fax Survey Request, form A-026 to the selected subcontractor. State in cover letter the reason and contingency for the survey requirements

TRAINING METRICS	Iss dt.	Rev dt.	Pg.
QUALITY OPERATING PROCEDURE	Sign	Rev #	13/15

REQUIREMENT: Control of Purchases (Internal PO), QSM 4.6.2

SUBJECT: Evaluation of Subcontractors

QOP 003 |DOCUMENT REQ'D: Form A-026, A-035

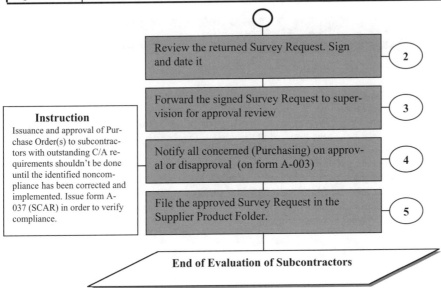

Instruction

Issuance and approval of Purchase Order(s) to subcontractors with outstanding C/A requirements shouldn't be done until the identified noncompliance has been corrected and implemented. Issue form A-037 (SCAR) in order to verify compliance.

Review the returned Survey Request. Sign and date it — 2

Forward the signed Survey Request to supervision for approval review — 3

Notify all concerned (Purchasing) on approval or disapproval (on form A-003) — 4

File the approved Survey Request in the Supplier Product Folder. — 5

End of Evaluation of Subcontractors

TRAINING METRICS	Iss dt.	Rev dt.	Pg.
QUALITY OPERATING PROCEDURE	Sign	Rev #	14/15
REQUIREMENT: Control of Purchases (Internal PO), QSM 4.6.2			
SUBJECT: Subcontractor Rating			
QOP 003 DOCUMENT REQ'D: Form A-003, and uncontrolled Vendor Rating form			

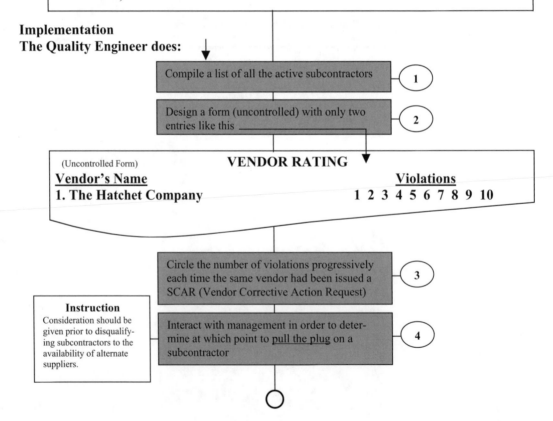

3.03

Subcontractor Rating

General

Subcontractors are expected to comply with the provisions of Purchase Orders issued to them. These Purchase Orders contain stipulated provisions relating to general and specific product related requirements. The Quality Provisions are just one component of the total requirements. In a rating process, the violations of the imposed PO provisions should be compared against a predetermined criteria (penalty points) made up according to the business interest of the Company. The predetermined criteria should be the standard rule against which subcontractors are rated and after <u>so many</u> violations disqualified. (In small businesses, subcontractors are relatively few and the rating system is based more on arbitrary decision than on an established rating system. Accordingly, the arbitrary decision making is heavily based on two factors: delivery and product quality capability. In tune with this approach, the following process instructions should be subject to individual consideration.)

Implementation
The Quality Engineer does:

Compile a list of all the active subcontractors   **1**

Design a form (uncontrolled) with only two entries like this _____   **2**

(Uncontrolled Form)   **VENDOR RATING**
Vendor's Name   **Violations**
1. The Hatchet Company   **1 2 3 4 5 6 7 8 9 10**

Circle the number of violations progressively each time the same vendor had been issued a SCAR (Vendor Corrective Action Request)   **3**

Instruction
Consideration should be given prior to disqualifying subcontractors to the availability of alternate suppliers.

Interact with management in order to determine at which point to <u>pull the plug</u> on a subcontractor   **4**

TRAINING METRICS	Iss dt.	Rev dt.		Pg.
QUALITY OPERATING PROCEDURE	Sign	Rev #		15/15
REQUIREMENT: Control of Purchases (Internal PO), QSM 4.6.2				
SUBJECT: Subcontractor Rating				
QOP 003	DOCUMENT REQ'D: Form A-003, and uncontrolled Vendor Rating form			

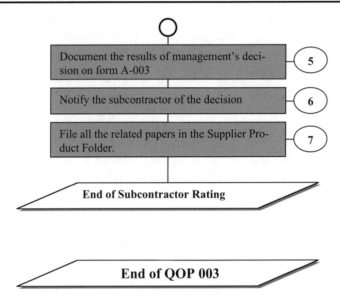

OVERVIEW TO QOP 004
(CONTROL OF CUSTOMER SUPPLIED PRODUCT)

Customer supplied products could be just about anything that an organization may need in supporting the contracted product's realization processes. The presumption that whatever the customer may provide in supporting product realization is already a qualified product, because the customer checked it before releasing it, is as far from the truth as catching a 10 pound bass on dry land. Of course, not every customer falls under this scenario. The customer's previously demonstrated capability should be the yardstick by which we should take any chances in accepting the supplied product(s). Without controlling customer supplied products, we are inviting the same type of problems as accepting unverified products from subcontractors. Just when we are going through the first production piece verification, we discover the problems. This is another area where any profit we wanted to make on the job has evaporated. Sometimes I wonder where customer satisfaction begins, now that I am buying product support from the customer. Anyway, QOP 004 is giving you the tools in order to control customer supplied products, just as you would buy products from anybody else. It's a fair deal.

DIAGRAMMATIC PROCEDURE QOP 004

CONTROL OF CUSTOMER SUPPLIED PRODUCTS

CONTENTS   **Page**

Control of Customer Supplied Product

TRAINING METRICS		Iss dt.		Rev dt.		Pg.
QUALITY OPERATING PROCEDURE		Sign		Rev #		2/8
REQUIREMENT: QUALITY SYSTEM MANUAL SEC: 4.7						
SUBJECT: CONTROL OF CUSTOMER SUPPLIED PRODUCT						
QOP 004	DOCUMENT REQ'D: QOP 004 (text) Form A-006, A-012, A-015, A-027					

REVISION HISTORY

Rev Date	Rev No	Description	Approval

TRAINING METRICS	Iss dt.	Rev dt.	Pg.
QUALITY OPERATING PROCEDURE	Sign	Rev #	3/8
REQUIREMENT: Control of Customer Supplied Product, QSM 4.7			
SUBJECT: Receiving and Verification			
QOP 004 |DOCUMENT REQ'D: QOP 004 (text) Form A-006, A-012, A-015, A-027			

1.0   PURPOSE

To control customer supplied products in accordance with the quality provisions of the Purchase Order and this procedure.

2.0   APPLICATION

Customer supplied product determines the processing schedule to furnish goods within delivery requirements of the contract. As such, the quality of the supplied product bears a substantial impact to comply with that requirement. The application of this procedure shall ensure the expedient identification, verification, documentation and release of the customer supplied product.

3.0   PROCEDURE

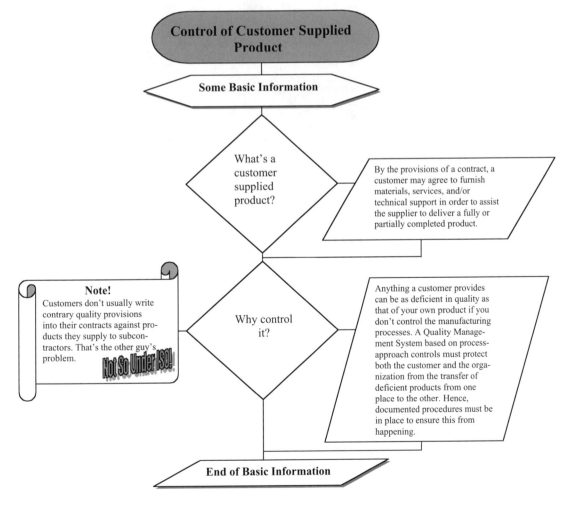

TRAINING METRICS	Iss dt.	Rev dt.	Pg.
QUALITY OPERATING PROCEDURE	Sign	Rev #	4/8
REQUIREMENT: Control of Customer Supplied Product, QSM 4.7			
SUBJECT: Receiving and Verification			
QOP 004 DOCUMENT REQ'D: QOP 004 (text) Form A-006, A-012, A-015, A-027			

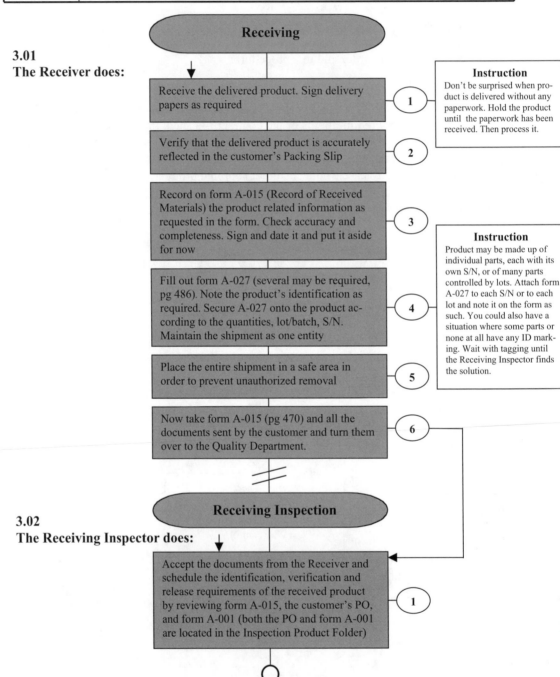

**3.01
The Receiver does:**

Receiving

1. Receive the delivered product. Sign delivery papers as required

   Instruction
   Don't be surprised when product is delivered without any paperwork. Hold the product until the paperwork has been received. Then process it.

2. Verify that the delivered product is accurately reflected in the customer's Packing Slip

3. Record on form A-015 (Record of Received Materials) the product related information as requested in the form. Check accuracy and completeness. Sign and date it and put it aside for now

   Instruction
   Product may be made up of individual parts, each with its own S/N, or of many parts controlled by lots. Attach form A-027 to each S/N or to each lot and note it on the form as such. You could also have a situation where some parts or none at all have any ID marking. Wait with tagging until the Receiving Inspector finds the solution.

4. Fill out form A-027 (several may be required, pg 486). Note the product's identification as required. Secure A-027 onto the product according to the quantities, lot/batch, S/N. Maintain the shipment as one entity

5. Place the entire shipment in a safe area in order to prevent unauthorized removal

6. Now take form A-015 (pg 470) and all the documents sent by the customer and turn them over to the Quality Department.

Receiving Inspection

**3.02
The Receiving Inspector does:**

1. Accept the documents from the Receiver and schedule the identification, verification and release requirements of the received product by reviewing form A-015, the customer's PO, and form A-001 (both the PO and form A-001 are located in the Inspection Product Folder)

TRAINING METRICS	Iss dt.	Rev dt.	Pg.
QUALITY OPERATING PROCEDURE	Sign	Rev #	5/8
REQUIREMENT: Control of Customer Supplied Product, QSM 4.7			
SUBJECT: Receiving and Verification			
QOP 004 |DOCUMENT REQ'D: QOP 004 (text) Form A-006, A-012, A-015, A-027			

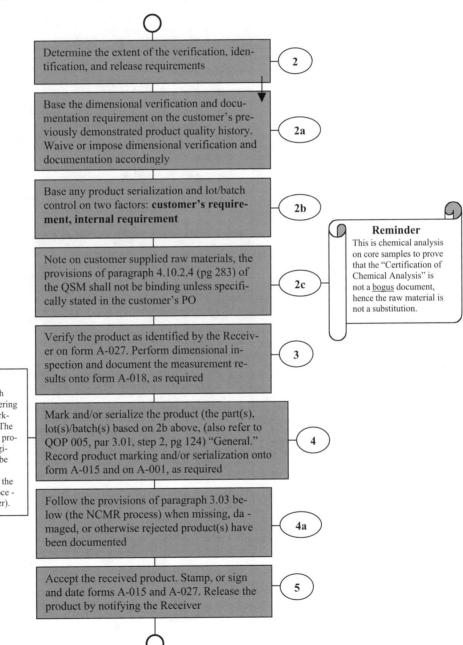

Determine the extent of the verification, identification, and release requirements — **2**

Base the dimensional verification and documentation requirement on the customer's previously demonstrated product quality history. Waive or impose dimensional verification and documentation accordingly — **2a**

Base any product serialization and lot/batch control on two factors: **customer's requirement, internal requirement** — **2b**

Note on customer supplied raw materials, the provisions of paragraph 4.10.2.4 (pg 283) of the QSM shall not be binding unless specifically stated in the customer's PO — **2c**

Reminder
This is chemical analysis on core samples to prove that the "Certification of Chemical Analysis" is not a bogus document, hence the raw material is not a substitution.

Verify the product as identified by the Receiver on form A-027. Perform dimensional inspection and document the measurement results onto form A-018, as required — **3**

Instruction
You must interact with Contracts and Engineering regarding product marking and serialization. The customer may control product serialization. Engineering also needs to be involved, for product identification impacts the processing related proce-dures (the Job Traveler).

Mark and/or serialize the product (the part(s), lot(s)/batch(s) based on 2b above, (also refer to QOP 005, par 3.01, step 2, pg 124) "General." Record product marking and/or serialization onto form A-015 and on A-001, as required — **4**

Follow the provisions of paragraph 3.03 below (the NCMR process) when missing, da-maged, or otherwise rejected product(s) have been documented — **4a**

Accept the received product. Stamp, or sign and date forms A-015 and A-027. Release the product by notifying the Receiver — **5**

TRAINING METRICS	Iss dt.	Rev dt.	Pg.
QUALITY OPERATING PROCEDURE	Sign	Rev #	6/8
REQUIREMENT: Control of Customer Supplied Product, QSM 4.7			
SUBJECT: Receiving and Verification			
QOP 004 |DOCUMENT REQ'D: QOP 004 (text) Form A-006, A-012, A-015, A-027			

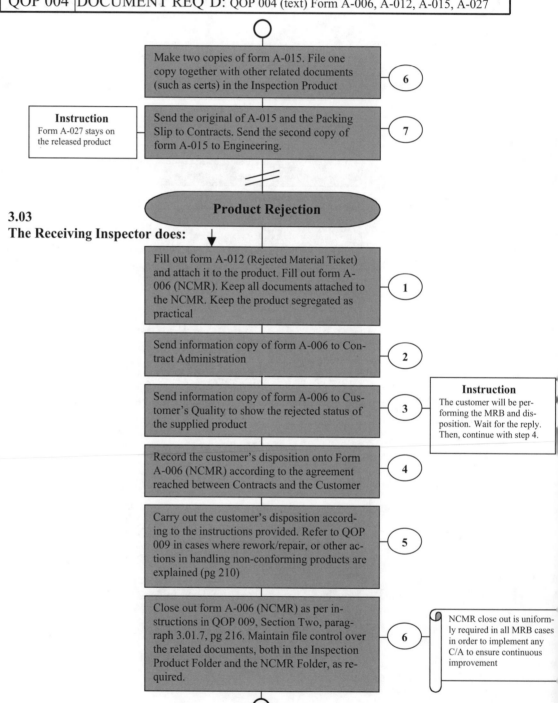

Make two copies of form A-015. File one copy together with other related documents (such as certs) in the Inspection Product — **6**

Instruction
Form A-027 stays on the released product

Send the original of A-015 and the Packing Slip to Contracts. Send the second copy of form A-015 to Engineering. — **7**

Product Rejection

3.03
The Receiving Inspector does:

Fill out form A-012 (Rejected Material Ticket) and attach it to the product. Fill out form A-006 (NCMR). Keep all documents attached to the NCMR. Keep the product segregated as practical — **1**

Send information copy of form A-006 to Contract Administration — **2**

Send information copy of form A-006 to Customer's Quality to show the rejected status of the supplied product — **3**

Instruction
The customer will be performing the MRB and disposition. Wait for the reply. Then, continue with step 4.

Record the customer's disposition onto Form A-006 (NCMR) according to the agreement reached between Contracts and the Customer — **4**

Carry out the customer's disposition according to the instructions provided. Refer to QOP 009 in cases where rework/repair, or other actions in handling non-conforming products are explained (pg 210) — **5**

Close out form A-006 (NCMR) as per instructions in QOP 009, Section Two, paragraph 3.01.7, pg 216. Maintain file control over the related documents, both in the Inspection Product Folder and the NCMR Folder, as required. — **6**

NCMR close out is uniformly required in all MRB cases in order to implement any C/A to ensure continuous improvement

TRAINING METRICS	Iss dt.	Rev dt.	Pg.
QUALITY OPERATING PROCEDURE	Sign	Rev #	7/8
REQUIREMENT: Control of Customer Supplied Product, QSM 4.7			
SUBJECT: Receiving and Verification			
QOP 004 |DOCUMENT REQ'D: QOP 004 (text) Form A-006, A-012, A-015, A-027			

3.04

Miscellaneous Products Supplied by the Customer

General

Customer supplied products may involve other than raw materials or semi-finished products. The customer's Purchase Order and/or amendments thereof should provide the necessary information regarding furnished products. Quality shall control and maintain documentation according to the type of product(s) supplied by the customer, which may involve:

a) production tooling;

b) fixtures;

c) inspection equipment;

d) software;

e) semi-finished product;

f) dropped-off product from the customer's other subcontractors;

g) other things (like shipping containers)

**3.04.1
The Receiver does:**

Follow the provisions of paragraph 3.01 above (114) from steps 1 through 6 in receiving miscellaneous products, as applicable.

**3.04.2
The Receiving Inspector does:**

Follow the provisions of paragraph 3.02 or 3.03 above (pg 114, 116) to the extent applicable in verifying, documenting, and releasing the miscellaneous products

(1)

Maintain file control over the related documents, both in the Inspection Product Folder and in the NCMR Folder, as required.

(2)

TRAINING METRICS	Iss dt.	Rev dt.	Pg.
QUALITY OPERATING PROCEDURE	Sign	Rev #	8/8
REQUIREMENT: Control of Customer Supplied Product, QSM 4.7			
SUBJECT: Receiving and Verification			
QOP 004 DOCUMENT REQ'D: QOP 004 (text) Form A-006, A-012, A-015, A-027			

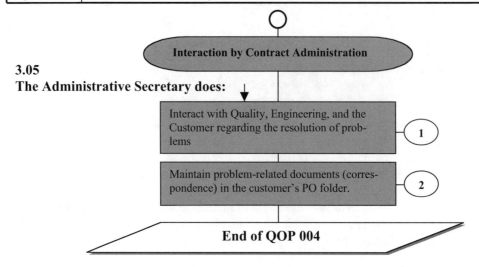

3.05
The Administrative Secretary does:

Interaction by Contract Administration

Interact with Quality, Engineering, and the Customer regarding the resolution of problems — 1

Maintain problem-related documents (correspondence) in the customer's PO folder. — 2

End of QOP 004

OVERVIEW TO QOP 005
(PRODUCT IDENTIFICATION AND TRACEABILITY)

Product identification is much like your birth certificate. It tells you something but not everything about yourself. Product-related documents do the same thing. This is the way we identify some very basic earmarks about the product we are putting through the various processes. The part, the serial, the lot or batch numbers, all serve to identify the product. Sometime, we don't mark the product, so we identify it on the document accompanying it. Then, we attach tags or labels to the product. We also identify the product on the Job Traveler and on the Operation Sheet. The customer does it on the Purchase Order and on the product specification. Now, we can not only match the product to its documents, but we can also trace it on the production floor, in the stockroom or even in the field. This is what product identification is all about. QOP 005 tells you how to control the process of product identification and traceability.

DIAGRAMMATIC PROCEDURE   QOP 005

PRODUCT IDENTIFICATION AND TRACEABILITY

CONTENTS   **PAGE**

Product Identification and Traceability

TRAINING METRICS	Iss dt.	Rev dt.	Pg.
QUALITY OPERATING PROCEDURE	Sign	Rev #	2/7
REQUIREMENT: QUALITY SYSTEM MANUAL SEC: 4.8			
SUBJECT: PRODUCT IDENTIFICATION AND TRACEABILITY			
QOP 005 DOCUMENT REQ'D: Forms A-001, A-008			

REVISION HISTORY

Rev Date	Rev No	Description	Approval

TRAINING METRICS	Iss dt.	Rev dt.	Pg.
QUALITY OPERATING PROCEDURE	Sign	Rev #	3/7
REQUIREMENT: Product Identification and Traceability, QSM 4.8			
SUBJECT: Basic Information			
QOP 005 |DOCUMENT REQ'D: Forms A-001, A-008			

1.0   PURPOSE

To maintain control over the process of product identification and traceability through the requirements of documented procedures.

2.0   APPLICATION

This procedure shall apply to product identification and traceability for the issuance of documents and the enforcement of marking.

3.0   PROCEDURE

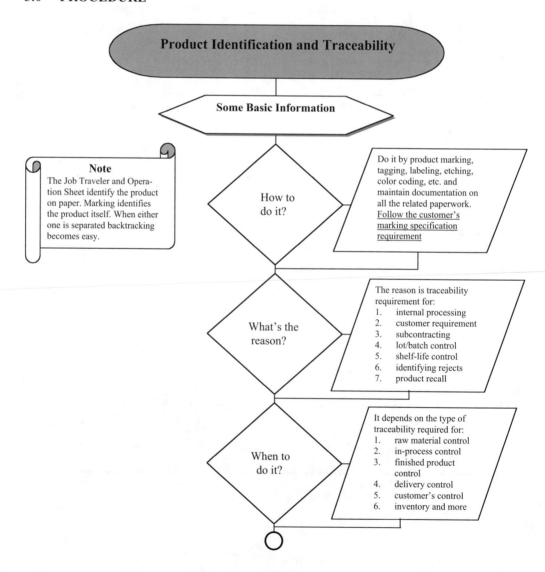

Product Identification and Traceability

Some Basic Information

Note
The Job Traveler and Operation Sheet identify the product on paper. Marking identifies the product itself. When either one is separated backtracking becomes easy.

How to do it?

Do it by product marking, tagging, labeling, etching, color coding, etc. and maintain documentation on all the related paperwork. Follow the customer's marking specification requirement

What's the reason?

The reason is traceability requirement for:
1.   internal processing
2.   customer requirement
3.   subcontracting
4.   lot/batch control
5.   shelf-life control
6.   identifying rejects
7.   product recall

When to do it?

It depends on the type of traceability required for:
1.   raw material control
2.   in-process control
3.   finished product control
4.   delivery control
5.   customer's control
6.   inventory and more

TRAINING METRICS	Iss dt.	Rev dt.	Pg.
QUALITY OPERATING PROCEDURE	Sign	Rev #	4/7
REQUIREMENT: Product Identification and Traceability, QSM 4.8			
SUBJECT: Internal Identification and Traceability (Basic Information)			
QOP 005 |DOCUMENT REQ'D: Forms A-001, A-006, A-012, A-013, A-014, A-018, A-027			

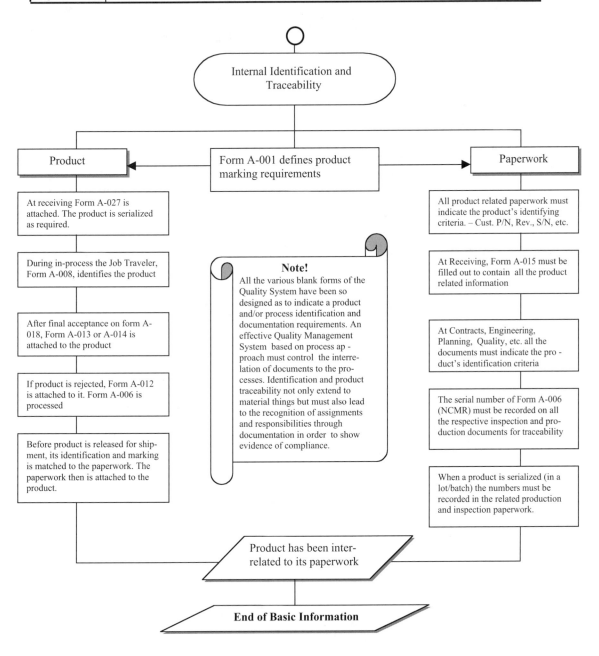

Internal Identification and Traceability

Form A-001 defines product marking requirements

Product

At receiving Form A-027 is attached. The product is serialized as required.

During in-process the Job Traveler, Form A-008, identifies the product

After final acceptance on form A-018, Form A-013 or A-014 is attached to the product

If product is rejected, Form A-012 is attached to it. Form A-006 is processed

Before product is released for shipment, its identification and marking is matched to the paperwork. The paperwork then is attached to the product.

Note!
All the various blank forms of the Quality System have been so designed as to indicate a product and/or process identification and documentation requirements. An effective Quality Management System based on process ap-proach must control the interrelation of documents to the processes. Identification and product traceability not only extend to material things but must also lead to the recognition of assignments and responsibilities through documentation in order to show evidence of compliance.

Paperwork

All product related paperwork must indicate the product's identifying criteria. – Cust. P/N, Rev., S/N, etc.

At Receiving, Form A-015 must be filled out to contain all the product related information

At Contracts, Engineering, Planning, Quality, etc. all the documents must indicate the pro-duct's identification criteria

The serial number of Form A-006 (NCMR) must be recorded on all the respective inspection and pro-duction documents for traceability

When a product is serialized (in a lot/batch) the numbers must be recorded in the related production and inspection paperwork.

Product has been inter-related to its paperwork

End of Basic Information

TRAINING METRICS	Iss dt.	Rev dt.	Pg.
QUALITY OPERATING PROCEDURE	Sign	Rev #	5/7
REQUIREMENT: Product Identification and Traceability, QSM 4.8			
SUBJECT: Internal Control			
QOP 005 DOCUMENT REQ'D: QOP 005 (text), Forms A-001, A-008			

3.01

Internal Control

General

1. The primary controlling method for document and product identification shall be the customer's Purchase Order imposed product part number and its revision level. Additional controls, such as individual serial numbering, batch or lot control, shall be imposed both in the related documents and on the product itself, as required.
2. Parts serialization, batch or lot control, when imposed, shall begin from the receipt of product(s), or as otherwise determined by the processing requirement. This shall form the basis also for identification later when splitting quantities for multiple processing is required.
3. Follow-on marking and documentation when, through processing, the original marking has been removed, shall be identically re-marked throughout processing.
4. Process owners' accountability for product identification and record keeping for the parts they have done shall be enforced and tracked by supervision.
5. Identification and documentation of the products and/or services, ordered on internally issued Purchase Orders, shall conform to the customer's Purchase Order requirements.
6. Departments issuing product-related documents shall enforce the identification and documentation requirements according to the customer's Purchase Order.
7. Quality shall flow-down the customer's specific product marking and traceability requirements in the Purchase Order Review Sheet, Form A-001.

3.01.1

Issuance of Product Related Documents

Contracts, Engineering, Purchasing, Planning, and Manufacturing do:

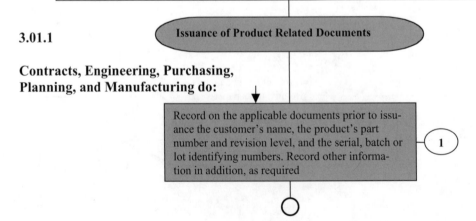

Record on the applicable documents prior to issuance the customer's name, the product's part number and revision level, and the serial, batch or lot identifying numbers. Record other information in addition, as required

1

TRAINING METRICS	Iss dt.	Rev dt.	Pg.
QUALITY OPERATING PROCEDURE	Sign	Rev #	6/7
REQUIREMENT: Product Identification and Traceability, QSM 4.8			
SUBJECT: Internal Control			
QOP 005 DOCUMENT REQ'D: QOP 005 (text), Forms A-001, A-008			

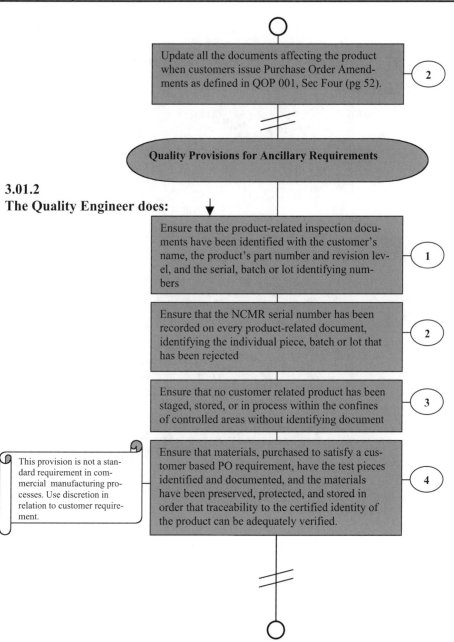

Update all the documents affecting the product when customers issue Purchase Order Amendments as defined in QOP 001, Sec Four (pg 52).

②

Quality Provisions for Ancillary Requirements

3.01.2
The Quality Engineer does:

Ensure that the product-related inspection documents have been identified with the customer's name, the product's part number and revision level, and the serial, batch or lot identifying numbers

①

Ensure that the NCMR serial number has been recorded on every product-related document, identifying the individual piece, batch or lot that has been rejected

②

Ensure that no customer related product has been staged, stored, or in process within the confines of controlled areas without identifying document

③

This provision is not a standard requirement in commercial manufacturing processes. Use discretion in relation to customer requirement.

Ensure that materials, purchased to satisfy a customer based PO requirement, have the test pieces identified and documented, and the materials have been preserved, protected, and stored in order that traceability to the certified identity of the product can be adequately verified.

④

TRAINING METRICS	Iss dt.	Rev dt.	Pg.
QUALITY OPERATING PROCEDURE	Sign	Rev #	7/7
REQUIREMENT: Product Identification and Traceability, QSM 4.8			
SUBJECT: Customer Flow-down Requirement			
QOP 005 |DOCUMENT REQ'D: QOP 005 (text), Forms A-001, A-008			

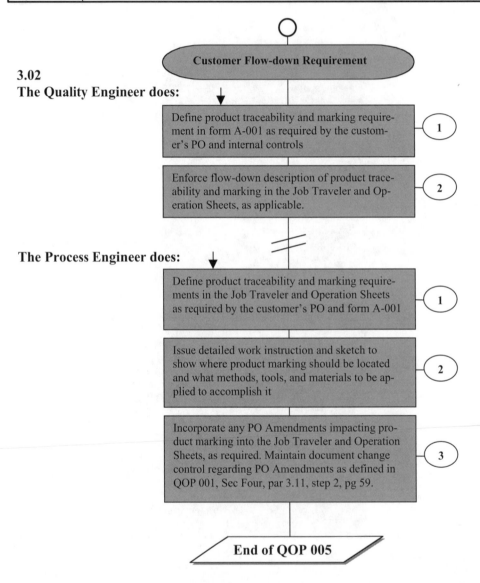

3.02
The Quality Engineer does:

Customer Flow-down Requirement

1. Define product traceability and marking requirement in form A-001 as required by the customer's PO and internal controls

2. Enforce flow-down description of product traceability and marking in the Job Traveler and Operation Sheets, as applicable.

The Process Engineer does:

1. Define product traceability and marking requirements in the Job Traveler and Operation Sheets as required by the customer's PO and form A-001

2. Issue detailed work instruction and sketch to show where product marking should be located and what methods, tools, and materials to be applied to accomplish it

3. Incorporate any PO Amendments impacting product marking into the Job Traveler and Operation Sheets, as required. Maintain document change control regarding PO Amendments as defined in QOP 001, Sec Four, par 3.11, step 2, pg 59.

End of QOP 005

OVERVIEW TO QOP 006
(INSPECTION AND TEST CONTROL)

This QOP is the longest of the quality operating procedures and it is by design that way. Instead of having a dozen or more procedures written separately on the different inspection assignments needed to support product quality enforcement, I have combined them all under one control. During my forty years in the field of Quality, I found that the various inspection functions were documented in such a fragmented, confusing manner that it took longer to find the procedure covering a specific type of inspection than to do the work itself. This was one reason why I combined them all under one directory. The second reason was to cut down on the volume of documents burdening the quality system, and the third was to maintain linkage to one QOP instead of a dozen. Now, anything you want to find regarding inspections will take you a second to find under its own index.

This consolidated QOP is the first most important controlling document of the quality management system, for it deals with product verification, acceptance, documentation and release of products to the customer. It has seven separate sections, each dedicated to control one or more inspection function, and it covers the entire processing cycle of a contracted product, including customer and subcontractor source inspection. Please look up the index, next page, to discover how extensively we cover the field of inspections.

Statistical process controls (SPC) may or may not be a determining factor, for the quality objective procedures, written for each assignment, have incorporated basic preventive instructions to monitor process variations and correct them when unacceptable levels arise. The use of control charts, commonly known as run charts, is strictly up to the customers, although the management representative could occasionally impose it on his own when it becomes necessary in the process of isolating the causes of problems.

DIAGRAMMATIC PROCEDURE   QOP 006

INSPECTION AND TEST CONTROL

CONTENTS   Page

TRAINING METRICS			Iss dt.		Rev dt.		Pg.
QUALITY OPERATING PROCEDURE			Sign		Rev #		3/52
REQUIREMENT: QUALITY SYSTEM MANUAL SEC: 4.10							
SUBJECT: INSPECTION AND TEST CONTROL							
QOP 006	DOCUMENT REQ'D: This procedure						

REVISION HISTORY

Rev Date	Rev No	Description	Approval

TRAINING METRICS	Iss dt.	Rev dt.	Pg.
QUALITY OPERATING PROCEDURE	Sign	Rev #	4/52
REQUIREMENT: Inspection and Test Control, QSM 4.10			
SUBJECT: General Requirements			
QOP 006	DOCUMENT REQ'D: This document		

GENERAL REQUIREMENTS

1.0   APPLICATION

This QOP 006 shall be applicable to enforce inspection and test requirements in the following areas:

1. First Piece Inspection
2. First Article Control (qualification process)
3. Process Control
4. Final Inspection
5. Receiving Inspection
6. Customer Source Inspection
7. Source Inspection at Subcontractors

2.0   DEFINITIONS   (used in controlling manufacturing processes)

1. **Controlled Process** – a defined activity undertaken to meet the requirement of a specification or work instruction after the approval of initial setup.
2. **Deviation** – a specific customer authorization issued prior to the processing of a product to allow departure from a defined design requirement for a specific number of units for a specific duration.
3. **First Article Inspection** – the complete inspection and test of a processed product made under process control wherein all measurements and test results, including special processes and workmanship, have been fully documented as required by a customer's purchase order.
4. **First Piece Inspection** – the complete inspection and test of a product's phase operation, which may be a single or multiple process, combined and carried out under one setup.
5. **Process Owner** – any employee trained to carry out a defined process in order to meet a specified requirement.
6. **Product non-conformance** – any condition that violates the requirement of a specification, process, or procedure.
7. **Repair** – relates to a non-conforming product, which cannot be further processed to meet a specified requirement without written approval from the customer.
8. **Rework** – relates to a non-conforming product that can be reprocessed under defined conditions to meet a specified drawing requirement.
9. **Waiver** – a written request sent to a customer to disposition of a nonconforming product.

TRAINING METRICS	Iss dt.	Rev dt.	Pg.
QUALITY OPERATING PROCEDURE	Sign	Rev #	5/52
REQUIREMENT: QUALITY SYSTEM MANUAL SEC: 4.10			
SUBJECT: General Requirements			
QOP 006 DOCUMENT REQ'D: This procedure			

3.0   PROCESS OWNER REQUIREMENTS

Process owners (and inspectors) performing production assignments shall observe the following:

1. that the measuring equipment being used is calibrated and periodically re-verified;
2. that the customer's product drawing number and revision level is verified against the Job Traveler and the program software. That notification to supervision is given when inconsistency is discovered;
3. that the first production piece is submitted for inspection verification and approval on all new setups identified in the Job Traveler;
4. that the production processes on customers' products are not carried out without work results verification and documentation.
5. that each completed operation as indicated in the Job Traveler is stamped, or signed and dated, and that all quantities have been accounted for.

End of General Requirements

TRAINING METRICS	Iss dt.	Rev dt.	Pg.
QUALITY OPERATING PROCEDURE	Sign	Rev #	6/52
REQUIREMENT: Inspection and Test Control, QSM 4.10			
SUBJECT: Some Basic Information			
QOP 006 DOCUMENT REQ'D: This document			

SECTION ONE

1.0 PURPOSE
To enforce quality provisions in controlling the verification, documentation, and approval of the first production unit.

2.0 APPLICATION
The provisions of this procedure shall apply to:
1. First production piece verification and approval;
2. First piece rework/repair verification and approval;
3. First piece verification and approval of subcontracted work, excluding special process.

3.0 PROCEDURE

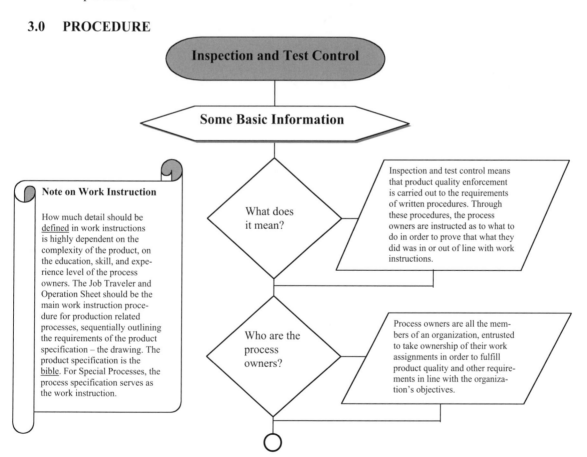

TRAINING METRICS	Iss dt.	Rev dt.	Pg.
QUALITY OPERATING PROCEDURE	Sign	Rev #	7/52
REQUIREMENT: Inspection and Test Control, QSM 4.10			
SUBJECT: Some Basic Information			
QOP 006	DOCUMENT REQ'D: This document			

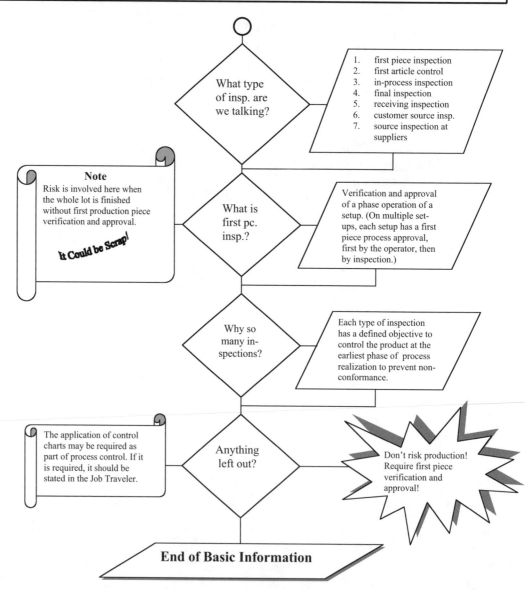

TRAINING METRICS	Iss dt.	Rev dt.	Pg.
QUALITY OPERATING PROCEDURE	Sign	Rev #	8/52
REQUIREMENT: Inspection and Test Control, QSM 4.10			
SUBJECT: First Piece Inspection (phase operation process control)			
QOP 006 DOCUMENT REQ'D: Forms A-006, A-008, A-012, A-013, A-017, A-018, A-023, A-033, A-037			

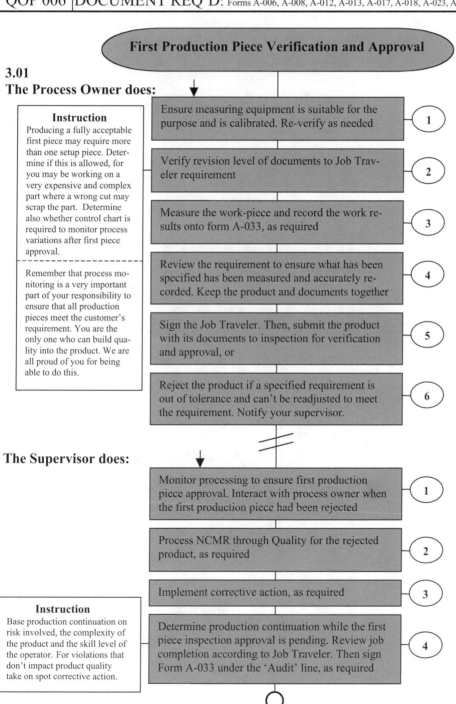

First Production Piece Verification and Approval

3.01
The Process Owner does:

Instruction

Producing a fully acceptable first piece may require more than one setup piece. Determine if this is allowed, for you may be working on a very expensive and complex part where a wrong cut may scrap the part. Determine also whether control chart is required to monitor process variations after first piece approval.

- - - - - - - - - - - - - -

Remember that process monitoring is a very important part of your responsibility to ensure that all production pieces meet the customer's requirement. You are the only one who can build quality into the product. We are all proud of you for being able to do this.

1. Ensure measuring equipment is suitable for the purpose and is calibrated. Re-verify as needed

2. Verify revision level of documents to Job Traveler requirement

3. Measure the work-piece and record the work results onto form A-033, as required

4. Review the requirement to ensure what has been specified has been measured and accurately recorded. Keep the product and documents together

5. Sign the Job Traveler. Then, submit the product with its documents to inspection for verification and approval, or

6. Reject the product if a specified requirement is out of tolerance and can't be readjusted to meet the requirement. Notify your supervisor.

The Supervisor does:

1. Monitor processing to ensure first production piece approval. Interact with process owner when the first production piece had been rejected

2. Process NCMR through Quality for the rejected product, as required

3. Implement corrective action, as required

Instruction

Base production continuation on risk involved, the complexity of the product and the skill level of the operator. For violations that don't impact product quality take on spot corrective action.

4. Determine production continuation while the first piece inspection approval is pending. Review job completion according to Job Traveler. Then sign Form A-033 under the 'Audit' line, as required

TRAINING METRICS	Iss dt.	Rev dt.	Pg.
QUALITY OPERATING PROCEDURE	Sign	Rev #	9/52
REQUIREMENT: Inspection and Test Control, QSM 4.10			
SUBJECT: First Piece Inspection (phase operation process control)			
QOP 006 |DOCUMENT REQ'D: Forms A-006, A-008, A-012, A-013, A-017, A-018, A-023, A-033, A-037			

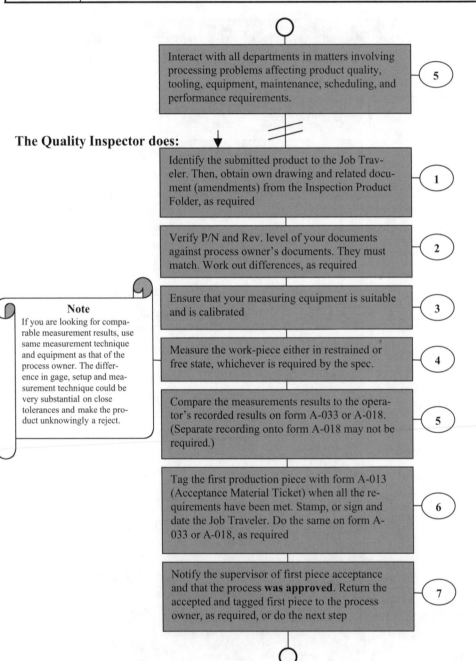

The Quality Inspector does:

Interact with all departments in matters involving processing problems affecting product quality, tooling, equipment, maintenance, scheduling, and performance requirements.   5

Identify the submitted product to the Job Traveler. Then, obtain own drawing and related document (amendments) from the Inspection Product Folder, as required   1

Verify P/N and Rev. level of your documents against process owner's documents. They must match. Work out differences, as required   2

Ensure that your measuring equipment is suitable and is calibrated   3

Note
If you are looking for comparable measurement results, use same measurement technique and equipment as that of the process owner. The difference in gage, setup and measurement technique could be very substantial on close tolerances and make the product unknowingly a reject.

Measure the work-piece either in restrained or free state, whichever is required by the spec.   4

Compare the measurements results to the operator's recorded results on form A-033 or A-018. (Separate recording onto form A-018 may not be required.)   5

Tag the first production piece with form A-013 (Acceptance Material Ticket) when all the requirements have been met. Stamp, or sign and date the Job Traveler. Do the same on form A-033 or A-018, as required   6

Notify the supervisor of first piece acceptance and that the process **was approved**. Return the accepted and tagged first piece to the process owner, as required, or do the next step   7

TRAINING METRICS	Iss dt.	Rev dt.	Pg.
QUALITY OPERATING PROCEDURE	Sign	Rev #	10/52
REQUIREMENT: Inspection and Test Control, QSM 4.10			
SUBJECT: First Piece Inspection (phase operation process control)			
QOP 006 DOCUMENT REQ'D: Forms A-006, A-008, A-012, A-013, A-017, A-018, A-023, A-033, A-037			

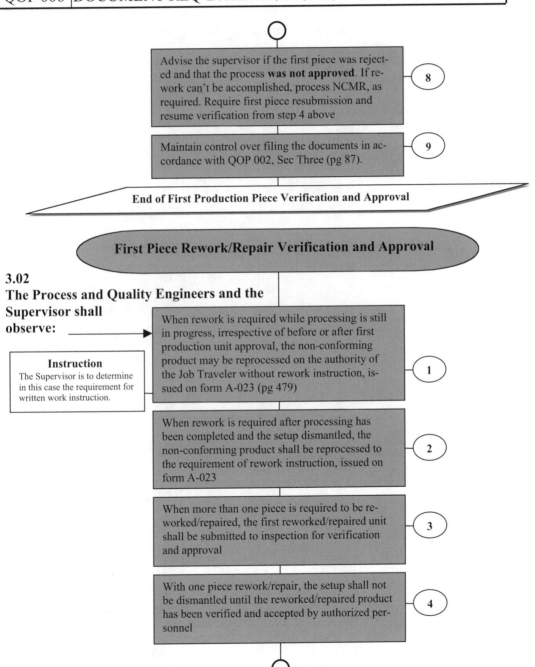

Advise the supervisor if the first piece was rejected and that the process **was not approved**. If rework can't be accomplished, process NCMR, as required. Require first piece resubmission and resume verification from step 4 above — **8**

Maintain control over filing the documents in accordance with QOP 002, Sec Three (pg 87). — **9**

End of First Production Piece Verification and Approval

First Piece Rework/Repair Verification and Approval

3.02
The Process and Quality Engineers and the Supervisor shall observe:

Instruction
The Supervisor is to determine in this case the requirement for written work instruction.

When rework is required while processing is still in progress, irrespective of before or after first production unit approval, the non-conforming product may be reprocessed on the authority of the Job Traveler without rework instruction, issued on form A-023 (pg 479) — **1**

When rework is required after processing has been completed and the setup dismantled, the non-conforming product shall be reprocessed to the requirement of rework instruction, issued on form A-023 — **2**

When more than one piece is required to be reworked/repaired, the first reworked/repaired unit shall be submitted to inspection for verification and approval — **3**

With one piece rework/repair, the setup shall not be dismantled until the reworked/repaired product has been verified and accepted by authorized personnel — **4**

| TRAINING METRICS | Iss dt. | Rev dt. | Pg. |
| QUALITY OPERATING PROCEDURE | Sign | Rev # | 11/52 |

REQUIREMENT: Inspection and Test Control, QSM 4.10

SUBJECT: First Piece Inspection (phase operation process control)

QOP 006 |DOCUMENT REQ'D: Forms A-006, A-008, A-012, A-013, A-017, A-018, A-023, A-033, A-037

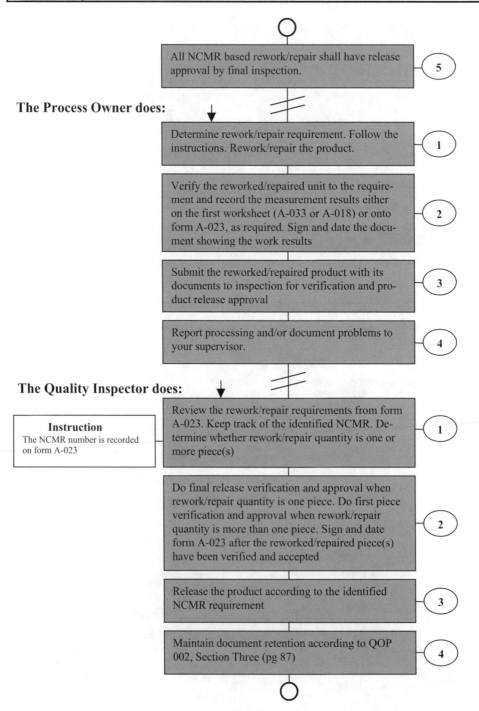

The Process Owner does:

5 — All NCMR based rework/repair shall have release approval by final inspection.

1 — Determine rework/repair requirement. Follow the instructions. Rework/repair the product.

2 — Verify the reworked/repaired unit to the requirement and record the measurement results either on the first worksheet (A-033 or A-018) or onto form A-023, as required. Sign and date the document showing the work results

3 — Submit the reworked/repaired product with its documents to inspection for verification and product release approval

4 — Report processing and/or document problems to your supervisor.

The Quality Inspector does:

Instruction
The NCMR number is recorded on form A-023

1 — Review the rework/repair requirements from form A-023. Keep track of the identified NCMR. Determine whether rework/repair quantity is one or more piece(s)

2 — Do final release verification and approval when rework/repair quantity is one piece. Do first piece verification and approval when rework/repair quantity is more than one piece. Sign and date form A-023 after the reworked/repaired piece(s) have been verified and accepted

3 — Release the product according to the identified NCMR requirement

4 — Maintain document retention according to QOP 002, Section Three (pg 87)

TRAINING METRICS	Iss dt.	Rev dt.	Pg.
QUALITY OPERATING PROCEDURE	Sign	Rev #	12/52

REQUIREMENT: Inspection and Test Control, QSM 4.10

SUBJECT: First Piece Inspection (phase operation process control)

QOP 006 | DOCUMENT REQ'D: Forms A-006, A-008, A-012, A-013, A-017, A-018, A-023, A-033, A-037

○

When the reworked/repaired product is rejected and it can't be corrected through additional processing, issue a new NCMR.   **5**

End of First Piece Rework/Repair Verification and Approval

3.03   First Piece Approval of Subcontracted Work, Excluding Special Process

The Receiver does:

Sign document(s) and receive the product sent by the subcontractor. (If the product had been hand carried, there may be nothing to sign.)   **1**

Turn the product and the attached document(s) over to the Quality Department.   **2**

The Quality Inspector does:

Identify the subcontractor and obtain the internally issued PO from the Supplier Product Folder   **1**

Determine the Quality Provisions from the PO as additional requirements to the terms and conditions of the PO   **2**

Verify the product to the PO and the specification (dwg) requirements. Record measurement results onto form A-018. Sign and date form A-018   **3**

Notify Purchasing (or Subcontractor's Quality) on verification results   **4**

If the product had been rejected, require resubmission of another first piece   **4a**

○

TRAINING METRICS	Iss dt.	Rev dt.	Pg.
QUALITY OPERATING PROCEDURE	Sign	Rev #	13/52
REQUIREMENT: Inspection and Test Control, QSM 4.10			
SUBJECT: First Piece Inspection (phase operation process control)			
QOP 006 DOCUMENT REQ'D: Forms A-006, A-008, A-012, A-013, A-017, A-018, A-023, A-033, A-037			

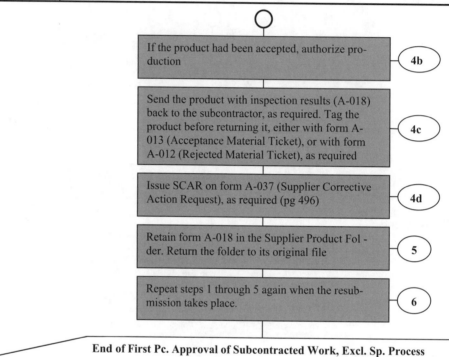

If the product had been accepted, authorize production — **4b**

Send the product with inspection results (A-018) back to the subcontractor, as required. Tag the product before returning it, either with form A-013 (Acceptance Material Ticket), or with form A-012 (Rejected Material Ticket), as required — **4c**

Issue SCAR on form A-037 (Supplier Corrective Action Request), as required (pg 496) — **4d**

Retain form A-018 in the Supplier Product Folder. Return the folder to its original file — **5**

Repeat steps 1 through 5 again when the resubmission takes place. — **6**

End of First Pc. Approval of Subcontracted Work, Excl. Sp. Process

End of Section One

TRAINING METRICS	Iss dt.	Rev dt.	Pg.
QUALITY OPERATING PROCEDURE	Sign	Rev #	14/52

REQUIREMENT: Inspection and Test Control, QSM 4.10

SUBJECT: First Article Control

QOP 006|DOCUMENT REQ'D: A-001, A-006, A-008, A-012, A-013, A-017, A-018, A-023, A-033, A-037

SECTION TWO

1.0 PURPOSE

Enforce the quality provisions in controlling First Article related assignments and documentation. (This is a first article qualification process control for a single unit. For multiunit, becoming an assembly, the same provisions apply, except in the plural.)

2.0 APPLICATION

Provisions of this procedure shall apply to controlling:
1. phases of the process assignments
2. inspection and documentation
3. customer source inspection
4. packaging and delivery
5. document retention

3.0 PROCEDURE

General Requirements
1. First Article quality requirements shall be detailed in the Purchase Order Review Sheet, Form A-001;
2. The Job Traveler (A-008) shall outline the processing steps in sequential order, including subcontracting, in order to control an orderly process realization;
3. The Quality Engineer shall enforce process controls in line with the requirements stated in the Job Traveler and the Purchase Order Review Sheet, A-001;
4. All the provisions of the quality system procedures shall be binding on all departments carrying out the First Article objectives, as required.

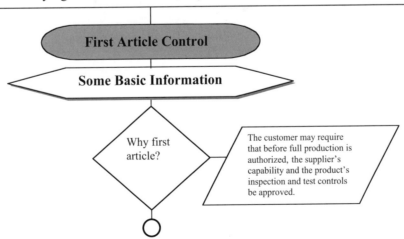

TRAINING METRICS	Iss dt.	Rev dt.	Pg.
QUALITY OPERATING PROCEDURE	Sign	Rev #	15/52
REQUIREMENT: Inspection and Test Control, QSM 4.10			
SUBJECT: First Article Control			
QOP 006|DOCUMENT REQ'D: A-001, A-006, A-008, A-012, A-013, A-017, A-018, A-023, A-033, A-037			

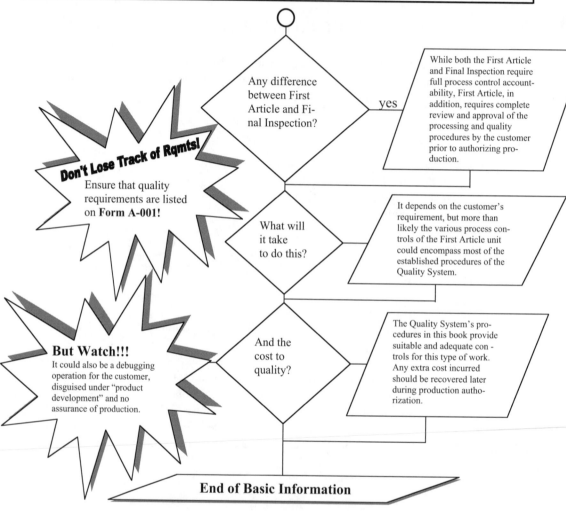

Any difference between First Article and Final Inspection?

yes

While both the First Article and Final Inspection require full process control accountability, First Article, in addition, requires complete review and approval of the processing and quality procedures by the customer prior to authorizing production.

Don't Lose Track of Rqmts!
Ensure that quality requirements are listed on **Form A-001!**

What will it take to do this?

It depends on the customer's requirement, but more than likely the various process controls of the First Article unit could encompass most of the established procedures of the Quality System.

But Watch!!!
It could also be a debugging operation for the customer, disguised under "product development" and no assurance of production.

And the cost to quality?

The Quality System's procedures in this book provide suitable and adequate controls for this type of work. Any extra cost incurred should be recovered later during production authorization.

End of Basic Information

TRAINING METRICS	Iss dt.	Rev dt.	Pg.
QUALITY OPERATING PROCEDURE	Sign	Rev #	16/52
REQUIREMENT: Inspection and Test Control, QSM 4.10			
SUBJECT: First Article Control			
QOP 006 DOCUMENT REQ'D: A-001, A-006, A-008, A-012, A-013, A-017, A-018, A-023, A-033, A-037			

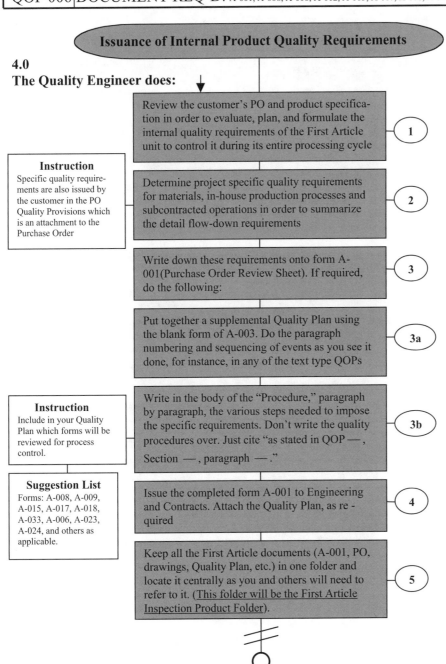

Issuance of Internal Product Quality Requirements

4.0

The Quality Engineer does:

1. Review the customer's PO and product specification in order to evaluate, plan, and formulate the internal quality requirements of the First Article unit to control it during its entire processing cycle

Instruction
Specific quality requirements are also issued by the customer in the PO Quality Provisions which is an attachment to the Purchase Order

2. Determine project specific quality requirements for materials, in-house production processes and subcontracted operations in order to summarize the detail flow-down requirements

3. Write down these requirements onto form A-001(Purchase Order Review Sheet). If required, do the following:

3a. Put together a supplemental Quality Plan using the blank form of A-003. Do the paragraph numbering and sequencing of events as you see it done, for instance, in any of the text type QOPs

Instruction
Include in your Quality Plan which forms will be reviewed for process control.

3b. Write in the body of the "Procedure," paragraph by paragraph, the various steps needed to impose the specific requirements. Don't write the quality procedures over. Just cite "as stated in QOP —, Section —, paragraph —."

Suggestion List
Forms: A-008, A-009, A-015, A-017, A-018, A-033, A-006, A-023, A-024, and others as applicable.

4. Issue the completed form A-001 to Engineering and Contracts. Attach the Quality Plan, as re-quired

5. Keep all the First Article documents (A-001, PO, drawings, Quality Plan, etc.) in one folder and locate it centrally as you and others will need to refer to it. (This folder will be the First Article Inspection Product Folder).

TRAINING METRICS	Iss dt.	Rev dt.	Pg.
QUALITY OPERATING PROCEDURE	Sign	Rev #	17/52
REQUIREMENT: Inspection and Test Control, QSM 4.10			
SUBJECT: First Article Control			
QOP 006|DOCUMENT REQ'D: A-001, A-006, A-008, A-012, A-013, A-017, A-018, A-023, A-033, A-037			

Issuance of the First Article Production Folder

5.0
The Project Engineer does:

Review the customer's PO and product specifications in order to evaluate, plan, and formulate the processing related requirements of the First Article unit to control it during its entire production cycle. Review form A-001, sent to you by the Quality Engineer, and incorporate the quality requirements into the processing related procedures (form A-008, pg 463), as required — **1**

Instruction
Ensure that workmanship requirement is defined for every processing phase impacting product realization.

Determine project related specific requirements for materials, in-house production processes and subcontracted operations in order to incorporate the detail flow-down requirements in your layout plans. Also determine the following: — **2**

Phase operational requirements for equipment, tools and fixtures, special gages, programming, and qualified personnel — **2a**

Processing capability for each phase operation. If required, invoke the 'developmental' process controls stated under QOP 012, Sec Three, pg 263, for proving out process techniques before issuing the final procedures — **2b**

Interact with the Quality Engineer in order that the processing related documents (A-008) and the product's inspection requirements (A-001) properly linked to control the First Article throughout its production cycle — **3**

Interact with the department heads to ensure adequate planning and coordination regarding purchasing, scheduling, production, maintenance, training and other related matters, as required — **4**

TRAINING METRICS	Iss dt.	Rev dt.	Pg.
QUALITY OPERATING PROCEDURE	Sign	Rev #	18/52

REQUIREMENT: Inspection and Test Control, QSM 4.10
SUBJECT: First Article Control
QOP 006|DOCUMENT REQ'D: A-001, A-006, A-008, A-012, A-013, A-017, A-018, A-023, A-033, A-037

Put together all the plans and formalize the pro-cessing related procedures (setup sheet, tool and gage sheets, cut sheet, programming, sketches, operation sheets and the job traveler) as required — **5**

Review and approve the controlling procedures. Issue the Production Folder, as required. — **6**

First Article Process Control

6.0
The Quality Engineer does:

Note
Check the header above. It contains the relevant form numbers.

Suggest you create a simple matrix on form A-003 showing the name of the process owner, the form(s) used, the com-pliance results yes – no –and C/A required yes – no -- .

Identify the process control specifics from the already issued Job Traveler and Purchase Order Review Sheet and write them down on a blank form of A-003. Title it 'In-process Audit and Follow-up.' Determine and note only the various forms on which work results have been docu-mented. Use form A-003 as your checklist — **1**

Go to Section Three of this QOP (pg 151) and determine the process control requirement for each form noted on A-003 — **2**

Carry out the monitoring in every department, as required. Review and verify whether each process owner had carried out the work results recording, as required. Note the results on form A-003 — **3**

Instruction
For major product quality violations, go to paragraph 7.0 below.

Carry out on-spot corrective action where in - fringements are minor. Note this also on A-003 — **4**

Sign and date each form under the 'Audit' line. Sign and date form A-003 also — **5**

Enter all the C/A related instructions you did onto form A-016 (Training Log). Refer to A-003 — **6**

TRAINING METRICS	Iss dt.	Rev dt.	Pg.
QUALITY OPERATING PROCEDURE	Sign	Rev #	19/52
REQUIREMENT: Inspection and Test Control, QSM 4.10			
SUBJECT: First Article Control			
QOP 006|DOCUMENT REQ'D: A-001, A-006, A-008, A-012, A-013, A-017, A-018, A-023, A-033, A-037			

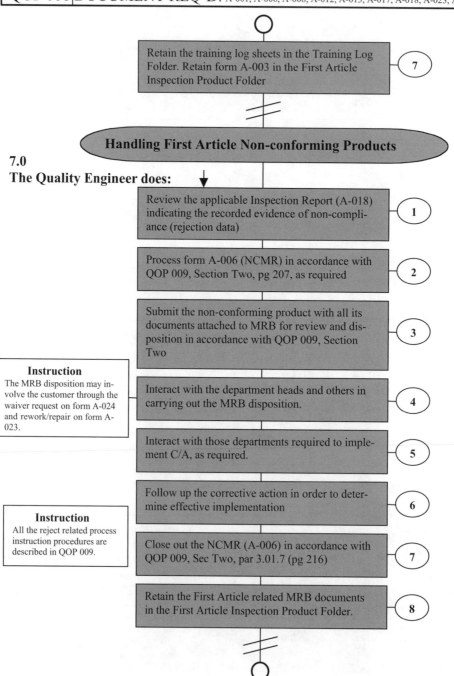

Retain the training log sheets in the Training Log Folder. Retain form A-003 in the First Article Inspection Product Folder — **7**

Handling First Article Non-conforming Products

7.0
The Quality Engineer does:

Review the applicable Inspection Report (A-018) indicating the recorded evidence of non-compliance (rejection data) — **1**

Process form A-006 (NCMR) in accordance with QOP 009, Section Two, pg 207, as required — **2**

Submit the non-conforming product with all its documents attached to MRB for review and disposition in accordance with QOP 009, Section Two — **3**

Instruction
The MRB disposition may involve the customer through the waiver request on form A-024 and rework/repair on form A-023.

Interact with the department heads and others in carrying out the MRB disposition. — **4**

Interact with those departments required to implement C/A, as required. — **5**

Follow up the corrective action in order to determine effective implementation — **6**

Instruction
All the reject related process instruction procedures are described in QOP 009.

Close out the NCMR (A-006) in accordance with QOP 009, Sec Two, par 3.01.7 (pg 216) — **7**

Retain the First Article related MRB documents in the First Article Inspection Product Folder. — **8**

TRAINING METRICS	Iss dt.	Rev dt.	Pg.
QUALITY OPERATING PROCEDURE	Sign	Rev #	20/52
REQUIREMENT: Inspection and Test Control, QSM 4.10			
SUBJECT: First Article Control			
QOP 006 DOCUMENT REQ'D: A-001, A-006, A-008, A-012, A-013, A-017, A-018, A-023, A-033, A-037			

Compiling the First Article Inspection Report

8.0

Note 1. The First Article Inspection Report shouldn't be put together until all the required processing steps have been completed and the product verified, approved, and released from final inspection.

Note 2. The First Article Inspection Report (FAIR) consists of all those documents that were issued to control the various processing operations including subcontracting, special processes and any MRB activity, proving work results documentation in compliance with requirements.

The Quality Engineer does:

Instruction
The First Article related processing and quality records should be located in these three folders.

1. Bring together the Production Folder, the First Article Inspection Product Folder, and the Supplier Product Folder.

Instruction
Look for the signatures after each operation number sequenced in the Job Traveler. No signature means the process owner forgot to sign off the specific operation or it wasn't done.

2. Find the Job Traveler and form A-001. Review both documents and ensure that all the requirements have been complied with. For any omission of the requirements, process an NCMR, as required, and execute it in accordance with QOP 009, Section One and Two. Begin gathering the documents needed to make up the First Article Inspection Report (FAIR)

You may omit this step if forms A-008 and A-001 were already reviewed under paragraph 6.0 above.

Instruction
When the customer supplies the material, it usually does not provide the certification with it. The customer's PO should specify what has been supplied.

When the First Article consists of more than one part, there should be a parts list notation in the PO.

3. Start with the Job Traveler. The first step should be 'Material Release.' Find the material's certification. This document should be your fourth record in the 'FAIR' stack. (The first one is the customer's PO. The second one is the Job Traveler. The third one is form A-001)

Stay neat in all document presentation. It's highly reflective of product quality!

4. Go, step by step, following the Job Traveler, pull together from the folders all those records substantiating work results documentation, including subcontracting and special processes. Place the documents in sequential order. Place any NCMR related documents where the non-conformance occurred. The last document should be the Final Inspection Report (A-018). (The C of C comes separately with the shipping documents)

Note
Any document missing is probably left behind somewhere in the processing area. Process owners are not the best organizers of job related papers.

TRAINING METRICS	Iss dt.	Rev dt.	Pg.
QUALITY OPERATING PROCEDURE	Sign	Rev #	21/52
REQUIREMENT: Inspection and Test Control, QSM 4.10			
SUBJECT: First Article Control			
QOP 006│DOCUMENT REQ'D: A-001, A-006, A-008, A-012, A-013, A-017, A-018, A-023, A-033, A-037			

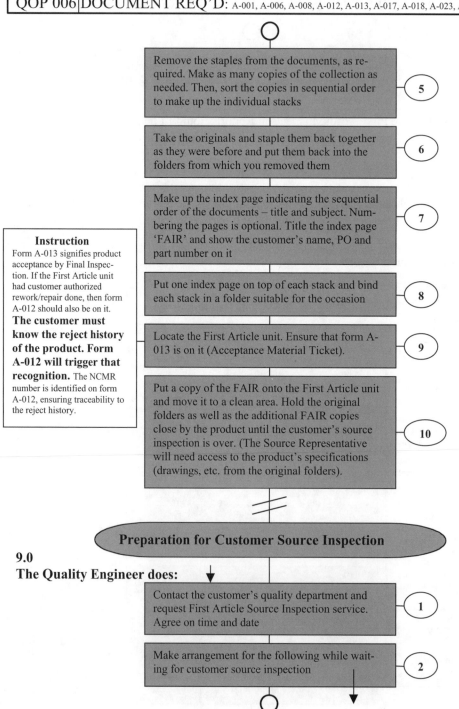

Instruction

Form A-013 signifies product acceptance by Final Inspection. If the First Article unit had customer authorized rework/repair done, then form A-012 should also be on it. **The customer must know the reject history of the product. Form A-012 will trigger that recognition.** The NCMR number is identified on form A-012, ensuring traceability to the reject history.

Remove the staples from the documents, as required. Make as many copies of the collection as needed. Then, sort the copies in sequential order to make up the individual stacks — 5

Take the originals and staple them back together as they were before and put them back into the folders from which you removed them — 6

Make up the index page indicating the sequential order of the documents – title and subject. Numbering the pages is optional. Title the index page 'FAIR' and show the customer's name, PO and part number on it — 7

Put one index page on top of each stack and bind each stack in a folder suitable for the occasion — 8

Locate the First Article unit. Ensure that form A-013 is on it (Acceptance Material Ticket). — 9

Put a copy of the FAIR onto the First Article unit and move it to a clean area. Hold the original folders as well as the additional FAIR copies close by the product until the customer's source inspection is over. (The Source Representative will need access to the product's specifications (drawings, etc. from the original folders). — 10

Preparation for Customer Source Inspection

9.0
The Quality Engineer does:

Contact the customer's quality department and request First Article Source Inspection service. Agree on time and date — 1

Make arrangement for the following while waiting for customer source inspection — 2

TRAINING METRICS	Iss dt.	Rev dt.	Pg.
QUALITY OPERATING PROCEDURE	Sign	Rev #	22/52

REQUIREMENT: Inspection and Test Control, QSM 4.10

SUBJECT: First Article Control

QOP 006|DOCUMENT REQ'D: A-001, A-006, A-008, A-012, A-013, A-017, A-018, A-023, A-033, A-037

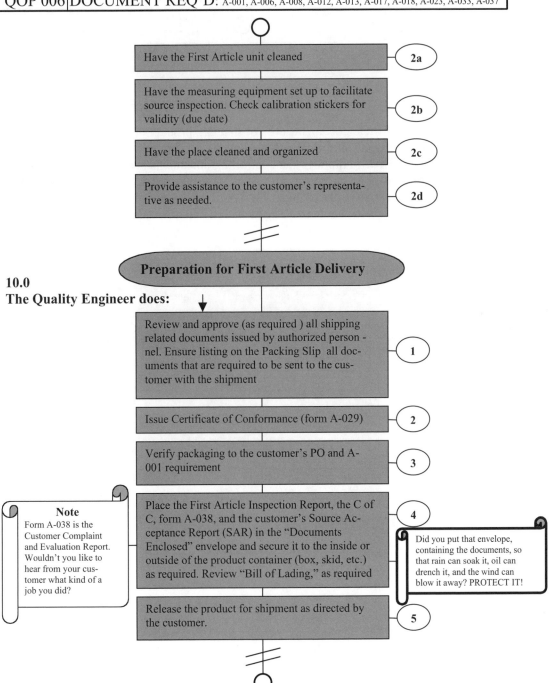

Have the First Article unit cleaned — **2a**

Have the measuring equipment set up to facilitate source inspection. Check calibration stickers for validity (due date) — **2b**

Have the place cleaned and organized — **2c**

Provide assistance to the customer's representative as needed. — **2d**

Preparation for First Article Delivery

10.0
The Quality Engineer does:

Review and approve (as required) all shipping related documents issued by authorized person-nel. Ensure listing on the Packing Slip all documents that are required to be sent to the customer with the shipment — **1**

Issue Certificate of Conformance (form A-029) — **2**

Verify packaging to the customer's PO and A-001 requirement — **3**

Note
Form A-038 is the Customer Complaint and Evaluation Report. Wouldn't you like to hear from your customer what kind of a job you did?

Place the First Article Inspection Report, the C of C, form A-038, and the customer's Source Acceptance Report (SAR) in the "Documents Enclosed" envelope and secure it to the inside or outside of the product container (box, skid, etc.) as required. Review "Bill of Lading," as required — **4**

Did you put that envelope, containing the documents, so that rain can soak it, oil can drench it, and the wind can blow it away? PROTECT IT!

Release the product for shipment as directed by the customer. — **5**

TRAINING METRICS	Iss dt.	Rev dt.	Pg.
QUALITY OPERATING PROCEDURE	Sign	Rev #	23/52
REQUIREMENT: Inspection and Test Control, QSM 4.10			
SUBJECT: First Article Control			
QOP 006|DOCUMENT REQ'D: A-001, A-006, A-008, A-012, A-013, A-017, A-018, A-023, A-033, A-037			

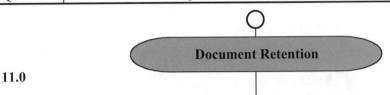

Document Retention

11.0

General

The First Article documents are not standard production documents due to the nature of project specific handling and controls imposed by the customer's Purchase Order. Although the controlling documents had been derived from the established quality system's procedures, the purpose for which they were applied made them isolated under First Article sectioning. Therefore, retention of these documents should not be intermixed with the standard production and quality records.

The Quality Engineer does:

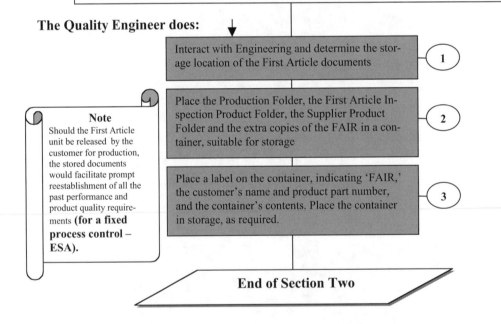

Interact with Engineering and determine the storage location of the First Article documents **1**

Place the Production Folder, the First Article Inspection Product Folder, the Supplier Product Folder and the extra copies of the FAIR in a container, suitable for storage **2**

Place a label on the container, indicating 'FAIR,' the customer's name and product part number, and the container's contents. Place the container in storage, as required. **3**

Note
Should the First Article unit be released by the customer for production, the stored documents would facilitate prompt reestablishment of all the past performance and product quality requirements **(for a fixed process control – ESA).**

End of Section Two

TRAINING METRICS	Iss dt.	Rev dt.	Pg.
QUALITY OPERATING PROCEDURE	Sign	Rev #	24/52
REQUIREMENT: Inspection and Test Control, QSM 4.9			
SUBJECT: Process Control (performance surveillance)			
QOP 006	DOCUMENT REQ'D: All the forms listed and controlled in the Quality System		

SECTION THREE

1.0   PURPOSE

Maintain performance surveillance over those activities affecting the quality system's objectives and requirements.

2.0   APPLICATION

Performance surveillance shall apply to the following document handling, recording, and processing related activities:

1. material control
2. tools, fixtures, and processing equipment
3. processing related documents
4. dimensional verification
5. special processes
6. final inspection
7. workmanship and other processing related requirements
8. non-conforming materials
9. product release
10. customer returns
11. corrective action and follow-up
12. contracts, engineering, and purchasing

3.0   PROCEDURE

General Requirements

1. Supervision shall be responsible to review performance results of the activities listed under paragraph 2.0 above in order to control those activities to ensure compliance to the requirements of documented procedures. The applicable forms shall be acknowledged by the reviewer by his/her signature and date at the time the performance surveillance was accomplished.

2. The supervisor shall have the authority to implement on spot corrective action with the process owner(s) who are assigned under his/her responsibility. For minor violations where only instructional methods have been used to correct documentation, the supervisor shall apply Form A-016 (Training Log Sheet) to record the activity. For major violations, impacting product quality, the supervisor shall apply the NCMR process (on Form A-006) in accordance with QOP 009, Section One and Two, as required. (pg 205, 207)

152   TRAINING METRICS 006

TRAINING METRICS	Iss dt.	Rev dt.	Pg.
QUALITY OPERATING PROCEDURE	Sign	Rev #	25/52
REQUIREMENT: Inspection and Test Control, QSM 4.9			
SUBJECT: Process Control (performance surveillance)			
QOP 006 DOCUMENT REQ'D: All the forms listed and controlled in the Quality System			

Process Control

Your assurance to ensure effective performance!

Some Basic Information

What's process control?

All activities that have been imposed in the progressive realization of a product need to be verified through a monitoring process, known as surveillance, in order to determine compliance to the imposed requirement.

Note!
We are only reviewing processing related assignments regarding performance requirements. Process reviews are not the "annual" quality system audits, which are scheduled according to the Audit Schedule established on Form A-021. Process surveillance has everything to do with preventive quality and the ongoing improvement practices to the operating system. This effort represents a major component of the Process Approach Quality Management System to enforce the established requirements.

How do we measure performance?

We measure performance through the recorded evidence of the work results. We evaluate this recorded evidence against the requirement stated in the applicable procedure (work instruction) in order to prove whether we did or did not comply with the stated requirement.

Where are the specified requirements?

For the Quality Management System, the specified requirements are stated in the Quality Manual. For controlling product quality, the specified requirements are stated in the Quality Operating Procedures. Together, these two tier documents regulate what we must do, step by step, to comply with those specified requirements needed to enforce product quality.

Note!
Linkage in-between the documents is maintained by references in order to indicate where to find the specified requirements.

Who needs all this detail?

The responsibility for product quality is no longer in the hands of one department – Quality. Other departments handling product related assignments must also be knowledgeable of the product quality requirements so that they don't pass on their mistakes from process to process. Being a process owner of a product related assignment, one must know how to comply with the rules of the quality system. Non-quality personnel need all this information.

You may have a better idea. Write it down! See book page 501, 'The Suggestion Path'

TRAINING METRICS	Iss dt.	Rev dt.	Pg.
QUALITY OPERATING PROCEDURE	Sign	Rev #	26/52
REQUIREMENT: Inspection and Test Control, QSM 4.9			
SUBJECT: Process Control (performance surveillance)			
QOP 006 |DOCUMENT REQ'D: All the forms listed and controlled in the Quality System			

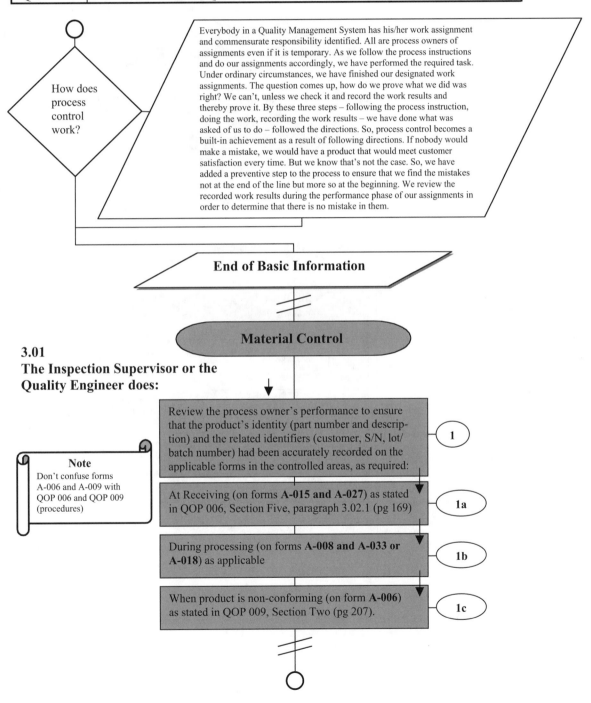

How does process control work?

Everybody in a Quality Management System has his/her work assignment and commensurate responsibility identified. All are process owners of assignments even if it is temporary. As we follow the process instructions and do our assignments accordingly, we have performed the required task. Under ordinary circumstances, we have finished our designated work assignments. The question comes up, how do we prove what we did was right? We can't, unless we check it and record the work results and thereby prove it. By these three steps – following the process instruction, doing the work, recording the work results – we have done what was asked of us to do – followed the directions. So, process control becomes a built-in achievement as a result of following directions. If nobody would make a mistake, we would have a product that would meet customer satisfaction every time. But we know that's not the case. So, we have added a preventive step to the process to ensure that we find the mistakes not at the end of the line but more so at the beginning. We review the recorded work results during the performance phase of our assignments in order to determine that there is no mistake in them.

End of Basic Information

Material Control

3.01
The Inspection Supervisor or the
Quality Engineer does:

Review the process owner's performance to ensure that the product's identity (part number and description) and the related identifiers (customer, S/N, lot/batch number) had been accurately recorded on the applicable forms in the controlled areas, as required: **1**

Note
Don't confuse forms A-006 and A-009 with QOP 006 and QOP 009 (procedures)

At Receiving (on forms **A-015 and A-027**) as stated in QOP 006, Section Five, paragraph 3.02.1 (pg 169) **1a**

During processing (on forms **A-008 and A-033 or A-018**) as applicable **1b**

When product is non-conforming (on form **A-006**) as stated in QOP 009, Section Two (pg 207). **1c**

TRAINING METRICS	Iss dt.	Rev dt.	Pg.
QUALITY OPERATING PROCEDURE	Sign	Rev #	27/52
REQUIREMENT: Inspection and Test Control, QSM 4.9			
SUBJECT: Process Control (performance surveillance)			
QOP 006	DOCUMENT REQ'D: All the forms listed and controlled in the Quality System		

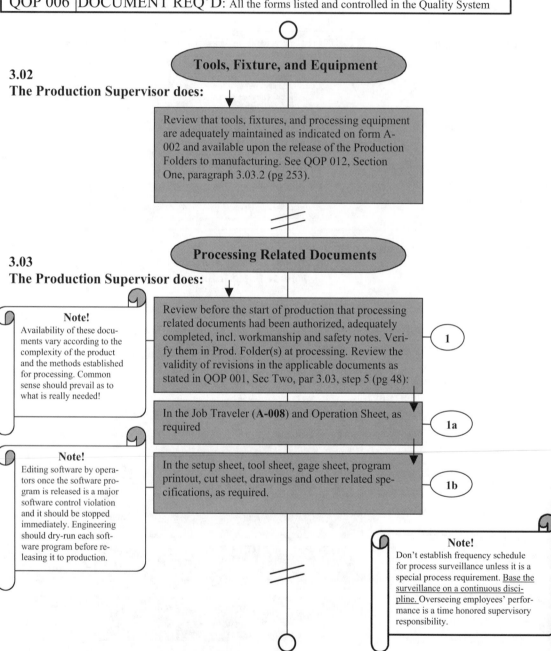

3.02
The Production Supervisor does:

Tools, Fixture, and Equipment

Review that tools, fixtures, and processing equipment are adequately maintained as indicated on form A-002 and available upon the release of the Production Folders to manufacturing. See QOP 012, Section One, paragraph 3.03.2 (pg 253).

3.03
The Production Supervisor does:

Processing Related Documents

Note!
Availability of these documents vary according to the complexity of the product and the methods established for processing. Common sense should prevail as to what is really needed!

Review before the start of production that processing related documents had been authorized, adequately completed, incl. workmanship and safety notes. Verify them in Prod. Folder(s) at processing. Review the validity of revisions in the applicable documents as stated in QOP 001, Sec Two, par 3.03, step 5 (pg 48): **1**

In the Job Traveler (**A-008**) and Operation Sheet, as required **1a**

In the setup sheet, tool sheet, gage sheet, program printout, cut sheet, drawings and other related specifications, as required. **1b**

Note!
Editing software by operators once the software program is released is a major software control violation and it should be stopped immediately. Engineering should dry-run each software program before releasing it to production.

Note!
Don't establish frequency schedule for process surveillance unless it is a special process requirement. Base the surveillance on a continuous discipline. Overseeing employees' performance is a time honored supervisory responsibility.

TRAINING METRICS	Iss dt.	Rev dt.	Pg.
QUALITY OPERATING PROCEDURE	Sign	Rev #	28/52
REQUIREMENT: Inspection and Test Control, QSM 4.9			
SUBJECT: Process Control (performance surveillance)			
QOP 006 |DOCUMENT REQ'D: All the forms listed and controlled in the Quality System			

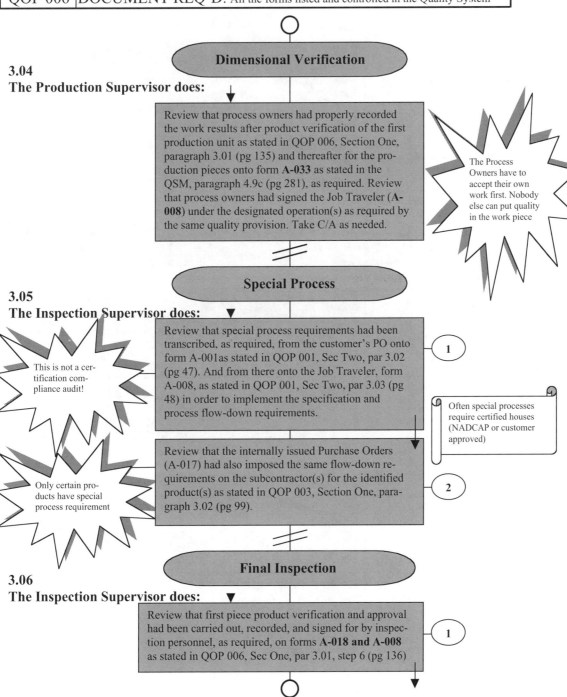

3.04
The Production Supervisor does:

Dimensional Verification

Review that process owners had properly recorded the work results after product verification of the first production unit as stated in QOP 006, Section One, paragraph 3.01 (pg 135) and thereafter for the production pieces onto form **A-033** as stated in the QSM, paragraph 4.9c (pg 281), as required. Review that process owners had signed the Job Traveler (**A-008**) under the designated operation(s) as required by the same quality provision. Take C/A as needed.

The Process Owners have to accept their own work first. Nobody else can put quality in the work piece

3.05
The Inspection Supervisor does:

Special Process

This is not a certification compliance audit!

Review that special process requirements had been transcribed, as required, from the customer's PO onto form A-001as stated in QOP 001, Sec Two, par 3.02 (pg 47). And from there onto the Job Traveler, form A-008, as stated in QOP 001, Sec Two, par 3.03 (pg 48) in order to implement the specification and process flow-down requirements.

1

Often special processes require certified houses (NADCAP or customer approved)

Review that the internally issued Purchase Orders (A-017) had also imposed the same flow-down requirements on the subcontractor(s) for the identified product(s) as stated in QOP 003, Section One, paragraph 3.02 (pg 99).

2

Only certain products have special process requirement

3.06
The Inspection Supervisor does:

Final Inspection

Review that first piece product verification and approval had been carried out, recorded, and signed for by inspection personnel, as required, on forms **A-018 and A-008** as stated in QOP 006, Sec One, par 3.01, step 6 (pg 136)

1

TRAINING METRICS	Iss dt.	Rev dt.	Pg.
QUALITY OPERATING PROCEDURE	Sign	Rev #	29/52
REQUIREMENT: Inspection and Test Control, QSM 4.9			
SUBJECT: Process Control (performance surveillance)			
QOP 006 DOCUMENT REQ'D: All the forms listed and controlled in the Quality System			

Note!
Form A-018 is the only document indicating the recording of data for product verification for final inspection.

Review final product verification and acceptance on forms **A-018, A-013, and A-008** as stated in QOP 006, Section Four. Review the inspectors' job performance as stated in QOP 006, Section Four (pg 161). All paragraphs apply

2

Note!
Final inspection may be carried out in several steps depending on processing requirements.

Review the tracking and recording of accepted and rejected quantities as described in the same quality provisions.

3

Workmanship and Other Processing Related Requirements

3.07
The Production Supervisor does:

Instruction
Other processing related requirements may involve product handling, stacking, preservation and storage, safety and protection gear, etc.

Review that workmanship and other processing related requirements had been completed, verified, and recorded as per work instructions in the Job Traveler and Operation Sheet. Refer to QOP 001, Section Two, paragraph 3.03, step 2 (pg 48)

1

Review that all the completed work pieces had been cleaned and the foreign objects (chips, dirt) removed and preservation is maintained to the same provisions of QOP 001, Section Two, paragraph 3.03, step 2.

2

Non-conforming Materials

3.08
The Inspection Supervisor does:

Advisement
Observe that MRB dispositions are carried out by the possible application of different forms.– document change on Form A-009; waivers on Form A-024; rework/ repair on Form A-023; subs C/A request on A-037. These are all part of an effective NCMR process control.

Review that non-conforming products had been identified and documented throughout the entire processing areas as stated in QOP 009, Section One and Two.

1

Review that the NCMR number had been transcribed onto the other processing related documents such as forms **A-008, A-009, A-018, A-012, A-024, A-023, and A-037** to ensure the recognition that the product had NCMR history. (See pg 207, step 3)

Instruction
Identification requires the application of form A-012 Documentation requires the application of form A-006

2

TRAINING METRICS	Iss dt.	Rev dt.	Pg.
QUALITY OPERATING PROCEDURE	Sign	Rev #	30/52
REQUIREMENT: Inspection and Test Control, QSM 4.9			
SUBJECT: Process Control (performance surveillance)			
QOP 006 |DOCUMENT REQ'D: All the forms listed and controlled in the Quality System			

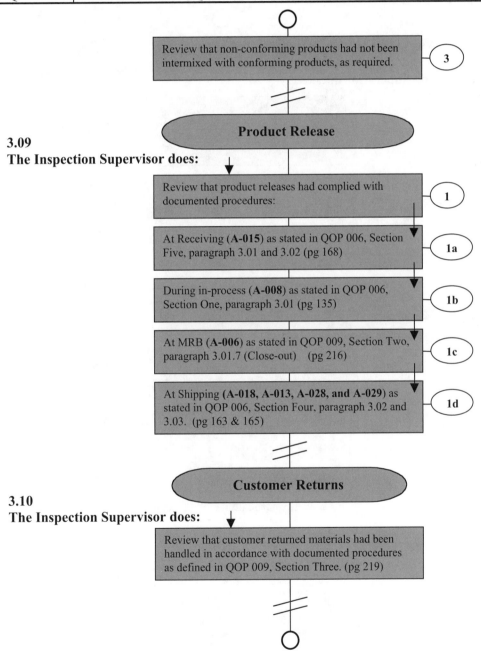

Review that non-conforming products had not been intermixed with conforming products, as required.   **3**

Product Release

3.09
The Inspection Supervisor does:

Review that product releases had complied with documented procedures: **1**

At Receiving (**A-015**) as stated in QOP 006, Section Five, paragraph 3.01 and 3.02 (pg 168)   **1a**

During in-process (**A-008**) as stated in QOP 006, Section One, paragraph 3.01 (pg 135)   **1b**

At MRB (**A-006**) as stated in QOP 009, Section Two, paragraph 3.01.7 (Close-out)   (pg 216)   **1c**

At Shipping (**A-018, A-013, A-028, and A-029**) as stated in QOP 006, Section Four, paragraph 3.02 and 3.03. (pg 163 & 165)   **1d**

Customer Returns

3.10
The Inspection Supervisor does:

Review that customer returned materials had been handled in accordance with documented procedures as defined in QOP 009, Section Three. (pg 219)

TRAINING METRICS	Iss dt.	Rev dt.	Pg.
QUALITY OPERATING PROCEDURE	Sign	Rev #	31/52
REQUIREMENT: Inspection and Test Control, QSM 4.9			
SUBJECT: Process Control (performance surveillance)			
QOP 006 |DOCUMENT REQ'D: All the forms listed and controlled in the Quality System			

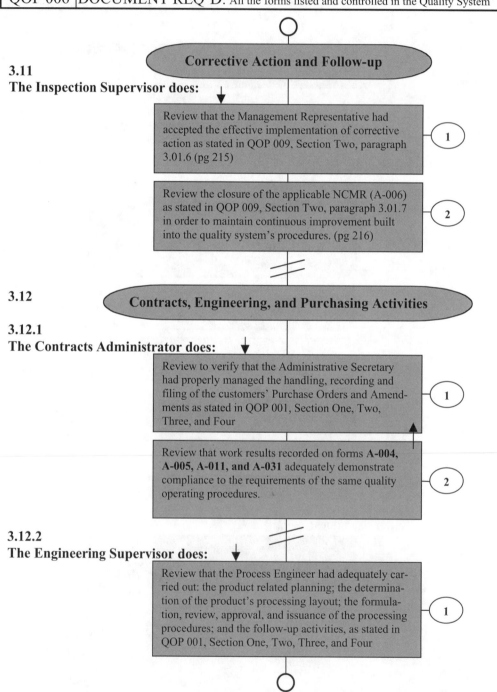

3.11
The Inspection Supervisor does:

Corrective Action and Follow-up

Review that the Management Representative had accepted the effective implementation of corrective action as stated in QOP 009, Section Two, paragraph 3.01.6 (pg 215) 1

Review the closure of the applicable NCMR (A-006) as stated in QOP 009, Section Two, paragraph 3.01.7 in order to maintain continuous improvement built into the quality system's procedures. (pg 216) 2

3.12 **Contracts, Engineering, and Purchasing Activities**

3.12.1
The Contracts Administrator does:

Review to verify that the Administrative Secretary had properly managed the handling, recording and filing of the customers' Purchase Orders and Amendments as stated in QOP 001, Section One, Two, Three, and Four 1

Review that work results recorded on forms **A-004, A-005, A-011, and A-031** adequately demonstrate compliance to the requirements of the same quality operating procedures. 2

3.12.2
The Engineering Supervisor does:

Review that the Process Engineer had adequately carried out: the product related planning; the determination of the product's processing layout; the formulation, review, approval, and issuance of the processing procedures; and the follow-up activities, as stated in QOP 001, Section One, Two, Three, and Four 1

TRAINING METRICS	Iss dt.	Rev dt.	Pg.
QUALITY OPERATING PROCEDURE	Sign	Rev #	32/52
REQUIREMENT: Inspection and Test Control, QSM 4.9			
SUBJECT: Process Control (performance surveillance)			
QOP 006 DOCUMENT REQ'D: All the forms listed and controlled in the Quality System			

◯

Review that the Process Engineer had effectively carried out and complied with the internal document change provisions of QOP 001, Section Four, in the application and execution of form **A-009**   (2)

Review that the Process Engineer had adequately carried out and complied with the provisions of QOP 009, Section One, Two, and Three regarding the review, disposition, and determination of corrective action on non-conforming products   (3)

Review that the Process Engineer had adequately complied with the provisions of QOP 002, Section Three, paragraph 3.03 in maintaining document retention   (4)

Review that the Process Engineer had carried out the engineering related performance requirements stated in QOP 012, Sections One, Two, and Three.   (5)

3.12.3
The Purchasing Supervisor does:

Review that the Purchasing Agent had adequately complied with the provisions of QOP 003, Section One and Two, in carrying out the preliminary investigation in the selection of subcontractors and the subsequent review, approval, and issuance of internal Purchase Orders in collaboration with the cognizant Quality Engineer.

3.13   **Audit Compliance Verification and Documentation**

3.13.1
The Management Representative does:

Instruction
The Management Representative shall not audit those areas for which he/she has direct work assignment responsibility as in Paragraph 3.11 above.

Audit the supervisory review requirements imposed in this Process Control procedure and those that are imposed in QOP 010 according to the Audit Schedule, form **A-021**. Record the audit results onto form **A-022**

◯

TRAINING METRICS	Iss dt.	Rev dt.	Pg.
QUALITY OPERATING PROCEDURE	Sign	Rev #	33/52
REQUIREMENT: Inspection and Test Control, QSM 4.9			
SUBJECT: Process Control (performance surveillance)			
QOP 006 |DOCUMENT REQ'D: All the forms listed and controlled in the Quality System			

3.13.2
Any Other Manager Except the Management
Representative does:

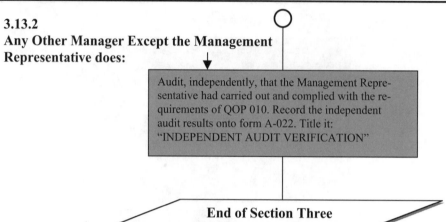

Audit, independently, that the Management Representative had carried out and complied with the requirements of QOP 010. Record the independent audit results onto form A-022. Title it:
"INDEPENDENT AUDIT VERIFICATION"

End of Section Three

TRAINING METRICS	Iss dt.	Rev dt.	Pg.
QUALITY OPERATING PROCEDURE	Sign	Rev #	34/52

REQUIREMENT: Inspection and Test Control, QSM 4.10.4

SUBJECT: Final Inspection and Testing

QOP 006|DOCUMENT REQ'D: A-001, A-006, A-008,A-012, A-013, A-014, A-018, A-028, A-029, A-033

SECTION FOUR

1.0   PURPOSE
Maintain control over final inspection and testing.

2.0   APPLICATION
This procedure shall apply to the verification, documentation, acceptance, and release of a completed product in compliance with the customer's Purchase Order requirements.

3.0   PROCEDURE

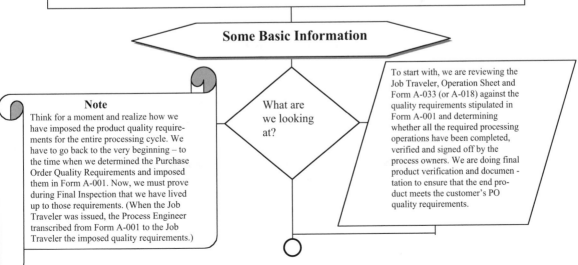

Final Inspection and Testing

General Requirement
1. Final inspection and testing entail the review of past performance regarding the product's processing, verification, acceptance, and documentation. Additional verification and documentation shall be done, as necessary, in order to demonstrate complete compliance to the customer's Purchase Order requirements, prior to delivery (or storage);
2. In the absence of PO quality provisions, the internally imposed and implemented quality requirements shall be reviewed and approved prior to releasing a completed product for delivery to the customer (or storage);
3. The above provisions equally apply to customer returned products.

Some Basic Information

Note
Think for a moment and realize how we have imposed the product quality requirements for the entire processing cycle. We have to go back to the very beginning – to the time when we determined the Purchase Order Quality Requirements and imposed them in Form A-001. Now, we must prove during Final Inspection that we have lived up to those requirements. (When the Job Traveler was issued, the Process Engineer transcribed from Form A-001 to the Job Traveler the imposed quality requirements.)

What are we looking at?

To start with, we are reviewing the Job Traveler, Operation Sheet and Form A-033 (or A-018) against the quality requirements stipulated in Form A-001 and determining whether all the required processing operations have been completed, verified and signed off by the process owners. We are doing final product verification and documentation to ensure that the end product meets the customer's PO quality requirements.

TRAINING METRICS	Iss dt.	Rev dt.	Pg.
QUALITY OPERATING PROCEDURE	Sign	Rev #	35/52
REQUIREMENT: Inspection and Test Control, QSM 4.10.4			
SUBJECT: Final Inspection and Testing			
QOP 006|DOCUMENT REQ'D: A-001, A-006, A-008,A-012, A-013, A-014, A-018, A-028, A-029, A-033			

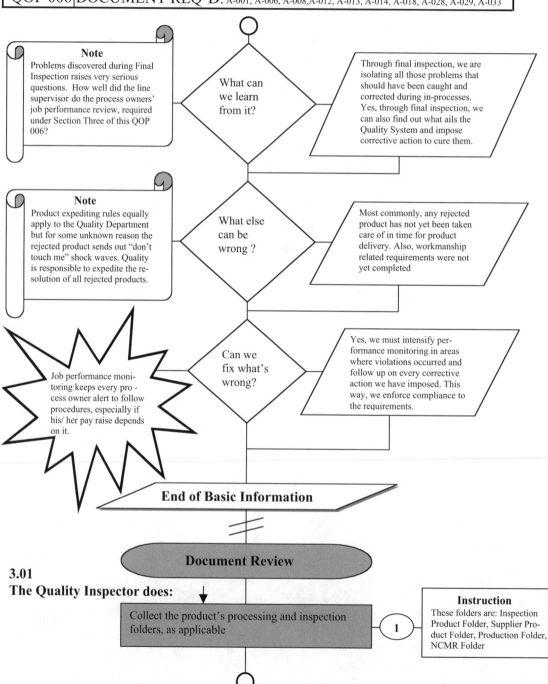

Note
Problems discovered during Final Inspection raises very serious questions. How well did the line supervisor do the process owners' job performance review, required under Section Three of this QOP 006?

What can we learn from it?

Through final inspection, we are isolating all those problems that should have been caught and corrected during in-processes. Yes, through final inspection, we can also find out what ails the Quality System and impose corrective action to cure them.

Note
Product expediting rules equally apply to the Quality Department but for some unknown reason the rejected product sends out "don't touch me" shock waves. Quality is responsible to expedite the re-solution of all rejected products.

What else can be wrong ?

Most commonly, any rejected product has not yet been taken care of in time for product delivery. Also, workmanship related requirements were not yet completed

Job performance monitoring keeps every process owner alert to follow procedures, especially if his/ her pay raise depends on it.

Can we fix what's wrong?

Yes, we must intensify performance monitoring in areas where violations occurred and follow up on every corrective action we have imposed. This way, we enforce compliance to the requirements.

End of Basic Information

Document Review

3.01
The Quality Inspector does:

Collect the product's processing and inspection folders, as applicable

1

Instruction
These folders are: Inspection Product Folder, Supplier Product Folder, Production Folder, NCMR Folder

TRAINING METRICS	Iss dt.	Rev dt.	Pg.
QUALITY OPERATING PROCEDURE	Sign	Rev #	36/52
REQUIREMENT: Inspection and Test Control, QSM 4.10.4			
SUBJECT: Final Inspection and Testing			
QOP 006|DOCUMENT REQ'D: A-001, A-006, A-008,A-012, A-013, A-014, A-018, A-028, A-029, A-033			

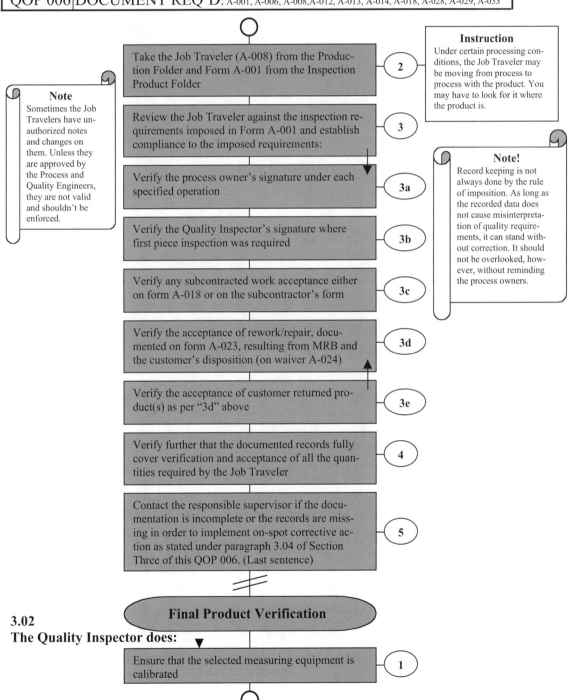

Instruction
Under certain processing conditions, the Job Traveler may be moving from process to process with the product. You may have to look for it where the product is.

Note
Sometimes the Job Travelers have un-authorized notes and changes on them. Unless they are approved by the Process and Quality Engineers, they are not valid and shouldn't be enforced.

Take the Job Traveler (A-008) from the Production Folder and Form A-001 from the Inspection Product Folder — 2

Review the Job Traveler against the inspection requirements imposed in Form A-001 and establish compliance to the imposed requirements: — 3

Verify the process owner's signature under each specified operation — 3a

Verify the Quality Inspector's signature where first piece inspection was required — 3b

Verify any subcontracted work acceptance either on form A-018 or on the subcontractor's form — 3c

Note!
Record keeping is not always done by the rule of imposition. As long as the recorded data does not cause misinterpretation of quality requirements, it can stand without correction. It should not be overlooked, however, without reminding the process owners.

Verify the acceptance of rework/repair, documented on form A-023, resulting from MRB and the customer's disposition (on waiver A-024) — 3d

Verify the acceptance of customer returned product(s) as per "3d" above — 3e

Verify further that the documented records fully cover verification and acceptance of all the quantities required by the Job Traveler — 4

Contact the responsible supervisor if the documentation is incomplete or the records are missing in order to implement on-spot corrective action as stated under paragraph 3.04 of Section Three of this QOP 006. (Last sentence) — 5

Final Product Verification

3.02
The Quality Inspector does:

Ensure that the selected measuring equipment is calibrated — 1

TRAINING METRICS	Iss dt.	Rev dt.	Pg.
QUALITY OPERATING PROCEDURE	Sign	Rev #	37/52
REQUIREMENT: Inspection and Test Control, QSM 4.10.4			
SUBJECT: Final Inspection and Testing			
QOP 006	DOCUMENT REQ'D: A-001, A-006, A-008,A-012, A-013, A-014, A-018, A-028, A-029, A-033		

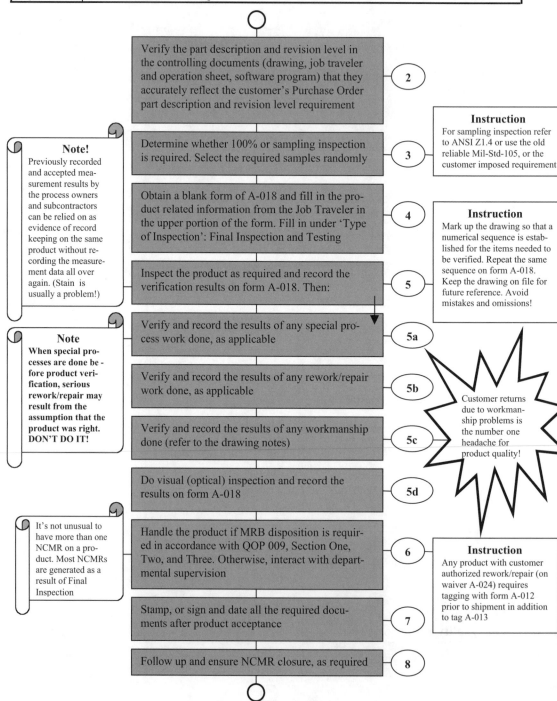

Verify the part description and revision level in the controlling documents (drawing, job traveler and operation sheet, software program) that they accurately reflect the customer's Purchase Order part description and revision level requirement — **2**

Determine whether 100% or sampling inspection is required. Select the required samples randomly — **3**

Instruction
For sampling inspection refer to ANSI Z1.4 or use the old reliable Mil-Std-105, or the customer imposed requirement

Obtain a blank form of A-018 and fill in the product related information from the Job Traveler in the upper portion of the form. Fill in under 'Type of Inspection': Final Inspection and Testing — **4**

Instruction
Mark up the drawing so that a numerical sequence is established for the items needed to be verified. Repeat the same sequence on form A-018. Keep the drawing on file for future reference. Avoid mistakes and omissions!

Note!
Previously recorded and accepted measurement results by the process owners and subcontractors can be relied on as evidence of record keeping on the same product without recording the measurement data all over again. (Stain is usually a problem!)

Inspect the product as required and record the verification results on form A-018. Then: — **5**

Verify and record the results of any special process work done, as applicable — **5a**

Note
When special processes are done before product verification, serious rework/repair may result from the assumption that the product was right. **DON'T DO IT!**

Verify and record the results of any rework/repair work done, as applicable — **5b**

Verify and record the results of any workmanship done (refer to the drawing notes) — **5c**

Customer returns due to workmanship problems is the number one headache for product quality!

Do visual (optical) inspection and record the results on form A-018 — **5d**

It's not unusual to have more than one NCMR on a product. Most NCMRs are generated as a result of Final Inspection

Handle the product if MRB disposition is required in accordance with QOP 009, Section One, Two, and Three. Otherwise, interact with departmental supervision — **6**

Instruction
Any product with customer authorized rework/repair (on waiver A-024) requires tagging with form A-012 prior to shipment in addition to tag A-013

Stamp, or sign and date all the required documents after product acceptance — **7**

Follow up and ensure NCMR closure, as required — **8**

TRAINING METRICS	Iss dt.	Rev dt.	Pg.
QUALITY OPERATING PROCEDURE	Sign	Rev #	38/52
REQUIREMENT: Inspection and Test Control, QSM 4.10.4			
SUBJECT: Final Inspection and Testing			
QOP 006	DOCUMENT REQ'D: A-001, A-006, A-008,A-012, A-013, A-014, A-018, A-028, A-029, A-033		

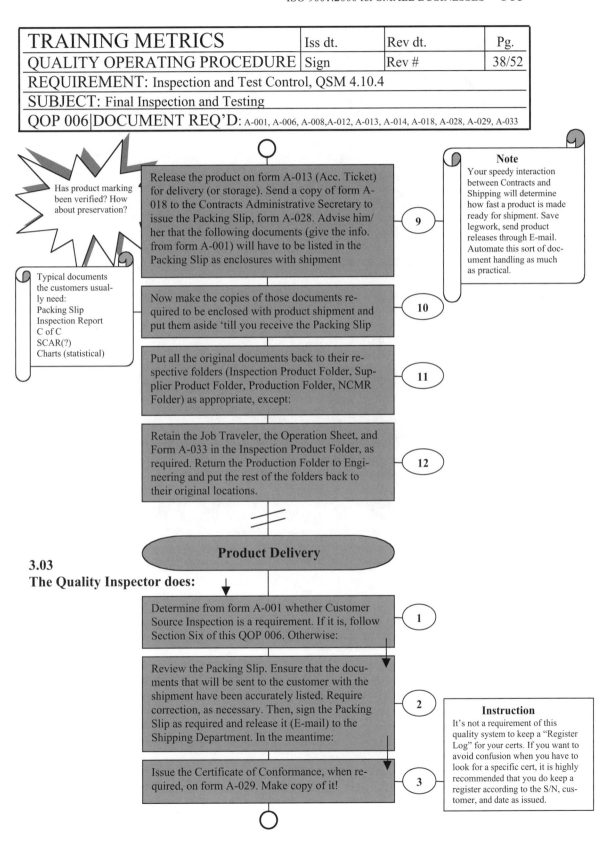

Has product marking been verified? How about preservation?

Release the product on form A-013 (Acc. Ticket) for delivery (or storage). Send a copy of form A-018 to the Contracts Administrative Secretary to issue the Packing Slip, form A-028. Advise him/her that the following documents (give the info. from form A-001) will have to be listed in the Packing Slip as enclosures with shipment

9

Note
Your speedy interaction between Contracts and Shipping will determine how fast a product is made ready for shipment. Save legwork, send product releases through E-mail. Automate this sort of document handling as much as practical.

Typical documents the customers usually need:
Packing Slip
Inspection Report
C of C
SCAR(?)
Charts (statistical)

Now make the copies of those documents required to be enclosed with product shipment and put them aside 'till you receive the Packing Slip

10

Put all the original documents back to their respective folders (Inspection Product Folder, Supplier Product Folder, Production Folder, NCMR Folder) as appropriate, except:

11

Retain the Job Traveler, the Operation Sheet, and Form A-033 in the Inspection Product Folder, as required. Return the Production Folder to Engineering and put the rest of the folders back to their original locations.

12

Product Delivery

3.03
The Quality Inspector does:

Determine from form A-001 whether Customer Source Inspection is a requirement. If it is, follow Section Six of this QOP 006. Otherwise:

1

Review the Packing Slip. Ensure that the documents that will be sent to the customer with the shipment have been accurately listed. Require correction, as necessary. Then, sign the Packing Slip as required and release it (E-mail) to the Shipping Department. In the meantime:

2

Instruction
It's not a requirement of this quality system to keep a "Register Log" for your certs. If you want to avoid confusion when you have to look for a specific cert, it is highly recommended that you do keep a register according to the S/N, customer, and date as issued.

Issue the Certificate of Conformance, when required, on form A-029. Make copy of it!

3

TRAINING METRICS	Iss dt.	Rev dt.	Pg.
QUALITY OPERATING PROCEDURE	Sign	Rev #	39/52
REQUIREMENT: Inspection and Test Control, QSM 4.10.4			
SUBJECT: Final Inspection and Testing			
QOP 006 DOCUMENT REQ'D: A-001, A-006, A-008, A-012, A-013, A-014, A-018, A-028, A-029, A-033			

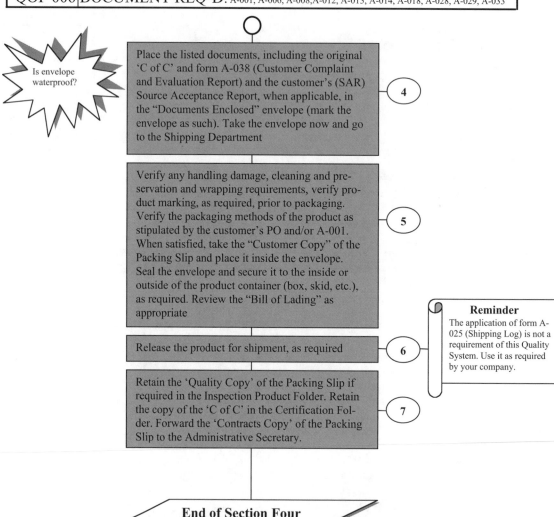

Is envelope waterproof?

Place the listed documents, including the original 'C of C' and form A-038 (Customer Complaint and Evaluation Report) and the customer's (SAR) Source Acceptance Report, when applicable, in the "Documents Enclosed" envelope (mark the envelope as such). Take the envelope now and go to the Shipping Department

4

Verify any handling damage, cleaning and pre-servation and wrapping requirements, verify product marking, as required, prior to packaging. Verify the packaging methods of the product as stipulated by the customer's PO and/or A-001. When satisfied, take the "Customer Copy" of the Packing Slip and place it inside the envelope. Seal the envelope and secure it to the inside or outside of the product container (box, skid, etc.), as required. Review the "Bill of Lading" as appropriate

5

Release the product for shipment, as required

6

Reminder
The application of form A-025 (Shipping Log) is not a requirement of this Quality System. Use it as required by your company.

Retain the 'Quality Copy' of the Packing Slip if required in the Inspection Product Folder. Retain the copy of the 'C of C' in the Certification Folder. Forward the 'Contracts Copy' of the Packing Slip to the Administrative Secretary.

7

End of Section Four

TRAINING METRICS	Iss dt.	Rev dt.	Pg.
QUALITY OPERATING PROCEDURE	Sign	Rev #	40/52
REQUIREMENT: Inspection and Test Control, 4.10.2			
SUBJECT: Receiving Inspection and Testing			
QOP 006 DOCUMENT REQ'D: Forms A-001, A-006. A-012, A-015, A-017, A-018, A-027			

SECTION FIVE

1.0   PURPOSE
Maintain control over received products.

2.0   APPLICATION
This procedure shall apply to Receiving and Receiving Inspection to control:
1. customer supplied materials, tools, gages, fixtures and other product related items;
2. internally purchased, product related materials, and subcontract services which may include special processes;
3. customer returns.

3.0   PROCEDURE

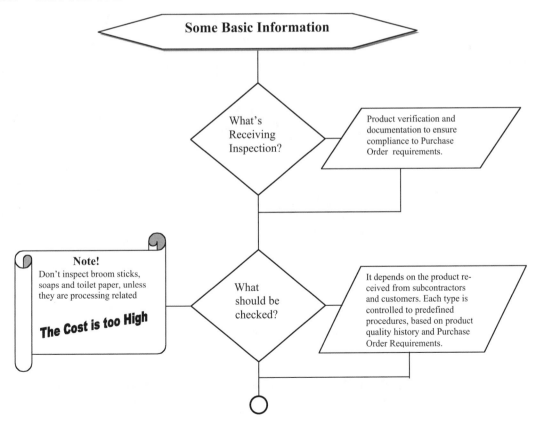

TRAINING METRICS	Iss dt.	Rev dt.	Pg.
QUALITY OPERATING PROCEDURE	Sign	Rev #	41/52
REQUIREMENT: Inspection and Test Control, 4.10.2			
SUBJECT: Receiving Inspection and Testing			
QOP 006|DOCUMENT REQ'D: Forms: A-001, A-006. A-012, A-015, A-017, A-018, A-027			

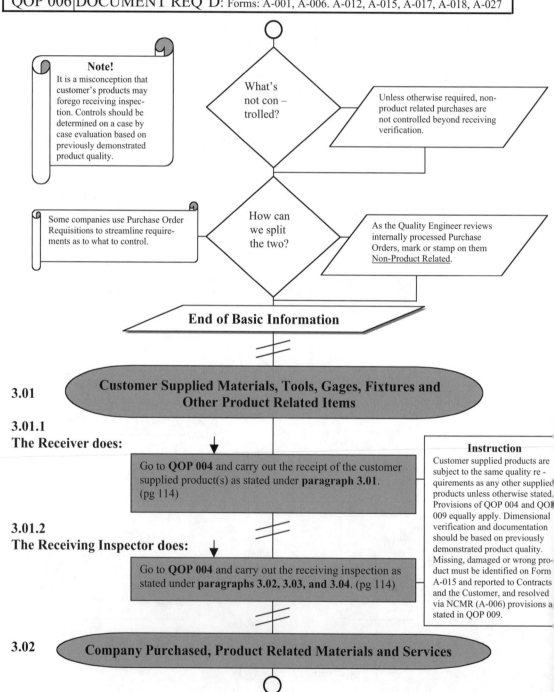

Note!
It is a misconception that customer's products may forego receiving inspection. Controls should be determined on a case by case evaluation based on previously demonstrated product quality.

What's not con‐trolled?

Unless otherwise required, non-product related purchases are not controlled beyond receiving verification.

Some companies use Purchase Order Requisitions to streamline require‐ments as to what to control.

How can we split the two?

As the Quality Engineer reviews internally processed Purchase Orders, mark or stamp on them Non-Product Related.

End of Basic Information

3.01   Customer Supplied Materials, Tools, Gages, Fixtures and Other Product Related Items

3.01.1
The Receiver does:

Go to **QOP 004** and carry out the receipt of the customer supplied product(s) as stated under **paragraph 3.01**. (pg 114)

3.01.2
The Receiving Inspector does:

Go to **QOP 004** and carry out the receiving inspection as stated under **paragraphs 3.02, 3.03, and 3.04**. (pg 114)

Instruction
Customer supplied products are subject to the same quality re‐quirements as any other supplied products unless otherwise stated. Provisions of QOP 004 and QOP 009 equally apply. Dimensional verification and documentation should be based on previously demonstrated product quality. Missing, damaged or wrong pro‐duct must be identified on Form A-015 and reported to Contracts and the Customer, and resolved via NCMR (A-006) provisions a stated in QOP 009.

3.02   Company Purchased, Product Related Materials and Services

TRAINING METRICS	Iss dt.	Rev dt.	Pg.
QUALITY OPERATING PROCEDURE	Sign	Rev #	42/52
REQUIREMENT: Inspection and Test Control, 4.10.2			
SUBJECT: Receiving Inspection and Testing			
QOP 006 DOCUMENT REQ'D: Forms: A-001, A-006. A-012, A-015, A-017, A-018, A-027			

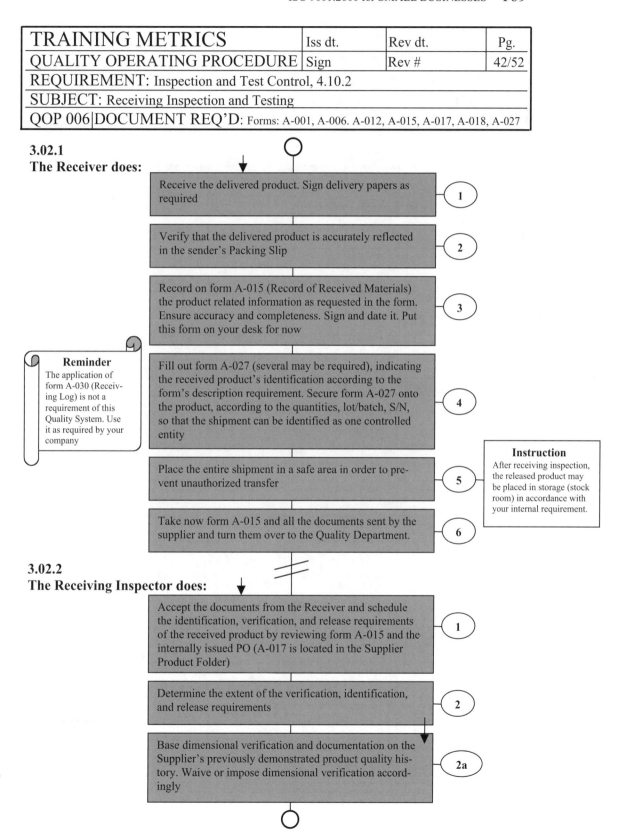

3.02.1
The Receiver does:

1. Receive the delivered product. Sign delivery papers as required

2. Verify that the delivered product is accurately reflected in the sender's Packing Slip

3. Record on form A-015 (Record of Received Materials) the product related information as requested in the form. Ensure accuracy and completeness. Sign and date it. Put this form on your desk for now

Reminder
The application of form A-030 (Receiving Log) is not a requirement of this Quality System. Use it as required by your company

4. Fill out form A-027 (several may be required), indicating the received product's identification according to the form's description requirement. Secure form A-027 onto the product, according to the quantities, lot/batch, S/N, so that the shipment can be identified as one controlled entity

5. Place the entire shipment in a safe area in order to prevent unauthorized transfer

Instruction
After receiving inspection, the released product may be placed in storage (stock room) in accordance with your internal requirement.

6. Take now form A-015 and all the documents sent by the supplier and turn them over to the Quality Department.

3.02.2
The Receiving Inspector does:

1. Accept the documents from the Receiver and schedule the identification, verification, and release requirements of the received product by reviewing form A-015 and the internally issued PO (A-017 is located in the Supplier Product Folder)

2. Determine the extent of the verification, identification, and release requirements

2a. Base dimensional verification and documentation on the Supplier's previously demonstrated product quality history. Waive or impose dimensional verification accordingly

TRAINING METRICS	Iss dt.	Rev dt.	Pg.
QUALITY OPERATING PROCEDURE	Sign	Rev #	43/52
REQUIREMENT: Inspection and Test Control, 4.10.2			
SUBJECT: Receiving Inspection and Testing			
QOP 006|DOCUMENT REQ'D: Forms: A-001, A-006. A-012, A-015, A-017, A-018, A-027			

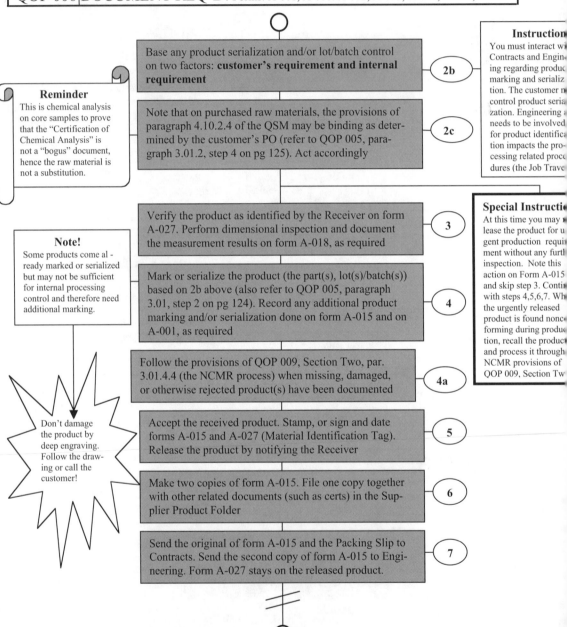

Instruction
You must interact wi
Contracts and Engine
ing regarding produc
marking and serializ
tion. The customer m
control product seria
zation. Engineering
needs to be involved
for product identifica
tion impacts the pro-
cessing related proce
dures (the Job Trave

Reminder
This is chemical analysis
on core samples to prove
that the "Certification of
Chemical Analysis" is
not a "bogus" document,
hence the raw material is
not a substitution.

2b Base any product serialization and/or lot/batch control on two factors: **customer's requirement and internal requirement**

2c Note that on purchased raw materials, the provisions of paragraph 4.10.2.4 of the QSM may be binding as determined by the customer's PO (refer to QOP 005, paragraph 3.01.2, step 4 on pg 125). Act accordingly

Special Instructi
At this time you may
lease the product for u
gent production requi
ment without any furth
inspection. Note this
action on Form A-015
and skip step 3. Conti
with steps 4,5,6,7. Wh
the urgently released
product is found nonc
forming during produc
tion, recall the produc
and process it through
NCMR provisions of
QOP 009, Section Tw

3 Verify the product as identified by the Receiver on form A-027. Perform dimensional inspection and document the measurement results on form A-018, as required

Note!
Some products come al -
ready marked or serialized
but may not be sufficient
for internal processing
control and therefore need
additional marking.

4 Mark or serialize the product (the part(s), lot(s)/batch(s)) based on 2b above (also refer to QOP 005, paragraph 3.01, step 2 on pg 124). Record any additional product marking and/or serialization done on form A-015 and on A-001, as required

4a Follow the provisions of QOP 009, Section Two, par. 3.01.4.4 (the NCMR process) when missing, damaged, or otherwise rejected product(s) have been documented

Don't damage the product by deep engraving. Follow the drawing or call the customer!

5 Accept the received product. Stamp, or sign and date forms A-015 and A-027 (Material Identification Tag). Release the product by notifying the Receiver

6 Make two copies of form A-015. File one copy together with other related documents (such as certs) in the Supplier Product Folder

7 Send the original of form A-015 and the Packing Slip to Contracts. Send the second copy of form A-015 to Engineering. Form A-027 stays on the released product.

TRAINING METRICS	Iss dt.	Rev dt.	Pg.
QUALITY OPERATING PROCEDURE	Sign	Rev #	44/52
REQUIREMENT: Inspection and Test Control, 4.10.2			
SUBJECT: Receiving Inspection and Testing			
QOP 006|DOCUMENT REQ'D: Forms: A-001, A-006. A-012, A-015, A-017, A-018, A-027			

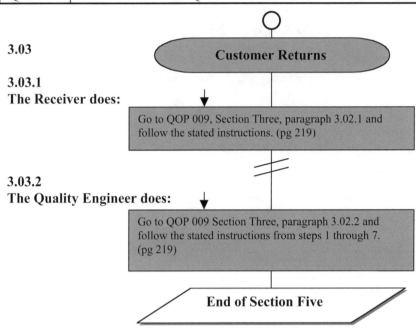

3.03

3.03.1
The Receiver does:

Customer Returns

Go to QOP 009, Section Three, paragraph 3.02.1 and follow the stated instructions. (pg 219)

3.03.2
The Quality Engineer does:

Go to QOP 009 Section Three, paragraph 3.02.2 and follow the stated instructions from steps 1 through 7. (pg 219)

End of Section Five

TRAINING METRICS	Iss dt.	Rev dt.	Pg.
QUALITY OPERATING PROCEDURE	Sign	Rev #	45/52
REQUIREMENT: Inspection and Test Control, QSM 4.10.4.1			
SUBJECT: Customer Source Inspection at the Supplier			
QOP 006 |DOCUMENT REQ'D: Inspection and Supplier Product Folders			

SECTION SIX

1.0   PURPOSE
Maintain control over how customer source inspection is carried out.

2.0   APPLICATION
This procedure shall apply to the presentation of a verified, documented, and accepted final product, made available for customer source inspection.

3.0   PROCEDURE
Customer source inspection may be imposed in two different ways:
1. at the supplier (the Company)
2. at the subcontractor (of the Company)

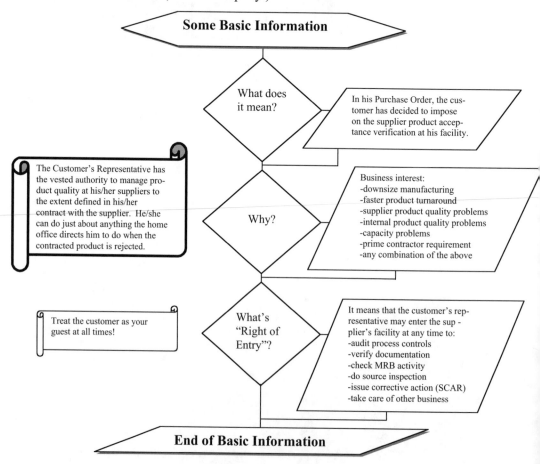

TRAINING METRICS	Iss dt.	Rev dt.	Pg.
QUALITY OPERATING PROCEDURE	Sign	Rev #	46/52
REQUIREMENT: Inspection and Test Control, QSM 4.10.4.1			
SUBJECT: Customer Source Inspection at the Supplier			
QOP 006 |DOCUMENT REQ'D: Inspection and Supplier Product Folders			

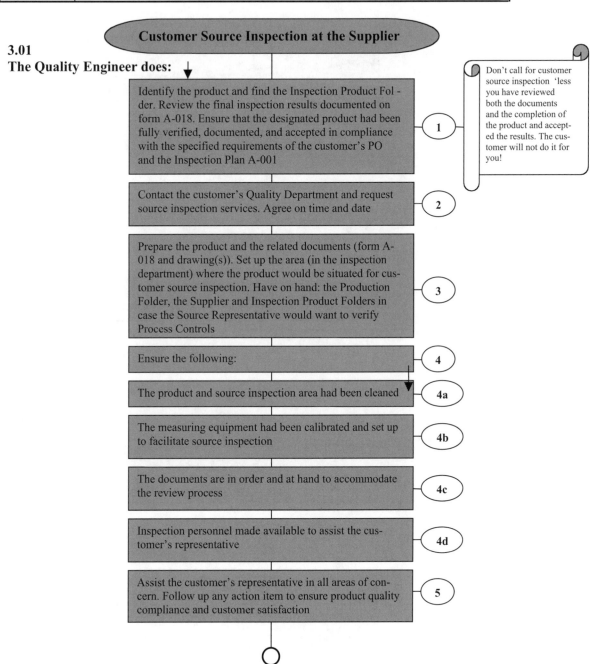

Customer Source Inspection at the Supplier

3.01
The Quality Engineer does:

1. Identify the product and find the Inspection Product Folder. Review the final inspection results documented on form A-018. Ensure that the designated product had been fully verified, documented, and accepted in compliance with the specified requirements of the customer's PO and the Inspection Plan A-001

Don't call for customer source inspection 'less you have reviewed both the documents and the completion of the product and accepted the results. The customer will not do it for you!

2. Contact the customer's Quality Department and request source inspection services. Agree on time and date

3. Prepare the product and the related documents (form A-018 and drawing(s)). Set up the area (in the inspection department) where the product would be situated for customer source inspection. Have on hand: the Production Folder, the Supplier and Inspection Product Folders in case the Source Representative would want to verify Process Controls

4. Ensure the following:

4a. The product and source inspection area had been cleaned

4b. The measuring equipment had been calibrated and set up to facilitate source inspection

4c. The documents are in order and at hand to accommodate the review process

4d. Inspection personnel made available to assist the customer's representative

5. Assist the customer's representative in all areas of concern. Follow up any action item to ensure product quality compliance and customer satisfaction

TRAINING METRICS	Iss dt.	Rev dt.	Pg.
QUALITY OPERATING PROCEDURE	Sign	Rev #	47/52
REQUIREMENT: Inspection and Test Control, QSM 4.10.4.1			
SUBJECT: Customer Source Inspection at the Supplier			
QOP 006 |DOCUMENT REQ'D: Inspection and Supplier Product Folders			

Arrange the final preparation to complete the source inspected product for delivery in accordance with the Job Traveler and/or Operation Sheet requirements, as applicable — 6

Release the accepted product for shipment in accordance with the "Product Delivery" requirements of this QOP 006, Section Four, paragraph 3.03, steps 2 through 7. (pg 165) — 7

Customer Source Inspection at the Subcontractor

3.02
The Quality Engineer does:

Interact between the customer's and the subcontractor's Quality Departments in setting up customer source inspection. Agree on time and date — 1

Prepare the subcontractor for the customer source inspection — 2

Verify that the subcontractor final inspected and documented the work results, and accepted the product — 2a

Is workmanship on the product completed?

Find out that the subcontractor had set up a clean area in which to do source inspection, had provided the appropriate measuring equipment, had organized the product related documents, had prepared (cleaned) the product for source inspection — 2b

Is Calibration OK?

Visit the subcontractor, as appropriate, and verify that 2a and 2b above had been in fact accomplished — 2c

Prepare yourself for the customer source inspection at the subcontractor — 3

Instruction
Don't let the Source Rep. end up waiting because the product not truly completed. **DO YOUR JOB UP FRONT!**

Review, as appropriate, the Inspection and Supplier Product Folders, the Production Folder, and the NCMR Folder in order to determine any in-house activity impacting the acceptance of the subcontracted product being readied for source inspection — 3a

TRAINING METRICS	Iss dt.	Rev dt.	Pg.
QUALITY OPERATING PROCEDURE	Sign	Rev #	48/52
REQUIREMENT: Inspection and Test Control, QSM 4.10.4.1			
SUBJECT: Customer Source Inspection at the Subcontractor			
QOP 006	DOCUMENT REQ'D: Inspection and Supplier (Subcontractor) Product Folders		

Reminder!
Documents you remove to take with you will have to be returned to their original file locations on your return from the subcontractor. FLAG THE FOLDERS, as required!

Organize to take all those documents that will be needed to substantiate any in-house processing related effort impacting the verification and acceptance of the presented product during source inspection — **3b**

Interact with Contracts, Purchasing, and Engineering, as appropriate, to take care of communication and possible product support liaison during and after customer source inspection — **3c**

Interact between the customer source representative and the subcontractor during source inspection in order to mediate inconsistencies and follow-up activities — **4**

Arrange the final preparation to complete the source accepted product for delivery in accordance with the Job Traveler and/or Operation Sheet requirements, as applicable — **5**

Release the accepted product for shipment in accordance with the 'Product Delivery' requirements of this QOP 006, Section Four, paragraph 3.03, steps 2 through 7, as applicable. (pg 165) — **6**

Instruction
Accepted product means that you have verified identification marking, preservation, packing and packaging to the specified requirements of the PO and A-001 prior to releasing the product.

End of Section Six

TRAINING METRICS	Iss dt.	Rev dt.	Pg.
QUALITY OPERATING PROCEDURE	Sign	Rev #	49/52
REQUIREMENT: Inspection and Test Control, QSM 4.10.4.2			
SUBJECT: Supplier Source Inspection at Subcontractor(s)			
QOP 006|DOCUMENT REQ'D: Inspection and Supplier Product Folders, Forms A-018, A-036			

SECTION SEVEN

1.0   PURPOSE

Maintain control over source inspection at the Company's subcontractor(s).

2.0   APPLICATION

Procedures in this section shall apply to source control activities of subcontracted product(s).

3.0   PROCEDURE

General Requirement

1. Internal Purchase Orders issued to subcontractors shall state source control requirements under the quality provisions imposed by the cognizant Quality Engineer as stipulated in QOP 003, Section One, par. 3.02, step 2. (pg 99)
2. Requests to schedule Source Inspection shall be directed to the Quality Department by all subcontractors. Purchasing shall be required to follow the same provision.

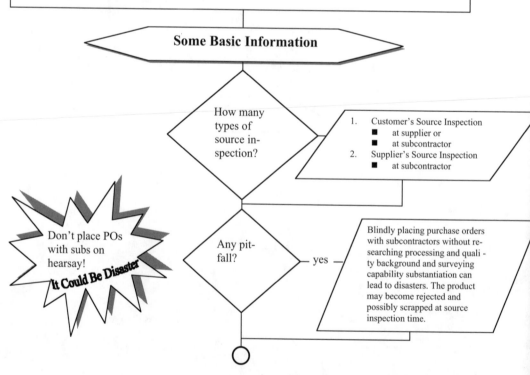

Some Basic Information

How many types of source inspection?

1. Customer's Source Inspection
    - at supplier or
    - at subcontractor
2. Supplier's Source Inspection
    - at subcontractor

Don't place POs with subs on hearsay! *It Could Be Disaster*

Any pitfall? — yes —

Blindly placing purchase orders with subcontractors without researching processing and quality background and surveying capability substantiation can lead to disasters. The product may become rejected and possibly scrapped at source inspection time.

TRAINING METRICS	Iss dt.	Rev dt.	Pg.
QUALITY OPERATING PROCEDURE	Sign	Rev #	50/52
REQUIREMENT: Inspection and Test Control, QSM 4.10.4.2			
SUBJECT: Supplier Source Inspection at Subcontractor(s)			
QOP 006|DOCUMENT REQ'D: Inspection and Supplier Product Folders, Forms A-018, A-036			

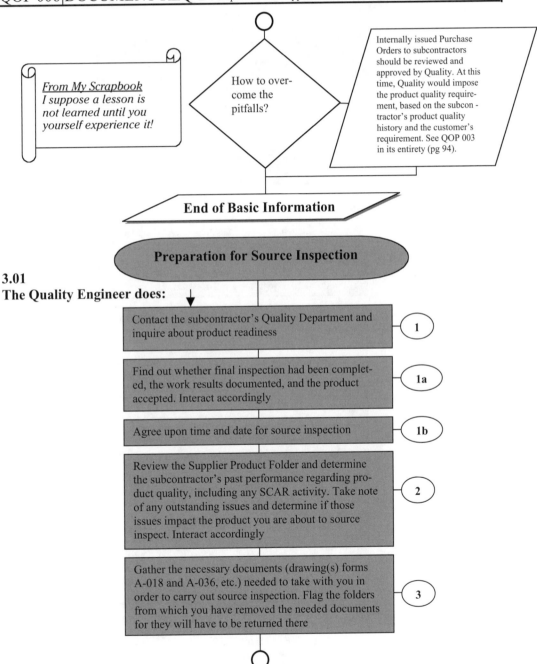

From My Scrapbook
I suppose a lesson is
not learned until you
yourself experience it!

How to over-
come the
pitfalls?

Internally issued Purchase
Orders to subcontractors
should be reviewed and
approved by Quality. At this
time, Quality would impose
the product quality requirement, based on the subcon -
tractor's product quality
history and the customer's
requirement. See QOP 003
in its entirety (pg 94).

End of Basic Information

Preparation for Source Inspection

3.01
The Quality Engineer does:

Contact the subcontractor's Quality Department and
inquire about product readiness — **1**

Find out whether final inspection had been completed, the work results documented, and the product
accepted. Interact accordingly — **1a**

Agree upon time and date for source inspection — **1b**

Review the Supplier Product Folder and determine
the subcontractor's past performance regarding product quality, including any SCAR activity. Take note
of any outstanding issues and determine if those
issues impact the product you are about to source
inspect. Interact accordingly — **2**

Gather the necessary documents (drawing(s) forms
A-018 and A-036, etc.) needed to take with you in
order to carry out source inspection. Flag the folders
from which you have removed the needed documents
for they will have to be returned there — **3**

TRAINING METRICS	Iss dt.	Rev dt.	Pg.
QUALITY OPERATING PROCEDURE	Sign	Rev #	51/52
REQUIREMENT: Inspection and Test Control, QSM 4.10.4.2			
SUBJECT: Supplier Source Inspection at Subcontractor(s)			
QOP 006 DOCUMENT REQ'D: Inspection and Supplier Product Folders, Forms A-018, A-036			

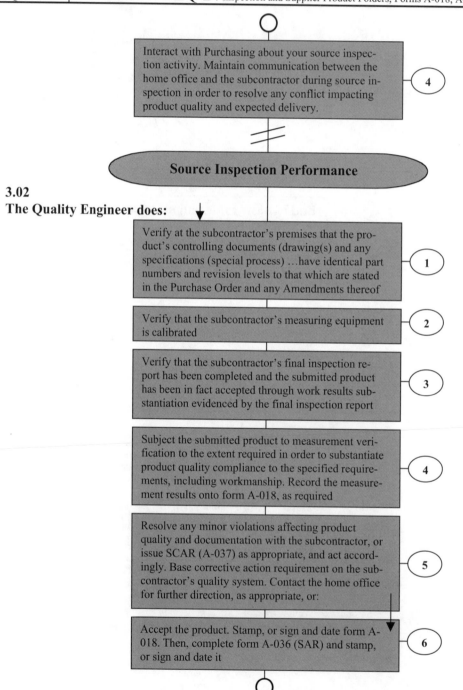

Interact with Purchasing about your source inspection activity. Maintain communication between the home office and the subcontractor during source inspection in order to resolve any conflict impacting product quality and expected delivery. **4**

Source Inspection Performance

3.02
The Quality Engineer does:

Verify at the subcontractor's premises that the product's controlling documents (drawing(s) and any specifications (special process) ...have identical part numbers and revision levels to that which are stated in the Purchase Order and any Amendments thereof **1**

Verify that the subcontractor's measuring equipment is calibrated **2**

Verify that the subcontractor's final inspection report has been completed and the submitted product has been in fact accepted through work results substantiation evidenced by the final inspection report **3**

Subject the submitted product to measurement verification to the extent required in order to substantiate product quality compliance to the specified requirements, including workmanship. Record the measurement results onto form A-018, as required **4**

Resolve any minor violations affecting product quality and documentation with the subcontractor, or issue SCAR (A-037) as appropriate, and act accordingly. Base corrective action requirement on the subcontractor's quality system. Contact the home office for further direction, as appropriate, or: **5**

Accept the product. Stamp, or sign and date form A-018. Then, complete form A-036 (SAR) and stamp, or sign and date it **6**

TRAINING METRICS	Iss dt.	Rev dt.	Pg.
QUALITY OPERATING PROCEDURE	Sign	Rev #	52/52
REQUIREMENT: Inspection and Test Control, QSM 4.10.4.2			
SUBJECT: Supplier Source Inspection at Subcontractor(s)			
QOP 006	DOCUMENT REQ'D: Inspection and Supplier Product Folders, Forms A-018, A-036		

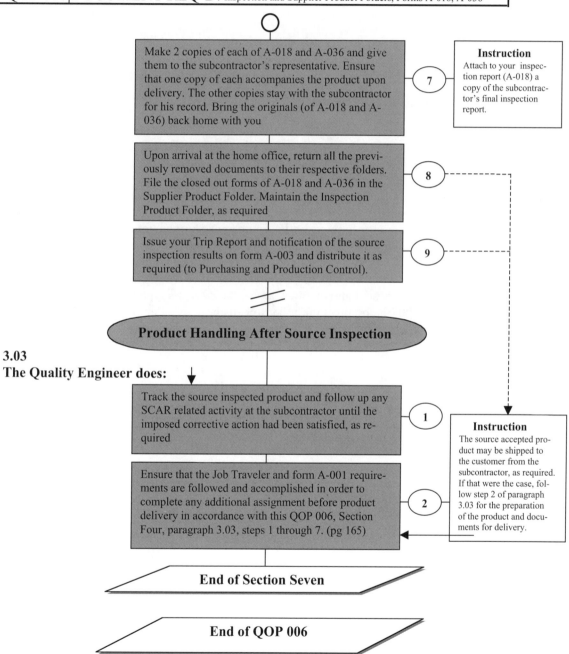

Make 2 copies of each of A-018 and A-036 and give them to the subcontractor's representative. Ensure that one copy of each accompanies the product upon delivery. The other copies stay with the subcontractor for his record. Bring the originals (of A-018 and A-036) back home with you

7

Instruction
Attach to your inspection report (A-018) a copy of the subcontractor's final inspection report.

Upon arrival at the home office, return all the previously removed documents to their respective folders. File the closed out forms of A-018 and A-036 in the Supplier Product Folder. Maintain the Inspection Product Folder, as required

8

Issue your Trip Report and notification of the source inspection results on form A-003 and distribute it as required (to Purchasing and Production Control).

9

Product Handling After Source Inspection

3.03
The Quality Engineer does:

Track the source inspected product and follow up any SCAR related activity at the subcontractor until the imposed corrective action had been satisfied, as required

1

Instruction
The source accepted product may be shipped to the customer from the subcontractor, as required. If that were the case, follow step 2 of paragraph 3.03 for the preparation of the product and documents for delivery.

Ensure that the Job Traveler and form A-001 requirements are followed and accomplished in order to complete any additional assignment before product delivery in accordance with this QOP 006, Section Four, paragraph 3.03, steps 1 through 7. (pg 165)

2

End of Section Seven

End of QOP 006

OVERVIEW TO QOP 007
(CONTROL OF INSPECTION, MEASURING AND TEST EQUIPMENT)

When we talk about product verification, what we really mean is to check that product with a measuring device to determine if it meets a determined tolerance limit. How often do we do this without questioning whether that measuring device we just used is good enough to do the job? So you see, that measuring device also has a tolerance limit. And then, when we confirm that measuring device to determine its accuracy, the validating equipment also has a tolerance limit. There must be someone in every shop, where measuring equipment is used, to take care of the gages, or very soon we be out of business, because we can't accurately measure what we made. All this, of course, is a reminder that we must always know, ahead of time, whether our measuring and test equipment was calibrated and that they are suitable for the type of measurement we want to make. QOP 007 is the procedure that tells us how to go about controlling measuring and test equipment. This procedure controls not only company owned equipment, but also personal and customer owned or loaned measuring and test equipment.

DIAGRAMMATIC PROCEDURE   QOP 007

CONTROL OF INSPECTION, MEASURING, AND TEST EQUIPMENT

CONTENTS   Page

TRAINING METRICS		Iss dt.	Rev dt.		Pg.
QUALITY OPERATING PROCEDURE		Sign	Rev #		2/11
REQUIREMENT: QUALITY SYSTEM MANUAL SEC: 4.11					
SUBJECT: Control of Inspection, Measuring, and Test Equipment					
QOP 007	DOCUMENT REQ'D: This procedure and Form A-020				

REVISION HISTORY

Rev Date	Rev No	Description	Approval

TRAINING METRICS	Iss dt.	Rev dt.	Pg.
QUALITY OPERATING PROCEDURE	Sign	Rev #	3/11
REQUIREMENT: Control of Inspection, Measuring, and Test Equipment, QSM 4.11			
SUBJECT: Basic Information			
QOP 007 |DOCUMENT REQ'D: This procedure and Form A-020			

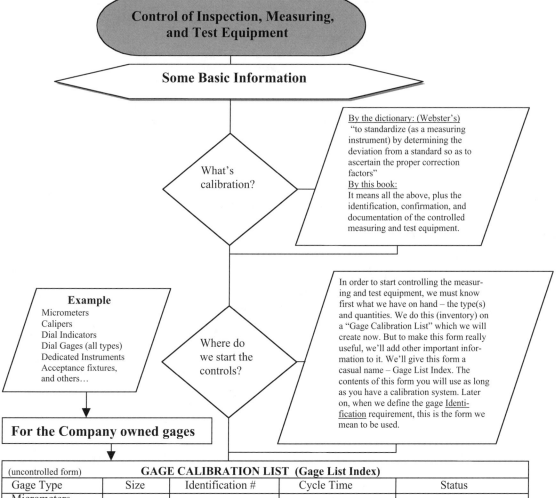

Control of Inspection, Measuring, and Test Equipment

Some Basic Information

What's calibration?

By the dictionary: (Webster's)
"to standardize (as a measuring instrument) by determining the deviation from a standard so as to ascertain the proper correction factors"
By this book:
It means all the above, plus the identification, confirmation, and documentation of the controlled measuring and test equipment.

Example
Micrometers
Calipers
Dial Indicators
Dial Gages (all types)
Dedicated Instruments
Acceptance fixtures,
and others…

Where do we start the controls?

In order to start controlling the measuring and test equipment, we must know first what we have on hand – the type(s) and quantities. We do this (inventory) on a "Gage Calibration List" which we will create now. But to make this form really useful, we'll add other important information to it. We'll give this form a casual name – Gage List Index. The contents of this form you will use as long as you have a calibration system. Later on, when we define the gage Identification requirement, this is the form we mean to be used.

For the Company owned gages

(uncontrolled form)	GAGE CALIBRATION LIST	(Gage List Index)		
Gage Type	Size	Identification #	Cycle Time	Status
Micrometers				
B&SH (Std)	0"- 1"	100001	3mo or less ☆	active
Starrett (Pin)	0"- 1"	100002	6mo or less	active
Federal (Blade)	0"- 3"	100003	3mo or less	inactive
Calipers				
Federal	0"- 6"	200001	3mo or less	active
B&SH (Dial)	0"- 12"	200002	3mo or less	active

(Keep going until you have listed all the Company owned measuring and test equipment.)

☆ "less" means dropped or damaged

TRAINING METRICS	Iss dt.	Rev dt.	Pg.
QUALITY OPERATING PROCEDURE	Sign	Rev #	4/11
REQUIREMENT: Control of Inspection, Measuring, and Test Equipment, QSM 4.11			
SUBJECT: Basic Information			
QOP 007 |DOCUMENT REQ'D: This procedure and Form A-020			

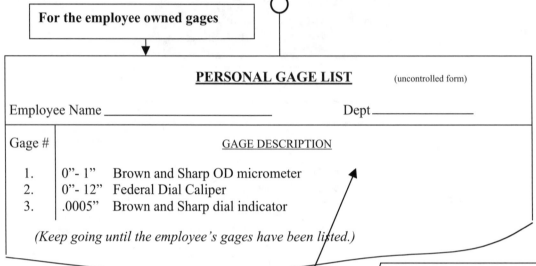

For the employee owned gages

PERSONAL GAGE LIST   (uncontrolled form)

Employee Name _____   Dept_____

Gage #	GAGE DESCRIPTION
1.	0"- 1" Brown and Sharp OD micrometer
2.	0"- 12" Federal Dial Caliper
3.	.0005" Brown and Sharp dial indicator

(Keep going until the employee's gages have been listed.)

No Gage ID
or Cycle
Time?

Employees' gages have to be
controlled only as long as they
stay with the company. So their
gages can't be controlled on a
permanent bases because that
would create unneeded and confusing paperwork with the Company's gages. As a result, we
have to create different forms to
show cycle time and calibration
record. Above is the "Gage List"
and below are: first, the "Cycle
Time" sheet, then, the "Calibration Record" sheet.

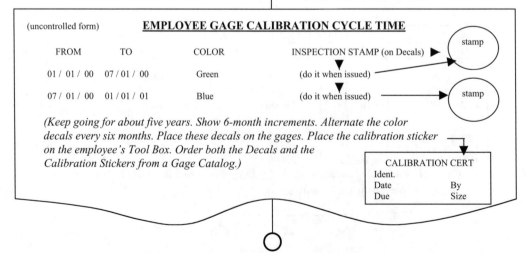

(uncontrolled form)   **EMPLOYEE GAGE CALIBRATION CYCLE TIME**

FROM	TO	COLOR	INSPECTION STAMP (on Decals) ▶	stamp
01 / 01 / 00	07 / 01 / 00	Green	(do it when issued)	
07 / 01 / 00	01 / 01 / 01	Blue	(do it when issued)	stamp

*(Keep going for about five years. Show 6-month increments. Alternate the color
decals every six months. Place these decals on the gages. Place the calibration sticker
on the employee's Tool Box. Order both the Decals and the
Calibration Stickers from a Gage Catalog.)*

CALIBRATION CERT
Ident.
Date   By
Due   Size

TRAINING METRICS	Iss dt.	Rev dt.	Pg.
QUALITY OPERATING PROCEDURE	Sign	Rev #	5/11
REQUIREMENT: Control of Inspection, Measuring, and Test Equipment, QSM 4.11			
SUBJECT: Basic Information			
QOP 007 DOCUMENT REQ'D: This procedure and Form A-020			

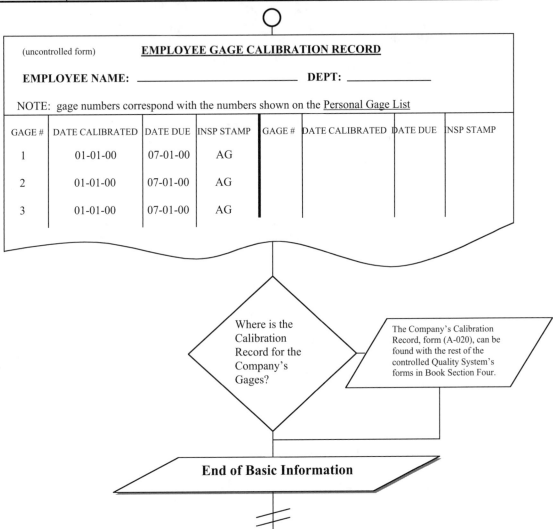

(uncontrolled form)   **EMPLOYEE GAGE CALIBRATION RECORD**

EMPLOYEE NAME: _____ **DEPT:** _____

NOTE: gage numbers correspond with the numbers shown on the <u>Personal Gage List</u>

GAGE #	DATE CALIBRATED	DATE DUE	INSP STAMP	GAGE #	DATE CALIBRATED	DATE DUE	INSP STAMP
1	01-01-00	07-01-00	AG				
2	01-01-00	07-01-00	AG				
3	01-01-00	07-01-00	AG				

Where is the Calibration Record for the Company's Gages?

The Company's Calibration Record, form (A-020), can be found with the rest of the controlled Quality System's forms in Book Section Four.

End of Basic Information

TRAINING METRICS	Iss dt.	Rev dt.	Pg.
QUALITY OPERATING PROCEDURE	Sign	Rev #	6/11
REQUIREMENT: Control of Inspection, Measuring, and Test Equipment, QSM 4.11			
SUBJECT: Work Instruction and Performance Requirement			
QOP 007 |DOCUMENT REQ'D: This procedure and Form A-020			

1.0   PURPOSE

Maintain control over inspection, measuring, and test equipment.

2.0   APPLICATION

This procedure shall apply to all types of Company and employee owned inspection, measuring, and test equipment, as well as to those under loan agreement (verbal or documented).

3.0   PROCEDURE

General Requirement

Record keeping shall be the acceptable means to demonstrate control of inspection, measuring and test equipment, including test software, and comparative references. The control of measuring and test equipment shall comply with ISO 10012 -1 and/or ANSI/NCSL Z540-1. Computerized data collection shall be allowed to maintain the records of calibration.

The built-in variability (gage maker's tolerance) of the selected measuring and test equipment shall be known in order to be able to assign that equipment to validate a specified requirement (dimension, weight, etc.) to the defined tolerance limits. (Match the equipment to the capability requirement.) Any software used in measuring applications shall be confirmed prior to its application.

The documented accuracy of the calibration masters shall be traceable to known standards (national or international, or even to a homemade standard in exceptional cases) in order to ensure suitability for the intended confirmation requirement. The documented accuracy and repeatability of the calibration masters and comparative references shall be certified by independent authority at predetermined intervals, based on the equipment manufacturer recommendations or as otherwise specified by internal requirement.

The calibration records shall demonstrate the frequency (cycle time) of confirmations of all the identified measuring and test equipment, including process owners' equipment and test software. Traceable record maintenance shall be a requirement as evidence of control. The calibration records shall be made available to customers or their representatives upon request.

TRAINING METRICS	Iss dt.	Rev dt.	Pg.
QUALITY OPERATING PROCEDURE	Sign	Rev #	7/11
REQUIREMENT: Control of Inspection, Measuring, and Test Equipment, QSM 4.11			
SUBJECT: Work Instruction and Performance Requirement			
QOP 007 |DOCUMENT REQ'D: This procedure and Form A-020			

Controlling the Identification of Gages

3.01

The Calibration Technician does:

Instruction

Check above under Basic Information the form called: "Gage Calibration List." The ID number shown for a gage will have to be transferred onto the Calibration Sticker for the specified Company's gage.

Check also "The Personal Gage List" for the same reason for the employee owned gages.

Caution

Use discretion when metal stamping or other types of identification means are used so as not to damage the internal mechanism of the equipment.

1. Uniquely identify (label) all inspection, measuring, and test equipment. Do this either on the instrument itself, or when that's not possible, on the container holding the instrument

2. Uniquely identify the employee owned measuring and test equipment which they use to verify and accept the Company's product(s)

3. Control the inventory of those gages that are on loan agreement or on borrowed bases. Identify them on form A-020 (Cal. Status Report) only. Don't calibrate these as the owners do the scheduling of calibration. Verify their accuracy and repeatability by similarity comparison and record the results on form A-020, as required

4. Ensure that the ID number assigned to each measuring and test equipment is always maintained on the specific equipment (Calibration Sticker) and on all the calibration related documents even if that equipment is removed from service

5. Don't reassign the once issued ID number to any other equipment (new or refurbished) unless the old calibration record clearly states the replacement requirements and can substantiate the <u>was</u> versus the <u>is</u> calibration accuracy and repeatability status

6. Identify all the equipment no longer suitable for measuring and test requirements as "No Longer in Service." Stamp, or sign and date the label. Or scrap the equipment.

TRAINING METRICS	Iss dt.	Rev dt.	Pg.
QUALITY OPERATING PROCEDURE	Sign	Rev #	8/11
REQUIREMENT: Control of Inspection, Measuring, and Test Equipment, QSM 4.11			
SUBJECT: Work Instruction and Performance Requirement			
QOP 007	DOCUMENT REQ'D: This procedure and Form A-020		

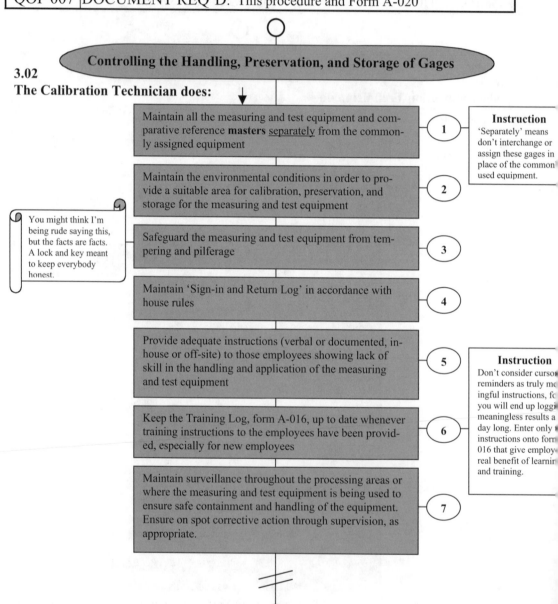

Controlling the Handling, Preservation, and Storage of Gages

3.02

The Calibration Technician does:

1 — Maintain all the measuring and test equipment and comparative reference **masters** separately from the commonly assigned equipment

Instruction
'Separately' means don't interchange or assign these gages in place of the commonly used equipment.

2 — Maintain the environmental conditions in order to provide a suitable area for calibration, preservation, and storage for the measuring and test equipment

You might think I'm being rude saying this, but the facts are facts. A lock and key meant to keep everybody honest.

3 — Safeguard the measuring and test equipment from tempering and pilferage

4 — Maintain 'Sign-in and Return Log' in accordance with house rules

5 — Provide adequate instructions (verbal or documented, in-house or off-site) to those employees showing lack of skill in the handling and application of the measuring and test equipment

Instruction
Don't consider cursor reminders as truly meaningful instructions, for you will end up logging meaningless results a day long. Enter only instructions onto form 016 that give employe real benefit of learning and training.

6 — Keep the Training Log, form A-016, up to date whenever training instructions to the employees have been provided, especially for new employees

7 — Maintain surveillance throughout the processing areas or where the measuring and test equipment is being used to ensure safe containment and handling of the equipment. Ensure on spot corrective action through supervision, as appropriate.

TRAINING METRICS	Iss dt.	Rev dt.	Pg.
QUALITY OPERATING PROCEDURE	Sign	Rev #	9/11
REQUIREMENT: Control of Inspection, Measuring, and Test Equipment, QSM 4.11			
SUBJECT: Work Instruction and Performance Requirement			
QOP 007 |DOCUMENT REQ'D: This procedure and Form A-020			

Controlling the Calibration of Gages

3.03
The Calibration Technician does:

Lack of gage readiness means stalling preparation and processing of the product

1 — Plan the calibration assignment of the measuring and test equipment in line with the requirement of production scheduling in order to ensure an uninterrupted production environment

2 — Review and follow the "Gage Calibration Technique Sheets" according to the requirement of the type of measuring and test equipment being calibrated. Do the calibration. Record the calibration results onto form A-020 simultaneously, as practical

3 — Calibrate the measuring and test equipment routinely as indicated by the cycle time (due date) posted on the Calibration Status Report, or on the Employee Gage Calibration Cycle time

4 — Calibrate the measuring and test equipment at other times as follows: when
θ   It's new,
θ   It has been repaired,
θ   It has been recalled,
θ   The accuracy is in question,
θ   Failure occurs (damage, etc.)

5 — Ensure, when calibrating the measuring and test equipment, that the previously recorded calibration results are reviewed, case by case, in order to assess the progressive deterioration to reclassify the equipment or to determine what's wrong in order to rework/repair it

6 — Lubricate the mechanical equipment as recommended by the manufacturer

7 — Maintain a safe work area and keep the place clean and organized.

Instruction
This Training Metrics, QOP 007, will show you the uncontrolled forms needed to manage the employee owned measuring and test equipment. Look these up under 'Basic Information.'

TRAINING METRICS	Iss dt.	Rev dt.	Pg.
QUALITY OPERATING PROCEDURE	Sign	Rev #	10/11
REQUIREMENT: Control of Inspection, Measuring, and Test Equipment, QSM 4.11			
SUBJECT: Work Instruction and Performance Requirement			
QOP 007	DOCUMENT REQ'D: This procedure and Form A-020		

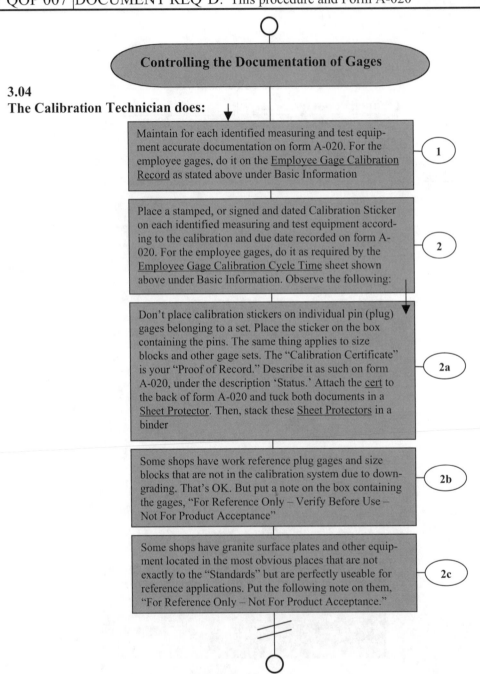

Controlling the Documentation of Gages

3.04
The Calibration Technician does:

1. Maintain for each identified measuring and test equipment accurate documentation on form A-020. For the employee gages, do it on the <u>Employee Gage Calibration Record</u> as stated above under Basic Information

2. Place a stamped, or signed and dated Calibration Sticker on each identified measuring and test equipment according to the calibration and due date recorded on form A-020. For the employee gages, do it as required by the <u>Employee Gage Calibration Cycle Time</u> sheet shown above under Basic Information. Observe the following:

2a. Don't place calibration stickers on individual pin (plug) gages belonging to a set. Place the sticker on the box containing the pins. The same thing applies to size blocks and other gage sets. The "Calibration Certificate" is your "Proof of Record." Describe it as such on form A-020, under the description 'Status.' Attach the <u>cert</u> to the back of form A-020 and tuck both documents in a <u>Sheet Protector</u>. Then, stack these <u>Sheet Protectors</u> in a binder

2b. Some shops have work reference plug gages and size blocks that are not in the calibration system due to downgrading. That's OK. But put a note on the box containing the gages, "For Reference Only – Verify Before Use – Not For Product Acceptance"

2c. Some shops have granite surface plates and other equipment located in the most obvious places that are not exactly to the "Standards" but are perfectly useable for reference applications. Put the following note on them, "For Reference Only – Not For Product Acceptance."

TRAINING METRICS	Iss dt.	Rev dt.	Pg.
QUALITY OPERATING PROCEDURE	Sign	Rev #	11/11
REQUIREMENT: Control of Inspection, Measuring, and Test Equipment, QSM 4.11			
SUBJECT: Work Instruction and Performance Requirement			
QOP 007 DOCUMENT REQ'D: This procedure and Form A-020			

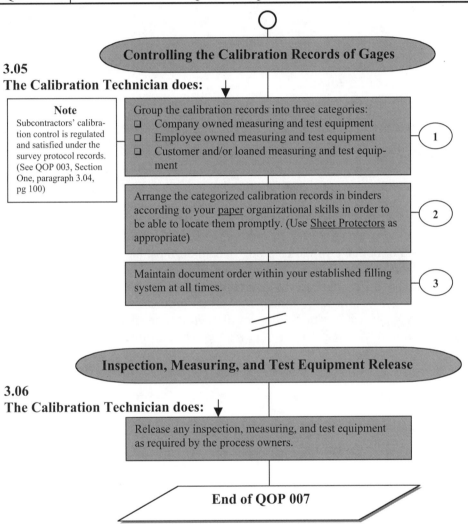

Controlling the Calibration Records of Gages

3.05
The Calibration Technician does:

Note
Subcontractors' calibration control is regulated and satisfied under the survey protocol records. (See QOP 003, Section One, paragraph 3.04, pg 100)

Group the calibration records into three categories:
❑   Company owned measuring and test equipment
❑   Employee owned measuring and test equipment
❑   Customer and/or loaned measuring and test equipment

1

Arrange the categorized calibration records in binders according to your <u>paper</u> organizational skills in order to be able to locate them promptly. (Use <u>Sheet Protectors</u> as appropriate)

2

Maintain document order within your established filling system at all times.

3

Inspection, Measuring, and Test Equipment Release

3.06
The Calibration Technician does:

Release any inspection, measuring, and test equipment as required by the process owners.

End of QOP 007

OVERVIEW TO QOP 008
(CONTROL OF INSPECTION AND TEST STATUS)

The control of inspection and test status means a lot more than just keeping track of inspection stamps and passwords. The meaning behind it has everything to do with the authorized use or application of them. The authorized use of this media signifies that a product has been verified, accepted, and released to go on to the next determined step in the product's finishing cycle. Without it, we would not know whether any phase operation was checked or not. We would be going on and on finishing a product only to find out at the end that from the first piece on to the last piece everything was rejected. Nobody indicated along the way that any processing step was controlled. What a way to learn a lesson! So, those who are authorized to release products from operation to operation bear a very significant responsibility in the application of inspection stamps and passwords. QOP 008 will define for you how to go about controlling the use of signatures, stamps, and passwords.

DIAGRAMMATIC PROCEDURE   QOP 008

CONTROL OF INSPECTION AND TEST STATUS

CONTENTS   **Page**

TRAINING METRICS		Iss dt.		Rev dt.		Pg.
QUALITY OPERATING PROCEDURE		Sign		Rev #		2/6
REQUIREMENT: QUALITY SYSTEM MANUAL SEC: 4.12						
SUBJECT: CONTROL OF INSPECTION AND TEST STATUS						
QOP 008	DOCUMENT REQ'D: This procedure and Forms A-003, A-022, A-032					

REVISION HISTORY

Rev Date	Rev No	Description	Approval

TRAINING METRICS	Iss dt.	Rev dt	Pg.
QUALITY OPERATING PROCEDURE	Sign	Rev #	3/6
REQUIREMENT: QUALITY SYSTEM MANUAL SEC: 4.12			
SUBJECT: CONTROL OF INSPECTION AND TEST STATUS			
QOP 008 DOCUMENT REQ'D: This procedure and Forms A-003, A-022, A-032			

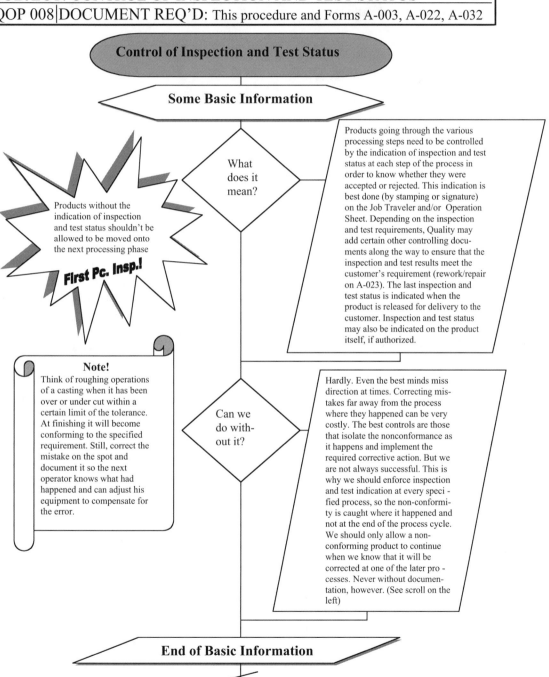

Control of Inspection and Test Status

Some Basic Information

Products without the indication of inspection and test status shouldn't be allowed to be moved onto the next processing phase

First Pc. Insp.!

What does it mean?

Products going through the various processing steps need to be controlled by the indication of inspection and test status at each step of the process in order to know whether they were accepted or rejected. This indication is best done (by stamping or signature) on the Job Traveler and/or Operation Sheet. Depending on the inspection and test requirements, Quality may add certain other controlling documents along the way to ensure that the inspection and test results meet the customer's requirement (rework/repair on A-023). The last inspection and test status is indicated when the product is released for delivery to the customer. Inspection and test status may also be indicated on the product itself, if authorized.

Note!
Think of roughing operations of a casting when it has been over or under cut within a certain limit of the tolerance. At finishing it will become conforming to the specified requirement. Still, correct the mistake on the spot and document it so the next operator knows what had happened and can adjust his equipment to compensate for the error.

Can we do without it?

Hardly. Even the best minds miss direction at times. Correcting mistakes far away from the process where they happened can be very costly. The best controls are those that isolate the nonconformance as it happens and implement the required corrective action. But we are not always successful. This is why we should enforce inspection and test indication at every specified process, so the non-conformity is caught where it happened and not at the end of the process cycle. We should only allow a nonconforming product to continue when we know that it will be corrected at one of the later processes. Never without documentation, however. (See scroll on the left)

End of Basic Information

TRAINING METRICS	Iss dt.	Rev dt	Pg.
QUALITY OPERATING PROCEDURE	Sign	Rev #	4/6
REQUIREMENT: QUALITY SYSTEM MANUAL SEC: 4.12			
SUBJECT: CONTROL OF INSPECTION AND TEST STATUS			
QOP 008 DOCUMENT REQ'D: This procedure and Forms A-003, A-022, A-032			

1.0   PURPOSE

Maintain control over inspection and test status indication, and computer access codes (passwords).

2.0   APPLICATION

This procedure shall apply to the application of inspection stamps and electronic passwords.

3.0   PROCEDURE

General Requirement

Inspection and test status indication shall be applied to product acceptance and release during the entire processing cycle, including products that are on hold and/or waiting for disposition. The indication of inspection and test status may be executed on the product itself, as required, but not without backup documentation to prove it. All products stored for further processing or shipment to customers shall have the inspection and test status indicated (on the Job Traveler or A-013).

Computerized (software based) document and data management for product related matters shall require access code (password) for all employees whose assignments affect product quality. Product acceptance and release indication executed through a software based documentation system shall be limited to designated personnel, responsible for the acceptance and release of the identified product(s). The issuance of access codes (passwords) shall be documented and controlled. Periodic monitoring by the issuing authority to ensure proper application of the access codes shall be enforced and adequately documented, as required.

3.01
The Quality Engineer does:

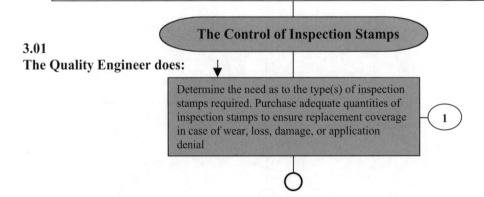

The Control of Inspection Stamps

Determine the need as to the type(s) of inspection stamps required. Purchase adequate quantities of inspection stamps to ensure replacement coverage in case of wear, loss, damage, or application denial

1

TRAINING METRICS	Iss dt.	Rev dt	Pg.
QUALITY OPERATING PROCEDURE	Sign	Rev #	5/6
REQUIREMENT: QUALITY SYSTEM MANUAL SEC: 4.12			
SUBJECT: CONTROL OF INSPECTION AND TEST STATUS			
QOP 008 DOCUMENT REQ'D: This procedure and Forms A-003, A-022, A-032			

Suggestion

Don't make QOPs or SOPs for the purpose of Quality Directives or other instructional communications for then you will have to control them similarly as the proce-dures are controlled in this book. Form A-003 was designed to take care of all internal communication requirements. Just fill in the header part of the form according to the type of direc-tive you wish to issue or control. Cite references as required and place "N/A" when things don't apply. Then, go and write your statements in the blank space below.

Determine who will be the stamp owners, autho-rized to use the stamps for specific applications, according to their assignments to accept and re-lease products. Issue and distribute to the stamp owners a Quality Directive on the application of stamps and the associated responsibility. Use form A-003 (blank) on which to issue this direc-tive, as appropriate — **2**

Instruction
Don't issue identical stamps to more than one person!

Assign the stamps according to documentation requirement stated on form A-032 — **3**

Instruction
The NCMR process for C/A is not limited for product handling only. Corrective Action ap-plies uniformly to pro-ducts, documents, and people, as required. (Our system in this book is meant to en-force continuous improvement in all aspects of the presented Quality Management System.)

Monitor the application of the issued stamps, as required, to prevent improper use. Record the mo-nitoring results onto form A-022. When finished, sign and date this form and hand it over to the Management Representative. Take corrective action, as required. In severe cases resort to the NCMR process (QOP 009) for C/A implementa-tion, as appropriate (pg 207) — **4**

STANDARD

The stamp owner's signa-ture on Form A-032 repre-sents the acceptance of re-sponsibility in the proper application of the issued stamp(s). The stamp(s) owner may apply his/her signature on the processing related documents instead of using the authorized stamp(s) in the acceptance and release of product(s).

Maintain security over the stamp inventory — **5**

Manage additional documentation requirement on form A-032 according to personnel changes — **6**

Instruction
Destroy all worn-out stamps and cross out their inventory status.

Retain forms A-003 (Quality Directive) and A-032 in one file folder. (When issuing stamps, issue the Quality Directive at the same time.) — **7**

The Control of Electronic Passwords

3.02

The Communications Manager does: ▼

Note
See 'Suggestion' scroll above under paragraph 3.01, step 2.

Determine the need as to methods used in the is-suance, limitation, and application of electronic access codes (passwords). Accordingly, issue and distribute a Communication Directive, using form A-003, as appropriate, informing the employees about basic requirements and conduct, regarding password ownership, application, and responsibi-lity — **1**

TRAINING METRICS	Iss dt.	Rev dt	Pg.
QUALITY OPERATING PROCEDURE	Sign	Rev #	6/6
REQUIREMENT: QUALITY SYSTEM MANUAL SEC: 4.12			
SUBJECT: CONTROL OF INSPECTION AND TEST STATUS			
QOP 008	DOCUMENT REQ'D: This procedure and Forms A-003, A-022, A-032		

Note!
The application of electronic passwords needs similar con–trol as inspection stamps do in order to prevent improper use.

Maintain internal registry on form A-032 on all password owners, including access limitations. Keep this registry secured. Provide access to it by upper management in case of your absence or termination

2

Note!
Refer to paragraph 3.01, step 4 above and read the 'Instruction' regarding the NCMR process. Also read the **STANDARD** under the same paragraph.

Monitor the application of the issued passwords, as required, to prevent improper use. Record the monitoring results onto form A-022. When finished, sign and date this form and send (E-mail) it over to the Management Representative. Take C/A implementation, as required. In severe cases resort to the NCMR process (QOP 009) for C/A implementation, as appropriate (pg 207)

3

Manage additional electronic documentation re-quirements on form A-032 according to personnel changes

4

Provide adequate system's backup to ensure nor-mal operations. Provide virus protection for all equipment engaged in product related assign-ments affecting product quality.

5

Quality should enforce account-ability for all issued stamps and pass-words regarding product quality!

End of QOP 008

OVERVIEW TO QOP 009
(CONTROL OF NON-CONFORMING PRODUCTS)

This procedure is about controlling non-conforming products and it is the third longest procedure by design. Don't be surprised to learn that in a process-approach quality management system, the control of non-conforming products also means controlling deficiencies in documents, processes, and methods. We cannot have an effectively run business by controlling hardware problems alone when the contributing causes of the non-conformance originated by deficiencies in documents, processes, and methods. Controlling deficiency in every process from planning to the delivery of the product is the quintessential driver in a process-approach system to effect on time continuous improvement to the operating management system. The lack of this type of control in businesses of all types is the hornet's nest from which everybody runs away instead of facing it and eliminating it through effective corrective action. This QOP has been linked to other QOPs in the quality management system, and the other QOPs in turn have been linked to this QOP, in order to bring about solving the festering problems in all the core departmental activities, contributing directly or indirectly to the deficiencies in documents, processes, methods and products. Once a non-conformance has been identified and written up on Form A-006, MRB is required to call for corrective action in line with the significance of the problem. The application of the type of statistical tools at this time will be determined by the Management Representative in the quest to identify the root cause(s) of the problem at hand, should supervision have failed to do so. The implementation of corrective action cannot be shoved aside, cannot be ignored, and cannot be turned over to Management Reviews as long as there are documented procedures in force to cure it. The Management Representative cannot close out any Form A-006 (Non-conforming Material Report) until he/she has followed up the effective implementation of corrective/preventive action. Under this system, the Management Representative is being independently audited for compliance. This is the significance of the stated provisions in QOP 009.

DIAGRAMMATIC PROCEDURE   QOP 009

CONTROL OF NON-CONFORMING PRODUCTS

CONTENTS

DIAGRAMMATIC PROCEDURE   QOP 009

CONTROL OF NON-CONFORMING PRODUCTS

CONTENTS (continued)   **Page**

TRAINING METRICS		Iss dt.		Rev dt.		Pg.
QUALITY OPERATING PROCEDURE		Sign		Rev #		3/ 24
REQUIREMENT: QUALITY SYSTEM MANUAL SEC: 4.13						
SUBJECT: CONTROL OF NON-CONFORMING PRODUCTS						
QOP 009	DOCUMENT REQ'D: This Quality Operating Procedure					

REVISION HISTORY

Rev Date	Rev No	Description	Approval

TRAINING METRICS	Iss dt.	Rev dt.	Pg.
QUALITY OPERATING PROCEDURE	Sign	Rev #	4/24
REQUIREMENT: Control of Non-conforming Products, QSM 4.13			
SUBJECT: Some Basic Information			
QOP 009 DOCUMENT REQ'D: Section One, Two, and Three of this procedure			

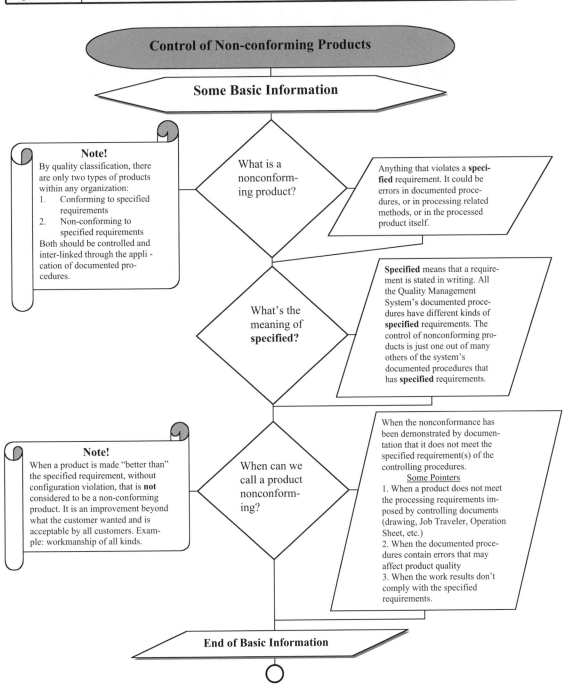

Control of Non-conforming Products

Some Basic Information

Note!
By quality classification, there are only two types of products within any organization:
1. Conforming to specified requirements
2. Non-conforming to specified requirements
Both should be controlled and inter-linked through the appli-cation of documented procedures.

What is a nonconform-ing product?

Anything that violates a **speci-fied** requirement. It could be errors in documented proce-dures, or in processing related methods, or in the processed product itself.

What's the meaning of **specified?**

Specified means that a require-ment is stated in writing. All the Quality Management System's documented proce-dures have different kinds of **specified** requirements. The control of nonconforming pro-ducts is just one out of many others of the system's documented procedures that has **specified** requirements.

Note!
When a product is made "better than" the specified requirement, without configuration violation, that is **not** considered to be a non-conforming product. It is an improvement beyond what the customer wanted and is acceptable by all customers. Exam-ple: workmanship of all kinds.

When can we call a product nonconform-ing?

When the nonconformance has been demonstrated by documen-tation that it does not meet the specified requirement(s) of the controlling procedures.
Some Pointers
1. When a product does not meet the processing requirements im-posed by controlling documents (drawing, Job Traveler, Operation Sheet, etc.)
2. When the documented proce-dures contain errors that may affect product quality
3. When the work results don't comply with the specified requirements.

End of Basic Information

TRAINING METRICS	Iss dt.	Rev dt.	Pg.
QUALITY OPERATING PROCEDURE	Sign	Rev #	5/24
REQUIREMENT: QUALITY SYSTEM MANUAL SEC: 4.13			
SUBJECT: CONTROL OF NON-CONFORMING PRODUCTS			
QOP 009	DOCUMENT REQ'D: Section One, Two, and Three of this procedure		

1.0 STANDARD

Maintain control over non-conforming products, documents, and actions affecting product quality.

2.0 APPLICATION

This procedure shall apply to the handling, reporting, and documentation of non-conforming products, documents, and actions affecting product quality.

3.0 Definitions

1. **Deviation** – a specific customer authorization issued prior to the manufacture of a product to allow departure from a defined design requirement for a specific number of units for a specific duration.
2. **Product non-conformance** – is any condition that violates the requirement of a specification, process, or procedure.
3. **Repair** -- a non-conforming product which cannot be further processed to meet a specified requirement without written approval from the customer.
4. **Rework** – a non-conforming product which can be reprocessed under defined conditions to meet a specified drawing requirement.
5. **Waiver** – a written request sent to a customer for disposition of a non-conforming product.

4.0 General Requirements

Non-conformance in product(s) and documented procedures shall be verified and documented in order to be processed as a reported non-conformance. Otherwise, it would lack the documented evidence to prove the indication of non-conformance (Forms A-018 and A-009).

Decisions made by the reviewing authority (MRB) shall be unanimous. In case of conflict, the management shall make the final decision.

A minimum of two (2) members of the reviewing authority shall be required to make a decision binding. One of the members shall always be a quality representative. The customer has final authority over non-reworkable products.

TRAINING METRICS	Iss dt.	Rev dt.	Pg.
QUALITY OPERATING PROCEDURE	Sign	Rev #	6/24
REQUIREMENT: QUALITY SYSTEM MANUAL SEC: 4.13			
SUBJECT: CONTROL OF NON-CONFORMING PRODUCTS			
QOP 009	DOCUMENT REQ'D: Section One, Two, and Three of this procedure		

SECTION ONE

1.0   PURPOSE
Maintain control over the reporting of non-conformance.

2.0   APPLICATION
This procedural section shall apply to the process of reporting non-conformance:
1. internally
2. to customers
3. to subcontractors

3.0   PROCEDURE

3.01

The Reporting Authority does:

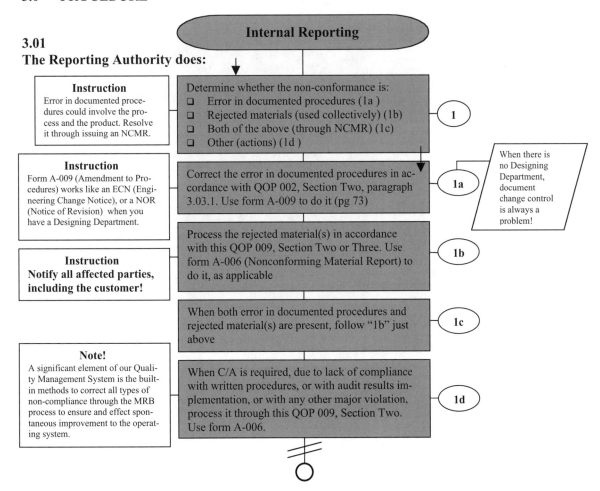

Internal Reporting

Instruction
Error in documented procedures could involve the process and the product. Resolve it through issuing an NCMR.

Determine whether the non-conformance is:
- ❑ Error in documented procedures (1a)
- ❑ Rejected materials (used collectively) (1b)
- ❑ Both of the above (through NCMR) (1c)
- ❑ Other (actions) (1d)

1

Instruction
Form A-009 (Amendment to Procedures) works like an ECN (Engineering Change Notice), or a NOR (Notice of Revision) when you have a Designing Department.

Correct the error in documented procedures in accordance with QOP 002, Section Two, paragraph 3.03.1. Use form A-009 to do it (pg 73)

1a

When there is no Designing Department, document change control is always a problem!

Instruction
Notify all affected parties, including the customer!

Process the rejected material(s) in accordance with this QOP 009, Section Two or Three. Use form A-006 (Nonconforming Material Report) to do it, as applicable

1b

When both error in documented procedures and rejected material(s) are present, follow "1b" just above

1c

Note!
A significant element of our Quality Management System is the built-in methods to correct all types of non-compliance through the MRB process to ensure and effect spontaneous improvement to the operating system.

When C/A is required, due to lack of compliance with written procedures, or with audit results implementation, or with any other major violation, process it through this QOP 009, Section Two. Use form A-006.

1d

TRAINING METRICS	Iss dt.	Rev dt.	Pg.
QUALITY OPERATING PROCEDURE	Sign	Rev #	7/24
REQUIREMENT: QUALITY SYSTEM MANUAL SEC: 4.13			
SUBJECT: CONTROL OF NON-CONFORMING PRODUCTS			
QOP 009	DOCUMENT REQ'D: Section One, Two, and Three of this procedure		

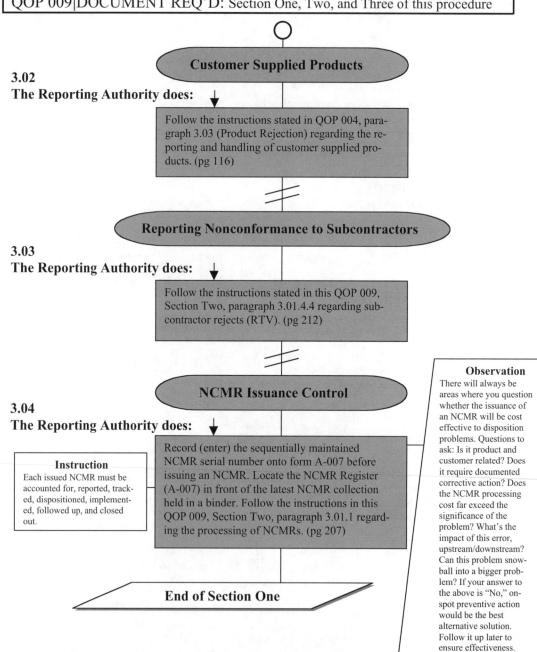

3.02
The Reporting Authority does:

Follow the instructions stated in QOP 004, paragraph 3.03 (Product Rejection) regarding the reporting and handling of customer supplied products. (pg 116)

Customer Supplied Products

Reporting Nonconformance to Subcontractors

3.03
The Reporting Authority does:

Follow the instructions stated in this QOP 009, Section Two, paragraph 3.01.4.4 regarding subcontractor rejects (RTV). (pg 212)

NCMR Issuance Control

3.04
The Reporting Authority does:

Instruction
Each issued NCMR must be accounted for, reported, tracked, dispositioned, implemented, followed up, and closed out.

Record (enter) the sequentially maintained NCMR serial number onto form A-007 before issuing an NCMR. Locate the NCMR Register (A-007) in front of the latest NCMR collection held in a binder. Follow the instructions in this QOP 009, Section Two, paragraph 3.01.1 regarding the processing of NCMRs. (pg 207)

Observation
There will always be areas where you question whether the issuance of an NCMR will be cost effective to disposition problems. Questions to ask: Is it product and customer related? Does it require documented corrective action? Does the NCMR processing cost far exceed the significance of the problem? What's the impact of this error, upstream/downstream? Can this problem snowball into a bigger problem? If your answer to the above is "No," on-spot preventive action would be the best alternative solution. Follow it up later to ensure effectiveness.

End of Section One

TRAINING METRICS	Iss dt.	Rev dt.	Pg.
QUALITY OPERATING PROCEDURE	Sign	Rev #	8/24
REQUIREMENT: QUALITY SYSTEM MANUAL SEC: 4.13			
SUBJECT: CONTROL OF NON-CONFORMING PRODUCTS			
QOP 009	DOCUMENT REQ'D: Forms A-003, A-006, A-007, A-009, A-012, A-018, A-023, A-024, A-033, A-037,A-039		

SECTION TWO

1.0   PURPOSE

Maintain control over the process of reporting, determining the cause, disposition, corrective/ preventive action, and follow-up of non-conforming products, processes, and work assignments.

2.0   APPLICATION

The provisions in this procedure shall apply to the processing of Non-conforming Material Report(s) (A-006) for the purpose of determining the cause, disposition, corrective/preventive action, and follow-up in respect to non-conforming products, processes, and work assignments.

3.0   PROCEDURE
3.01   Performance Requirement

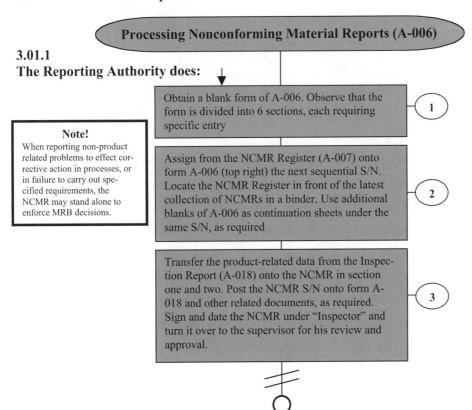

Processing Nonconforming Material Reports (A-006)

3.01.1
The Reporting Authority does:

Note!
When reporting non-product related problems to effect corrective action in processes, or in failure to carry out specified requirements, the NCMR may stand alone to enforce MRB decisions.

1. Obtain a blank form of A-006. Observe that the form is divided into 6 sections, each requiring specific entry

2. Assign from the NCMR Register (A-007) onto form A-006 (top right) the next sequential S/N. Locate the NCMR Register in front of the latest collection of NCMRs in a binder. Use additional blanks of A-006 as continuation sheets under the same S/N, as required

3. Transfer the product-related data from the Inspection Report (A-018) onto the NCMR in section one and two. Post the NCMR S/N onto form A-018 and other related documents, as required. Sign and date the NCMR under "Inspector" and turn it over to the supervisor for his review and approval.

TRAINING METRICS	Iss dt.	Rev dt.	Pg.
QUALITY OPERATING PROCEDURE	Sign	Rev #	9/24
REQUIREMENT: QUALITY SYSTEM MANUAL SEC: 4.13			
SUBJECT: CONTROL OF NON-CONFORMING PRODUCTS			
QOP 009	DOCUMENT REQ'D: Forms A-003, A-006, A-007, A-009, A-012, A-018, A-023, A-024, A-033, A-037, A-039		

Cause Determination

3.01.2
The Supervisor does:

Instruction
Rely on technical support when facing the unknown. Don't guess! The management representative is your best guide in this process. He may need to look deeper in order to isolate the contributing causes to the problem at hand that you were unable to do from the available data.

1. Review the reported non-conformance and determine the underlying reason(s) (the root cause) as to why this problem happened

2. Describe the contributing reasons for the nonconformance in the third block on form A-006. Don't explain the end result. We all know that from the Inspection Report (A-018)

The effectiveness of corrective action depends on your explanation of what the cause was. So don't guess or write something silly!

3. Sign and date under 'Supervisor' as indicated and send this NCMR together with all related documents (A-018, etc.) and the product to the Material Review Board (MRB), as practical.

MRB Disposition

3.01.3
The Material Review Board does:
(Quality and Engineering)

Note!
Quality is a permanent member of MRB. "Engineering" may be substituted, as required.
_ _ _ _ _ _ _ _ _ _ _ _ _ _
Observe that there may be cases where MRB authority, other than "REWORK," is not granted by the customer. A question may arise: Should I need an "Engineering" signature on the NCMR (Form A-006) in order to submit a "Waiver" to the customer? The answer is: Yes. While the waiver shows only a partial information of the rejected product, the NCMR shows all of it. It serves as a cross-functional activity indicator, incorporating the required actions needed to implement improvements. The closure of NCMR should ensure that all requirements have been satisfied including waiver requirement(s). See paragraph 3.01.7 below.

1. Review and evaluate the inspection results (from A-018) and the cause (from A-006) of the nonconformance in order to determine the appropriate steps to take in dispositioning the subject nonconformance

2. Make the disposition and indicate under 'MRB Disposition' the appropriate action required. Do this by encircling the activity indicator(s) (RWK, RPR, ACC, RTV, SCR, Waiver. Next, add under 'Instruction/Comment,' in line with the affected item(s), any relevant information required to be followed by those who will be implementing your disposition(s)

3. Determine whether any C/A is required. Enter this requirement as indicated on A-006. Then, sign and date as shown, authorizing enforcement

TRAINING METRICS	Iss dt.	Rev dt.	Pg.
QUALITY OPERATING PROCEDURE	Sign	Rev #	10/24
REQUIREMENT: QUALITY SYSTEM MANUAL SEC: 4.13			
SUBJECT: CONTROL OF NON-CONFORMING PRODUCTS			
QOP 009	DOCUMENT REQ'D: Forms A-003, A-006, A-007, A-009, A-012, A-018, A-023, A-024, A-033, A-037, A-039		

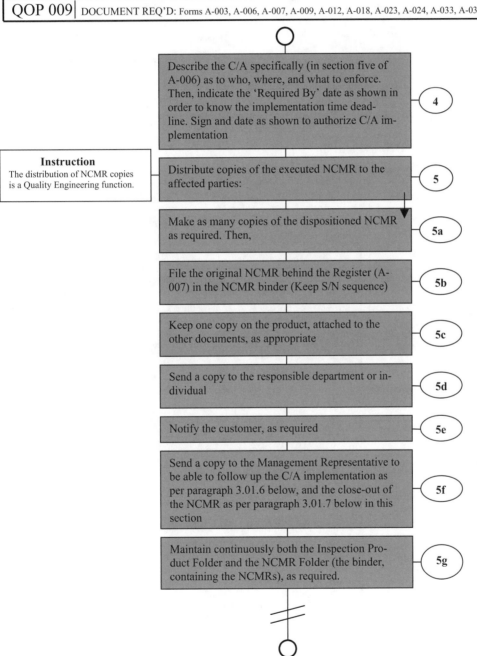

Describe the C/A specifically (in section five of A-006) as to who, where, and what to enforce. Then, indicate the 'Required By' date as shown in order to know the implementation time deadline. Sign and date as shown to authorize C/A implementation — **4**

Instruction
The distribution of NCMR copies is a Quality Engineering function.

Distribute copies of the executed NCMR to the affected parties: — **5**

Make as many copies of the dispositioned NCMR as required. Then, — **5a**

File the original NCMR behind the Register (A-007) in the NCMR binder (Keep S/N sequence) — **5b**

Keep one copy on the product, attached to the other documents, as appropriate — **5c**

Send a copy to the responsible department or individual — **5d**

Notify the customer, as required — **5e**

Send a copy to the Management Representative to be able to follow up the C/A implementation as per paragraph 3.01.6 below, and the close-out of the NCMR as per paragraph 3.01.7 below in this section — **5f**

Maintain continuously both the Inspection Product Folder and the NCMR Folder (the binder, containing the NCMRs), as required. — **5g**

TRAINING METRICS	Iss dt.	Rev dt.	Pg.
QUALITY OPERATING PROCEDURE	Sign	Rev #	11/24
REQUIREMENT: QUALITY SYSTEM MANUAL SEC: 4.13			
SUBJECT: CONTROL OF NON-CONFORMING PRODUCTS			
QOP 009 DOCUMENT REQ'D: Forms A-003, A-006, A-007, A-009, A-012, A-018, A-023, A-024, A-033, A-037,A-039			

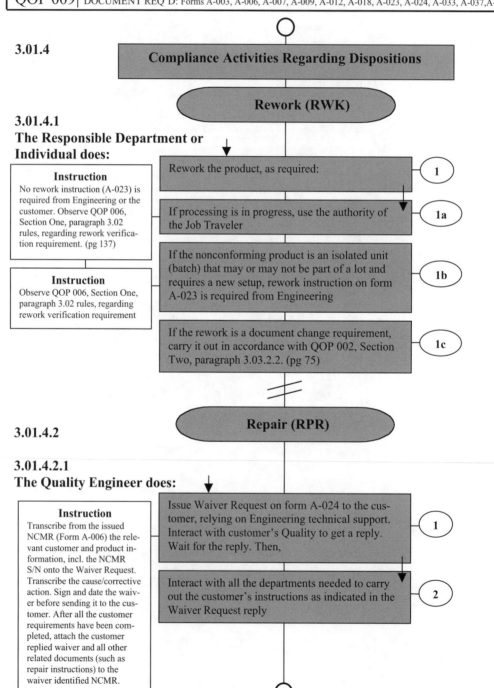

3.01.4

Compliance Activities Regarding Dispositions

Rework (RWK)

3.01.4.1
The Responsible Department or
Individual does:

Instruction
No rework instruction (A-023) is required from Engineering or the customer. Observe QOP 006, Section One, paragraph 3.02 rules, regarding rework verification requirement. (pg 137)

Rework the product, as required: **1**

If processing is in progress, use the authority of the Job Traveler   **1a**

Instruction
Observe QOP 006, Section One, paragraph 3.02 rules, regarding rework verification requirement

If the nonconforming product is an isolated unit (batch) that may or may not be part of a lot and requires a new setup, rework instruction on form A-023 is required from Engineering   **1b**

If the rework is a document change requirement, carry it out in accordance with QOP 002, Section Two, paragraph 3.03.2.2. (pg 75)   **1c**

Repair (RPR)

3.01.4.2

3.01.4.2.1
The Quality Engineer does:

Instruction
Transcribe from the issued NCMR (Form A-006) the relevant customer and product information, incl. the NCMR S/N onto the Waiver Request. Transcribe the cause/corrective action. Sign and date the waiver before sending it to the customer. After all the customer requirements have been completed, attach the customer replied waiver and all other related documents (such as repair instructions) to the waiver identified NCMR.

Issue Waiver Request on form A-024 to the customer, relying on Engineering technical support. Interact with customer's Quality to get a reply. Wait for the reply. Then,   **1**

Interact with all the departments needed to carry out the customer's instructions as indicated in the Waiver Request reply   **2**

TRAINING METRICS	Iss dt.	Rev dt.	Pg.
QUALITY OPERATING PROCEDURE	Sign	Rev #	12/24
REQUIREMENT: QUALITY SYSTEM MANUAL SEC: 4.13			
SUBJECT: CONTROL OF NON-CONFORMING PRODUCTS			
QOP 009	DOCUMENT REQ'D: Forms A-003, A-006, A-007, A-009, A-012, A-018, A-023, A-024, A-033, A-037, A-039		

If the disposition by the customer is to repair the product, interact with the Process Engineer to issue the Rework/Repair Procedure on form A-023 — **2a**

If the disposition by the customer is to send the product for his evaluation, follow the instructions as stated in QOP 006, Section Four, paragraph 3.03, as appropriate. Observe "Instruction" under step 6 in paragraph 3.02 of the same procedure, regarding tagging (pg 164 & 165) — **2b**

If the disposition by the customer is to scrap the product, follow the instruction as stated below in paragraph 3.01.4.5 in this section. — **2c**

3.01.4.2.2 The Process Engineer does:

Issue the Rework/Repair Procedure on form A-023 according to instructions given by the customer in the Waiver Request reply. Interact with the customer's Process Engineer in order to define the repair procedure, as appropriate — **1**

Monitor the rework/repair process and ensure adequate technical support. — **2**

3.01.4.2.3 The Production Supervisor does:

Assign the rework/repair job to the appropriate process owner who will carry out the rework/repair according to the work instruction issued by the Process Engineer on form A-023 — **1**

Monitor the process owner's work performance. Observe product verification requirements as stated under QOP 006, Section One, paragraph 3.02. (pg 137) — **2**

TRAINING METRICS	Iss dt.	Rev dt.	Pg.
QUALITY OPERATING PROCEDURE	Sign	Rev #	13/24
REQUIREMENT: QUALITY SYSTEM MANUAL SEC: 4.13			
SUBJECT: CONTROL OF NON-CONFORMING PRODUCTS			
QOP 009	DOCUMENT REQ'D: Forms A-003, A-006, A-007, A-009, A-012, A-018, A-023, A-024, A-033, A-037,A-039		

3.01.4.2.4 The Quality Engineer does:

Monitor the completion of the reworked/repaired product through final release verification by the Quality Inspector — **1**

Instruction
Keep the released product tagged with form A-012 throughout other processes. Ship the product tagged (A-012) to the customer with his/ her concession agreement (the Waiver, A-024) added to the shipping documents.

Work together with the Quality Inspector in line with instructions given under QOP 006, Section One, paragraph 3.02, as appropriate. (pg 137) — **2**

Accept (ACC)
(USE AS IS CONDITION)

3.01.4.3
The Process and Quality Engineers do:

Observe that acceptance of non-conforming product(s), except to rework it, is not permitted on customers' POs. Issue rework instructions on form A-023, as required, or resort to the 'Repair' process as outlined above under paragraph 3.01.4.2

Return to Vendor (RTV)

3.01.4.4
The Quality Engineer does:

Notify the Purchasing and Planning Departments. (A copy of the dispositioned NCMR will do it.) See paragraph 3.01.3, step 5 above — **1**

Request Packing Slip (form A-028) — **2**

Attach to the Packing Slip the Inspection Report (A-018), a copy of the dispositioned NCMR (A-006), and the completed Supplier Corrective Action Request (SCAR, form A-037) — **3**

Release the product to the Shipping Department to return it to the subcontractor — **4**

TRAINING METRICS	Iss dt.	Rev dt.	Pg.
QUALITY OPERATING PROCEDURE	Sign	Rev #	14/24
REQUIREMENT: QUALITY SYSTEM MANUAL SEC: 4.13			
SUBJECT: CONTROL OF NON-CONFORMING PRODUCTS			
QOP 009	DOCUMENT REQ'D: Forms A-003, A-006, A-007, A-009, A-012, A-018, A-023, A-024, A-033, A-037, A-039		

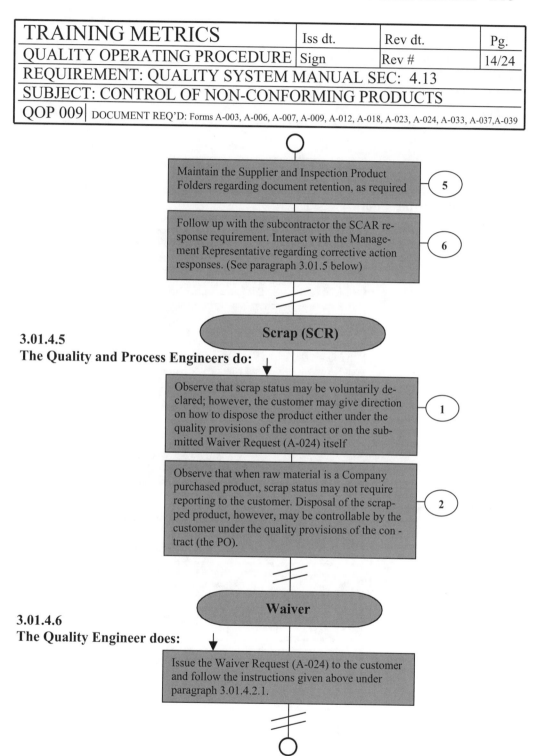

Maintain the Supplier and Inspection Product Folders regarding document retention, as required — 5

Follow up with the subcontractor the SCAR response requirement. Interact with the Management Representative regarding corrective action responses. (See paragraph 3.01.5 below) — 6

Scrap (SCR)

3.01.4.5
The Quality and Process Engineers do:

Observe that scrap status may be voluntarily declared; however, the customer may give direction on how to dispose the product either under the quality provisions of the contract or on the submitted Waiver Request (A-024) itself — 1

Observe that when raw material is a Company purchased product, scrap status may not require reporting to the customer. Disposal of the scrapped product, however, may be controllable by the customer under the quality provisions of the contract (the PO). — 2

Waiver

3.01.4.6
The Quality Engineer does:

Issue the Waiver Request (A-024) to the customer and follow the instructions given above under paragraph 3.01.4.2.1.

TRAINING METRICS	Iss dt.	Rev dt.	Pg.
QUALITY OPERATING PROCEDURE	Sign	Rev #	15/24
REQUIREMENT: QUALITY SYSTEM MANUAL SEC: 4.13			
SUBJECT: CONTROL OF NON-CONFORMING PRODUCTS			
QOP 009	DOCUMENT REQ'D: Forms A-003, A-006, A-007, A-009, A-012, A-018, A-023, A-024, A-033, A-037, A-039		

Regrading Material

3.01.4.7
The Quality Engineer does:

1 Observe that regrading is a customer imposed requirement resulting from his/her dispositioning of a non-conforming product, which will be identified either on the Waiver Request or on the customer's processed NCMR (on his own form)

2 Observe that product identification, traceability, and documentation requirements to handle regrading may be imposed by the customer. Carry this out according to the customer's requirements and in line with the internal documentation requirement as described in QOP 005, paragraph 3.02, as appropriate. (pg 126)

Handling Corrective and Preventive Action

3.01.5
The Management Representative does:

Instruction
Corrective action implementation may not be an easy process when more than one problem-one cause is at hand. You may need to do *cause and effect* diagramming and *failure mode effect analysis* (FMEA) as well as some serious brainstorming and benchmarking to pin down the actual causes of problems before deciding on cross-functional corrective action implementation. While the implementation process is fully described here for you, the knowledge and skill in the application of statistical tools you should possess as part of your job description as Management Representative.

1 Determine from the dispositioned NCMR the extent of the corrective action requirement

Instruction
Ensure timely response from subcontractors on all issued SCARs

2 Interact with the responsible department head, or the individual, that has to implement the corrective action and work out the implementation plan with him/her in line with the required date noted on the specific NCMR

3 Summarize the implementation plan on form A-003 (blank) and issue it to the responsible agent. Maintain a copy of it for follow-up verification

4 Observe that corrective action may involve change to the Quality System documents. Refer to QOP 002, Section Two (all paragraphs may apply) to implement it. (pg 71)

TRAINING METRICS	Iss dt.	Rev dt.	Pg.
QUALITY OPERATING PROCEDURE	Sign	Rev #	16/24
REQUIREMENT: QUALITY SYSTEM MANUAL SEC: 4.13			
SUBJECT: CONTROL OF NON-CONFORMING PRODUCTS			
QOP 009	DOCUMENT REQ'D: Forms A-003, A-006, A-007, A-009, A-012, A-018, A-023, A-024, A-033, A-037, A-039		

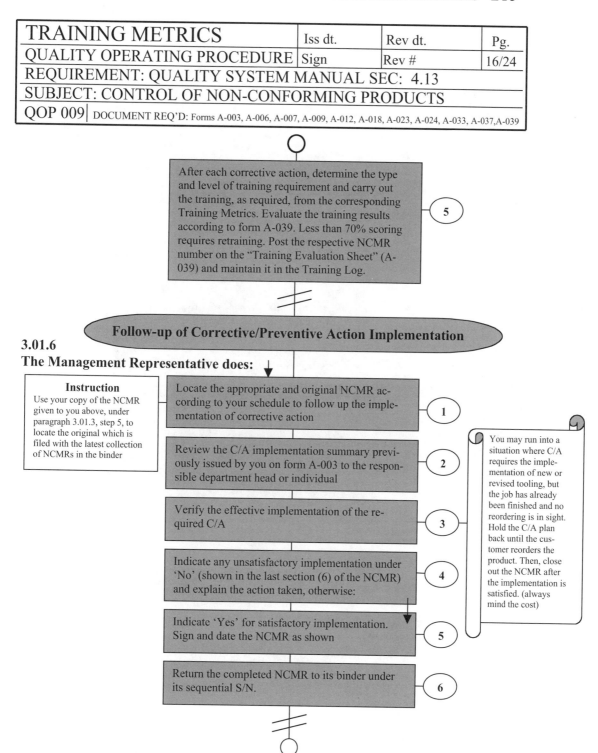

After each corrective action, determine the type and level of training requirement and carry out the training, as required, from the corresponding Training Metrics. Evaluate the training results according to form A-039. Less than 70% scoring requires retraining. Post the respective NCMR number on the "Training Evaluation Sheet" (A-039) and maintain it in the Training Log.

5

Follow-up of Corrective/Preventive Action Implementation

3.01.6

The Management Representative does:

Instruction
Use your copy of the NCMR given to you above, under paragraph 3.01.3, step 5, to locate the original which is filed with the latest collection of NCMRs in the binder

Locate the appropriate and original NCMR according to your schedule to follow up the implementation of corrective action

1

Review the C/A implementation summary previously issued by you on form A-003 to the responsible department head or individual

2

Verify the effective implementation of the required C/A

3

Indicate any unsatisfactory implementation under 'No' (shown in the last section (6) of the NCMR) and explain the action taken, otherwise:

4

Indicate 'Yes' for satisfactory implementation. Sign and date the NCMR as shown

5

Return the completed NCMR to its binder under its sequential S/N.

6

You may run into a situation where C/A requires the implementation of new or revised tooling, but the job has already been finished and no reordering is in sight. Hold the C/A plan back until the customer reorders the product. Then, close out the NCMR after the implementation is satisfied. (always mind the cost)

TRAINING METRICS	Iss dt.	Rev dt.	Pg.
QUALITY OPERATING PROCEDURE	Sign	Rev #	17/24
REQUIREMENT: QUALITY SYSTEM MANUAL SEC: 4.13			
SUBJECT: CONTROL OF NON-CONFORMING PRODUCTS			
QOP 009	DOCUMENT REQ'D: Forms A-003, A-006, A-007, A-009, A-012, A-018, A-023, A-024, A-033, A-037, A-039		

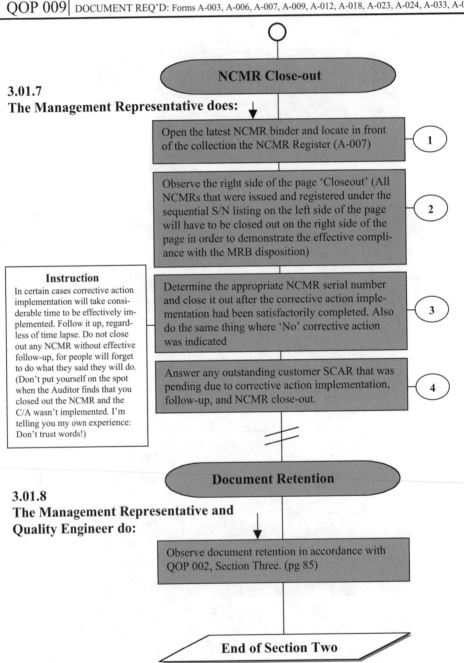

NCMR Close-out

3.01.7
The Management Representative does:

1. Open the latest NCMR binder and locate in front of the collection the NCMR Register (A-007)

2. Observe the right side of the page 'Closeout' (All NCMRs that were issued and registered under the sequential S/N listing on the left side of the page will have to be closed out on the right side of the page in order to demonstrate the effective compliance with the MRB disposition)

Instruction
In certain cases corrective action implementation will take considerable time to be effectively implemented. Follow it up, regardless of time lapse. Do not close out any NCMR without effective follow-up, for people will forget to do what they said they will do. (Don't put yourself on the spot when the Auditor finds that you closed out the NCMR and the C/A wasn't implemented. I'm telling you my own experience: Don't trust words!)

3. Determine the appropriate NCMR serial number and close it out after the corrective action implementation had been satisfactorily completed. Also do the same thing where 'No' corrective action was indicated

4. Answer any outstanding customer SCAR that was pending due to corrective action implementation, follow-up, and NCMR close-out.

Document Retention

3.01.8
The Management Representative and
Quality Engineer do:

Observe document retention in accordance with QOP 002, Section Three. (pg 85)

End of Section Two

TRAINING METRICS	Iss dt.	Rev dt.		Pg.
QUALITY OPERATING PROCEDURE	Sign	Rev #		18/24
REQUIREMENT: Control of Non-conforming Products, QSM 4.13				
SUBJECT: Customer Complaints, Product Return, and Satisfaction Reports				
QOP 009	DOCUMENT REQ'D: Forms A-003, A-006, A-007, A-009, A-012, A-018, A-023, A-024, A-033, A-038			

Customer Complaints, Product Return, and Satisfaction Reports

SECTION THREE

1.0   PURPOSE

Maintain control over customer complaints, product returns, and customer satisfaction reporting.

2.0   APPLICATION

The provisions of this procedure shall apply to the processing of customer complaints, product returns, and customer satisfaction reporting.

3.0   PROCEDURE

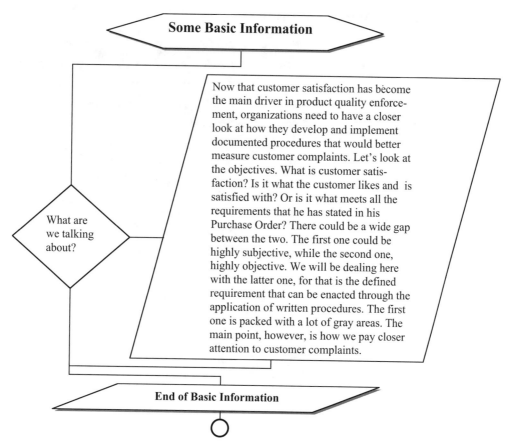

Some Basic Information

Now that customer satisfaction has become the main driver in product quality enforcement, organizations need to have a closer look at how they develop and implement documented procedures that would better measure customer complaints. Let's look at the objectives. What is customer satisfaction? Is it what the customer likes and is satisfied with? Or is it what meets all the requirements that he has stated in his Purchase Order? There could be a wide gap between the two. The first one could be highly subjective, while the second one, highly objective. We will be dealing here with the latter one, for that is the defined requirement that can be enacted through the application of written procedures. The first one is packed with a lot of gray areas. The main point, however, is how we pay closer attention to customer complaints.

What are we talking about?

End of Basic Information

TRAINING METRICS	Iss dt.	Rev dt.	Pg.
QUALITY OPERATING PROCEDURE	Sign	Rev #	19/24
REQUIREMENT: Control of Non-conforming Products, QSM 4.13			
SUBJECT: Customer Complaints, Product Return, and Satisfaction Reports			
QOP 009 DOCUMENT REQ'D: Forms A-003, A-006, A-007, A-009, A-012, A-018, A-023, A-024, A-033, A-038			

3.01

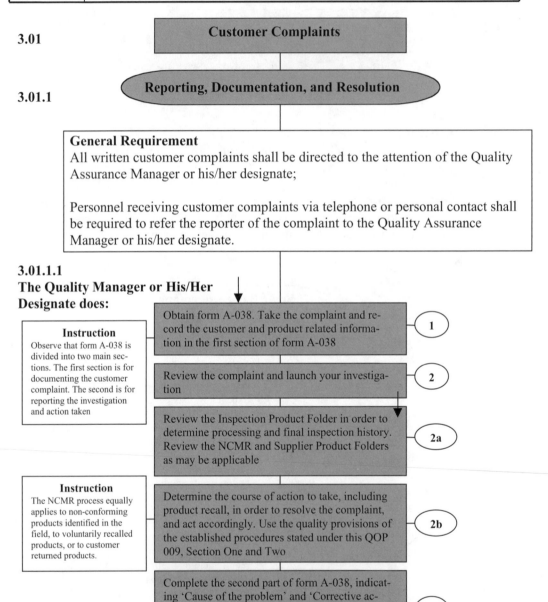

Customer Complaints

3.01.1

Reporting, Documentation, and Resolution

General Requirement

All written customer complaints shall be directed to the attention of the Quality Assurance Manager or his/her designate;

Personnel receiving customer complaints via telephone or personal contact shall be required to refer the reporter of the complaint to the Quality Assurance Manager or his/her designate.

3.01.1.1
The Quality Manager or His/Her Designate does:

Instruction
Observe that form A-038 is divided into two main sections. The first section is for documenting the customer complaint. The second is for reporting the investigation and action taken

1 — Obtain form A-038. Take the complaint and record the customer and product related information in the first section of form A-038

2 — Review the complaint and launch your investigation

2a — Review the Inspection Product Folder in order to determine processing and final inspection history. Review the NCMR and Supplier Product Folders as may be applicable

Instruction
The NCMR process equally applies to non-conforming products identified in the field, to voluntarily recalled products, or to customer returned products.

2b — Determine the course of action to take, including product recall, in order to resolve the complaint, and act accordingly. Use the quality provisions of the established procedures stated under this QOP 009, Section One and Two

3 — Complete the second part of form A-038, indicating 'Cause of the problem' and 'Corrective action taken.' Sign and date form A-038. Send a copy of it to the customer (to the person reported the complaint)

TRAINING METRICS	Iss dt.	Rev dt.	Pg.
QUALITY OPERATING PROCEDURE	Sign	Rev #	20/24
REQUIREMENT: Control of Non-conforming Products, QSM 4.13			
SUBJECT: Customer Complaints, Product Return, and Satisfaction Reports			
QOP 009	DOCUMENT REQ'D: Forms A-003, A-006, A-007, A-009, A-012, A-018, A-023, A-024, A-033, A-038		

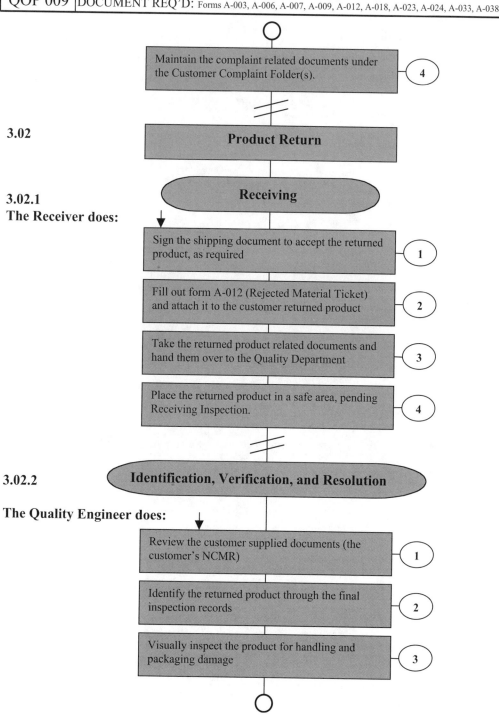

Maintain the complaint related documents under the Customer Complaint Folder(s). ④

3.02 **Product Return**

3.02.1
The Receiver does:

Receiving

Sign the shipping document to accept the returned product, as required ①

Fill out form A-012 (Rejected Material Ticket) and attach it to the customer returned product ②

Take the returned product related documents and hand them over to the Quality Department ③

Place the returned product in a safe area, pending Receiving Inspection. ④

3.02.2 Identification, Verification, and Resolution

The Quality Engineer does:

Review the customer supplied documents (the customer's NCMR) ①

Identify the returned product through the final inspection records ②

Visually inspect the product for handling and packaging damage ③

TRAINING METRICS	Iss dt.	Rev dt.	Pg.
QUALITY OPERATING PROCEDURE	Sign	Rev #	21/24
REQUIREMENT: Control of Non-conforming Products, QSM 4.13			
SUBJECT: Customer Complaints, Product Return, and Satisfaction Reports			
QOP 009	DOCUMENT REQ'D: Forms A-003, A-006, A-007, A-009, A-012, A-018, A-023, A-024, A-033, A-038		

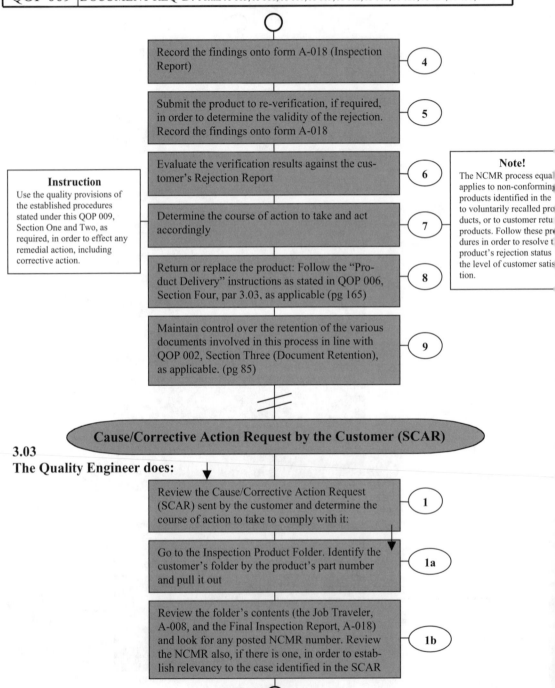

Record the findings onto form A-018 (Inspection Report)   **4**

Submit the product to re-verification, if required, in order to determine the validity of the rejection. Record the findings onto form A-018   **5**

Evaluate the verification results against the customer's Rejection Report   **6**

Instruction
Use the quality provisions of the established procedures stated under this QOP 009, Section One and Two, as required, in order to effect any remedial action, including corrective action.

Determine the course of action to take and act accordingly   **7**

Return or replace the product: Follow the "Product Delivery" instructions as stated in QOP 006, Section Four, par 3.03, as applicable (pg 165)   **8**

Maintain control over the retention of the various documents involved in this process in line with QOP 002, Section Three (Document Retention), as applicable. (pg 85)   **9**

Note!
The NCMR process equal[ly] applies to non-conforming products identified in the [house,] to voluntarily recalled products, or to customer retu[rned] products. Follow these pr[oce-]dures in order to resolve t[he] product's rejection status [to] the level of customer satis[fac-]tion.

Cause/Corrective Action Request by the Customer (SCAR)

3.03
The Quality Engineer does:

Review the Cause/Corrective Action Request (SCAR) sent by the customer and determine the course of action to take to comply with it: **1**

Go to the Inspection Product Folder. Identify the customer's folder by the product's part number and pull it out   **1a**

Review the folder's contents (the Job Traveler, A-008, and the Final Inspection Report, A-018) and look for any posted NCMR number. Review the NCMR also, if there is one, in order to establish relevancy to the case identified in the SCAR   **1b**

TRAINING METRICS	Iss dt.	Rev dt.	Pg.
QUALITY OPERATING PROCEDURE	Sign	Rev #	22/24
REQUIREMENT: Control of Non-conforming Products, QSM 4.13			
SUBJECT: Customer Complaints, Product Return, and Satisfaction Reports			
QOP 009 DOCUMENT REQ'D: Forms A-003, A-006, A-007, A-009, A-012, A-018, A-023, A-024, A-033, A-038			

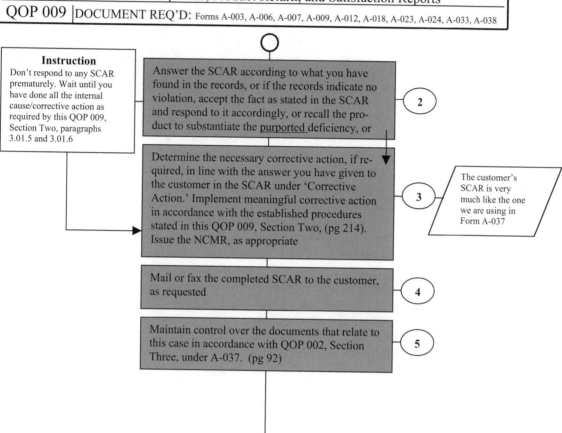

Instruction
Don't respond to any SCAR prematurely. Wait until you have done all the internal cause/corrective action as required by this QOP 009, Section Two, paragraphs 3.01.5 and 3.01.6

Answer the SCAR according to what you have found in the records, or if the records indicate no violation, accept the fact as stated in the SCAR and respond to it accordingly, or recall the product to substantiate the purported deficiency, or — 2

Determine the necessary corrective action, if required, in line with the answer you have given to the customer in the SCAR under 'Corrective Action.' Implement meaningful corrective action in accordance with the established procedures stated in this QOP 009, Section Two, (pg 214). Issue the NCMR, as appropriate — 3

The customer's SCAR is very much like the one we are using in Form A-037

Mail or fax the completed SCAR to the customer, as requested — 4

Maintain control over the documents that relate to this case in accordance with QOP 002, Section Three, under A-037. (pg 92) — 5

TRAINING METRICS	Iss dt.	Rev dt.	Pg.
QUALITY OPERATING PROCEDURE	Sign	Rev #	23/24
REQUIREMENT: Control of Non-conforming Products, QSM 4.13			
SUBJECT: Customer Complaints, Product Return, and Satisfaction Reports			
QOP 009 DOCUMENT REQ'D: Forms A-003, A-006, A-007, A-009, A-012, A-018, A-023, A-024, A-033, A-038			

3.04

Customer Satisfaction Reporting

General Requirement

Product acceptance by the customers through the "Customer Complaint and Evaluation Report" (Form A-038) is not a contractual requirement unless it is specifically stated in the customer's Purchase Order. Product deficiencies between the customer and the supplier's organization in contractual relationships are handled through the Quality Provisions of the customer's Purchase Orders (SCAR provisions) and the Company's documented procedures, the "Control of Non-conforming Products" stipulated in QOP 009. Therefore, the application of Form A-038 is a self-imposed requirement to document and implement those areas of product concerns that fall outside the established documented system. The purpose of the application of Form A-038 is to ensure ultimate customer satisfaction in product quality beyond those provisions stipulated in customers' Purchase Orders. Failure to define by the customers the Quality Provisions in their Purchase Orders shall not be construed by the customers as accepting responsibility by the supplier's organization through the application of Form A-038.

Form A-038 is a multipurpose document: 1. designed to handle customer complaints reported outside of the SCAR provisions of the customer's Purchase Order. (See paragraph 3.01 above) 2. designed to handle the unsolicited product evaluation responses that customers may wish to express when they receive a delivered product. (Covered under this paragraph 3.04)

3.04.1   **Handling and Controlling Customer Evaluation Reports (A-038)**

3.04.1.1
The Final Inspector does:

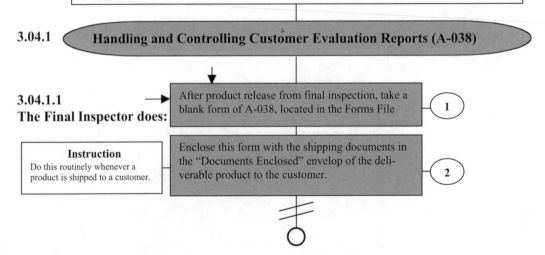

After product release from final inspection, take a blank form of A-038, located in the Forms File   ①

Instruction
Do this routinely whenever a product is shipped to a customer.

Enclose this form with the shipping documents in the "Documents Enclosed" envelop of the deliverable product to the customer.   ②

TRAINING METRICS	Iss dt.	Rev dt.	Pg.
QUALITY OPERATING PROCEDURE	Sign	Rev #	24/24
REQUIREMENT: Control of Non-conforming Products, QSM 4.13			
SUBJECT: Customer Complaints, Product Return, and Satisfaction Reports			
QOP 009 DOCUMENT REQ'D: Forms A-003, A-006, A-007, A-009, A-012, A-018, A-023, A-024, A-033, A-038			

3.04.1.2
The Management Representative does:

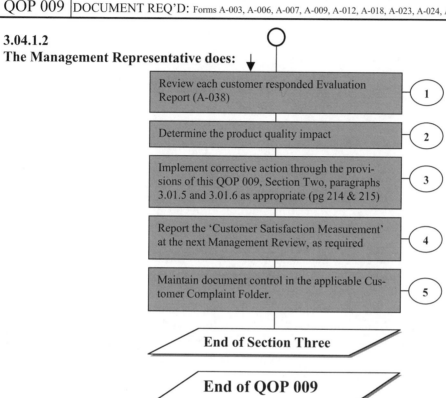

Review each customer responded Evaluation Report (A-038) — 1

Determine the product quality impact — 2

Implement corrective action through the provisions of this QOP 009, Section Two, paragraphs 3.01.5 and 3.01.6 as appropriate (pg 214 & 215) — 3

Report the 'Customer Satisfaction Measurement' at the next Management Review, as required — 4

Maintain document control in the applicable Customer Complaint Folder. — 5

End of Section Three

End of QOP 009

OVERVIEW TO QOP 010
(INTERNAL QUALITY AUDIT)

The quality system procedures in this process-approach system don't resemble the standard operating procedures (SOP) in conventional quality systems. The quality procedures in this system have been designed to be user-friendly in all areas as to the identification of requirement, the definition of responsibility, work assignment, accountability, and process flow. For an auditor to spend more time on compiling the checklist than on doing the audit itself is absolutely unacceptable in economic terms. This, of course, is not the auditor's problem. How the quality system's procedures have been laid out and integrated must lend a user-friendly flow process in which what you need is right in front of you. Since internal audits are done mostly by in-house employees, it makes a big difference economically whether the auditor spends a few hours or a few days in accomplishing internal audits. I paid serious attention to this also when I planned the layout of my procedures, for I used to do audits at many companies in my forty years service and could not easily find where and how responsibilities were linked to work assignments in the conventionally written SOPs. Every quality operating procedure in this process-approach system is also a checklist, not only for the auditor, but also for the process owner and his supervisor as well. Note the dramatic shift from conventional methods as to how we maintain continuous improvement in training employees whenever corrective action requirement points to the operators' lack of understanding of their work assignment. The details of the audit process are the subject of this QOP.

DIAGRAMMATIC PROCEDURE   QOP 010

INTERNAL QUALITY AUDITS

CONTENTS	Internal Quality Audits	Page

TRAINING METRICS			Iss dt.	Rev dt.		Pg.
QUALITY OPERATING PROCEDURE			Sign	Rev #		2/9
REQUIREMENT: QUALITY SYSTEM MANUAL SEC: 4.17						
SUBJECT: Internal Quality Audits						
QOP 010	DOCUMENT REQ'D: This QOP 010					

REVISION HISTORY

Rev Date	Rev No	Description	Approval

TRAINING METRICS	Iss dt.	Rev dt.	Pg.
QUALITY OPERATING PROCEDURE	Sign	Rev #	3/9
REQUIREMENT: Internal Quality Audits, QSM 4.17			
SUBJECT: Some Basic Information			
QOP 010 |DOCUMENT REQ'D: Forms A-006, A-007, A-021, A-022			

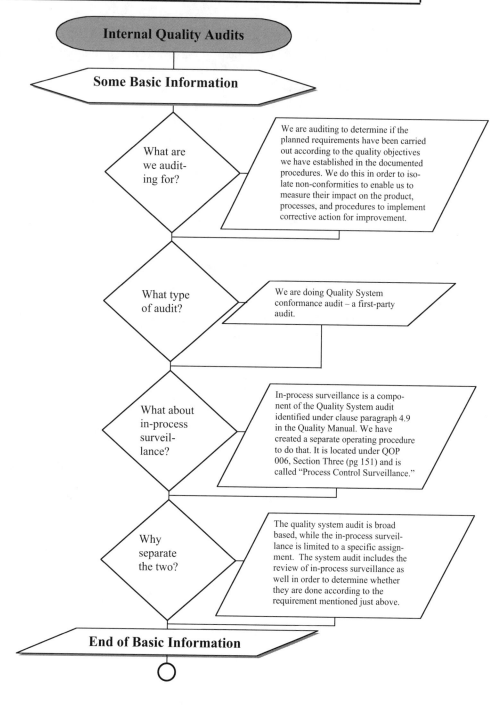

Internal Quality Audits

Some Basic Information

What are we auditing for?

We are auditing to determine if the planned requirements have been carried out according to the quality objectives we have established in the documented procedures. We do this in order to isolate non-conformities to enable us to measure their impact on the product, processes, and procedures to implement corrective action for improvement.

What type of audit?

We are doing Quality System conformance audit – a first-party audit.

What about in-process surveillance?

In-process surveillance is a component of the Quality System audit identified under clause paragraph 4.9 in the Quality Manual. We have created a separate operating procedure to do that. It is located under QOP 006, Section Three (pg 151) and is called "Process Control Surveillance."

Why separate the two?

The quality system audit is broad based, while the in-process surveillance is limited to a specific assignment. The system audit includes the review of in-process surveillance as well in order to determine whether they are done according to the requirement mentioned just above.

End of Basic Information

TRAINING METRICS	Iss dt.	Rev dt.	Pg.
QUALITY OPERATING PROCEDURE	Sign	Rev #	4/9
REQUIREMENT: Internal Quality Audits, QSM 4.17			
SUBJECT: Performing the Quality System Audit			
QOP 010 DOCUMENT REQ'D: Forms A-006, A-007, A-021, A-022			

1.0   PURPOSE

Maintain control over the process of carrying out internal quality system audits.

2.0   APPLICATION

The provisions of this operating procedure shall apply to: determining internal audit requirements, carrying out internal quality system audits and documentation of those activities.

3.0   PROCEDURE

3.01   Definition

Audit – a planned, independent review and verification process to determine compliance to the approved procedures of the Quality Management System.

3.02

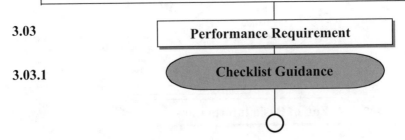

General Requirement

The manual defines in policy statements the quality system's requirements and gives linkage by reference or inclusion to the applicable process instructions, known as Quality Operating Procedures (QOPs). The manual follows the layout of ISO 9002/1994 and within these provisions incorporates the requirements of ISO 9001/2000. It reflects every clause of the Standard according to its numerical layout, except clauses 4.4 and 4.19.

Auditors must first understand the specific paragraph requirement in the quality manual and have familiarity with the referenced or included process instructions (QOPs) before undertaking audits. This is a prerequisite for auditors to enable them to carry out any planned audit(s) effectively. Auditors shall not impose their own set of requirements to contradict the provisions of the established and implemented Quality Management System.

3.03   **Performance Requirement**

3.03.1   **Checklist Guidance**

TRAINING METRICS	Iss dt.	Rev dt.	Pg.
QUALITY OPERATING PROCEDURE	Sign	Rev #	5/9
REQUIREMENT: Internal Quality Audits, QSM 4.17			
SUBJECT: Performing the Quality System Audit			
QOP 010 DOCUMENT REQ'D: Forms A-006, A-007, A-021, A-022			

The Auditor does:

Don't commit audit scheduling to exact timing! It doesn't happen that way!

1 **Observe:** that a QOP outlines the specific work compliance requirement stated in the Quality Manual: that each page of a QOP has been page plated to guide you to understand linkage and document requirement

2 **Observe:** that the process instruction procedures embodied in QOPs guide you forward to understand the task at hand and at the same time refers you to the specific forms used by the process owners to record work results: that each QOP gives you indication of linkage to other QOPs, enabling you to find the exact enforcement requirement for the identified task

Example
QOP 006 (Inspection and Test Control) contains 7 interrelated work assignments:
- first piece inspection
- first article qualification
- process control
- final inspection
- receiving inspection
- customer source insp.
- source inspection at subs.

3 **Observe:** that each QOP is a stand-alone document that may contain several interrelated work assignments, step by step, requiring work performance accomplishment and documentation of work results by the process owners

Instruction
Form A-022 is both, your checklist and evaluation record for reporting

4 **Observe:** that when auditing is carried out, the QOP sections and the related paragraphs should be identified and documented in the Audit Plan (Reporting), form A-022, to indicate the specific audit effort

5 **Observe:** that in order to determine conformance or nonconformance to the specified requirements, the applicable procedures and forms will have to be reviewed and understood ahead of time

6 **Observe:** that the Management Representative may perform the quality system audit(s) provided that he/she is not the direct process owner of the work assignment being audited

Reminder!
While internal system audits are often done by the Management Representative, there must be someone else appointed also who would audit his audit performance to ensure independent assessment. It sounds like no end to the audit process, but the Registrar will look for results in this area.

Reminder
The same Audit Schedule (A-021) should be used year after year, even if changes have been made to it.

7 **Observe:** that when the Management Representative performs a scheduled (A-021) quality system audit, his/her audit performance shall be verified (audited) by any other independent agent to establish definite audit compliance. (Use the same QOP 010, including Audit Reporting (A-022))

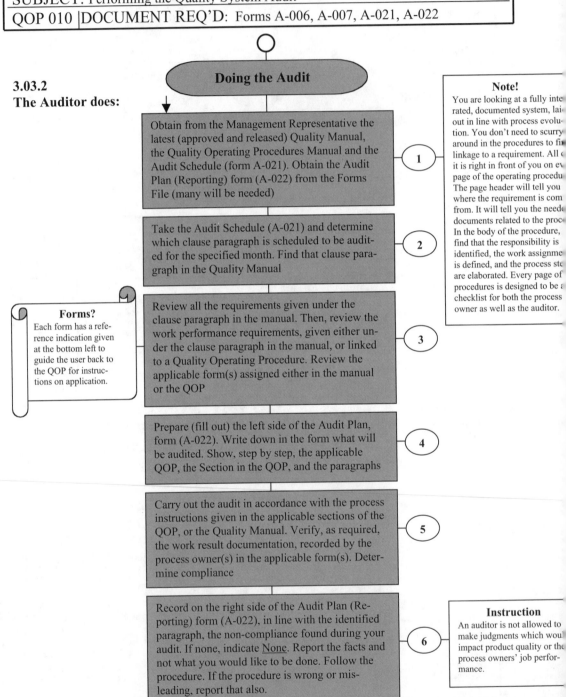

TRAINING METRICS	Iss dt.	Rev dt.	Pg.
QUALITY OPERATING PROCEDURE	Sign	Rev #	6/9
REQUIREMENT: Internal Quality Audits, QSM 4.17			
SUBJECT: Performing the Quality System Audit			
QOP 010 |DOCUMENT REQ'D: Forms A-006, A-007, A-021, A-022			

Doing the Audit

3.03.2
The Auditor does:

Note!
You are looking at a fully interated, documented system, laiout in line with process evolution. You don't need to scurry around in the procedures to filinkage to a requirement. All it is right in front of you on evpage of the operating proceduThe page header will tell you where the requirement is comfrom. It will tell you the neede documents related to the proceIn the body of the procedure, find that the responsibility is identified, the work assignmeis defined, and the process steare elaborated. Every page of procedures is designed to be a checklist for both the process owner as well as the auditor.

1. Obtain from the Management Representative the latest (approved and released) Quality Manual, the Quality Operating Procedures Manual and the Audit Schedule (form A-021). Obtain the Audit Plan (Reporting) form (A-022) from the Forms File (many will be needed)

2. Take the Audit Schedule (A-021) and determine which clause paragraph is scheduled to be audited for the specified month. Find that clause paragraph in the Quality Manual

Forms?
Each form has a reference indication given at the bottom left to guide the user back to the QOP for instructions on application.

3. Review all the requirements given under the clause paragraph in the manual. Then, review the work performance requirements, given either under the clause paragraph in the manual, or linked to a Quality Operating Procedure. Review the applicable form(s) assigned either in the manual or the QOP

4. Prepare (fill out) the left side of the Audit Plan, form (A-022). Write down in the form what will be audited. Show, step by step, the applicable QOP, the Section in the QOP, and the paragraphs

5. Carry out the audit in accordance with the process instructions given in the applicable sections of the QOP, or the Quality Manual. Verify, as required, the work result documentation, recorded by the process owner(s) in the applicable form(s). Determine compliance

6. Record on the right side of the Audit Plan (Reporting) form (A-022), in line with the identified paragraph, the non-compliance found during your audit. If none, indicate None. Report the facts and not what you would like to be done. Follow the procedure. If the procedure is wrong or misleading, report that also.

Instruction
An auditor is not allowed to make judgments which woulimpact product quality or the process owners' job performance.

TRAINING METRICS	Iss dt.	Rev dt.	Pg.
QUALITY OPERATING PROCEDURE	Sign	Rev #	7/9
REQUIREMENT: Internal Quality Audits, QSM 4.17			
SUBJECT: Performing the Quality System Audit			
QOP 010 DOCUMENT REQ'D: Forms A-006, A-007, A-021, A-022			

7 — Write down your observations (the facts) and not your recommendations under 'Comments' in the Audit Reporting form

8 — Sign and date the Audit Plan (Reporting) form as shown. Turn in the Audit Plan (Reporting) form, the Quality Manual and the Quality Operating Procedures Manual to the Management Representative. Put the remainder blanks (A-022) back to the Forms File under the correct serial number.

Reviewing the Audit Reporting (A-022)

3.03.3
The Management Representative does: ↓

1 — Receive the Audit Plan (Reporting), one or several, from the Auditor

Instruction

Corrective action imple - mentation may not be an easy process when more than one problem-one cause is at hand. You may need to do *cause and effect* diagram- ming and *failure mode effect analysis* (FMEA) as well as some serious brainstorming and benchmarking to pin down the actual causes of problems before deciding on cross-functional corrective action implementation. While the implementation process is fully described here for you, the knowledge and skill in the application of statistical tools you should possess as part of your job description as Man- agement Representative.

2 — Review what had been audited and evaluate any nonconformance written down by the Auditor. Evaluate 'Comments' made, as appropriate

3 — Determine any negative impact to product quali- ty, to processes, and to procedures

4 — Implement corrective action in accordance with the following:

4a — If performance had not been carried out and doc- umented as required by the work instruction, im- plement corrective action through supervision in accordance with QOP 006, Section Three, par 3.0, step 2 (pg 151), as appropriate, in line with the identified supervisory responsibility. If NCMR enforcement is required, do it in accor- dance with QOP 009, Section Two (pg 207)

TRAINING METRICS	Iss dt.	Rev dt.	Pg.
QUALITY OPERATING PROCEDURE	Sign	Rev #	8/9
REQUIREMENT: Internal Quality Audits, QSM 4.17			
SUBJECT: Performing the Quality System Audit			
QOP 010 DOCUMENT REQ'D: Forms A-006, A-007, A-021, A-022			

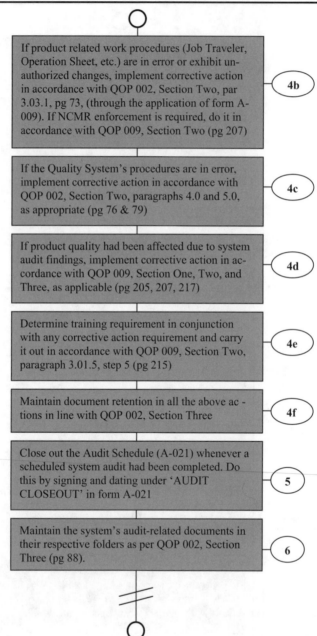

TRAINING METRICS	Iss dt.	Rev dt.	Pg.
QUALITY OPERATING PROCEDURE	Sign	Rev #	9/9
REQUIREMENT: Internal Quality Audits, QSM 4.17			
SUBJECT: Performing the Quality System Audit			
QOP 010 DOCUMENT REQ'D: Forms A-006, A-007, A-021, A-022			

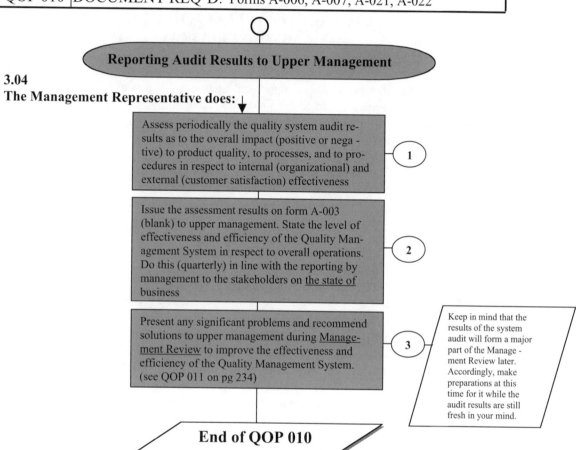

Reporting Audit Results to Upper Management

3.04

The Management Representative does:

1. Assess periodically the quality system audit results as to the overall impact (positive or negative) to product quality, to processes, and to procedures in respect to internal (organizational) and external (customer satisfaction) effectiveness

2. Issue the assessment results on form A-003 (blank) to upper management. State the level of effectiveness and efficiency of the Quality Management System in respect to overall operations. Do this (quarterly) in line with the reporting by management to the stakeholders on the state of business

3. Present any significant problems and recommend solutions to upper management during Management Review to improve the effectiveness and efficiency of the Quality Management System. (see QOP 011 on pg 234)

Keep in mind that the results of the system audit will form a major part of the Management Review later. Accordingly, make preparations at this time for it while the audit results are still fresh in your mind.

End of QOP 010

OVERVIEW TO QOP 011
(MANAGEMENT REVIEW)

Management Review in a process-approach system is not the forum to present a stupendous pile of papers to impress upper management that lots of things were done but there are still numerous unmanageable problems outstanding, making the operating system ineffective. This sort of thing happens when companies implement ISO compliant systems according to the way they always ran their businesses. They have documented all their weaknesses and not much had been improved in product quality. They have now little more to show except a mountain of records and lots of bills, awaiting review. You see, under a process-approach system the responsibilities for process owners are defined from the beginning to the end in the product's processing cycle and work assignment accountability to requirement is demonstrated at every process in the chain of events. The product is only an object, moving through the cycle, controlled by the process and not by the problems. Every process owner checks his/her work, because nobody else can build quality in the product. At the end you are ready to ship that product to the customer. This is how customer satisfaction is also built into the processes.

Controlling non-conformance in documents, processes, methods, and products is also managed as a process. The problems are not pushed aside to be forgotten until they strike back again. So, management is given all the good news how problems were solved as part of process control and not as pending management review to look for improvements where improvements would do nothing more than fanning the wind – too late to add value to the process. Yet, management reviews are very important to assess successes and through it set new business plans for growth in an effectively managed production environment. In a process-approach system, management makes improvements on the basis of successes – a very significant change from the ISO philosophy.

DIAGRAMMATIC PROCEDURE   QOP 011

MANAGEMENT REVIEW

CONTENTS   **Page**

TRAINING METRICS		Iss dt.		Rev dt.		Pg.
QUALITY OPERATING PROCEDURE		Sign		Rev #		2/9
REQUIREMENT: QUALITY SYSTEM MANUAL SEC: 4.1.3						
SUBJECT: MANAGEMENT REVIEW						
QOP 011	DOCUMENT REQ'D: Forms A-003, A-006, A-007, A-034					

REVISION HISTORY

Rev Date	Rev No	Description	Approval

TRAINING METRICS	Iss dt.	Rev dt.	Pg.
QUALITY OPERATING PROCEDURE	Sign	Rev #	3/9
REQUIREMENT: Management Review, QSM 4.1.3			
SUBJECT: Some Basic Information			
QOP 011 DOCUMENT REQ'D: Forms A-003, A-006, A-007, A-034			

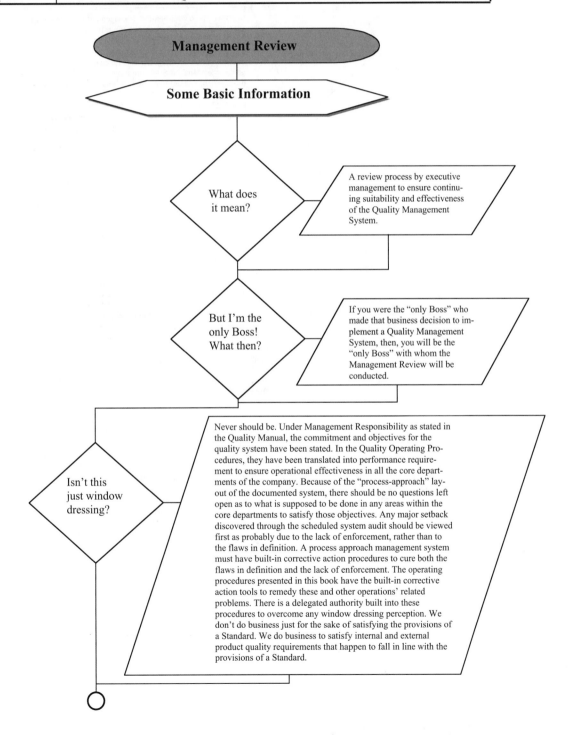

Management Review

Some Basic Information

What does it mean?

A review process by executive management to ensure continuing suitability and effectiveness of the Quality Management System.

But I'm the only Boss! What then?

If you were the "only Boss" who made that business decision to implement a Quality Management System, then, you will be the "only Boss" with whom the Management Review will be conducted.

Isn't this just window dressing?

Never should be. Under Management Responsibility as stated in the Quality Manual, the commitment and objectives for the quality system have been stated. In the Quality Operating Procedures, they have been translated into performance requirement to ensure operational effectiveness in all the core departments of the company. Because of the "process-approach" layout of the documented system, there should be no questions left open as to what is supposed to be done in any areas within the core departments to satisfy those objectives. Any major setback discovered through the scheduled system audit should be viewed first as probably due to the lack of enforcement, rather than to the flaws in definition. A process approach management system must have built-in corrective action procedures to cure both the flaws in definition and the lack of enforcement. The operating procedures presented in this book have the built-in corrective action tools to remedy these and other operations' related problems. There is a delegated authority built into these procedures to overcome any window dressing perception. We don't do business just for the sake of satisfying the provisions of a Standard. We do business to satisfy internal and external product quality requirements that happen to fall in line with the provisions of a Standard.

TRAINING METRICS	Iss dt.	Rev dt.	Pg.
QUALITY OPERATING PROCEDURE	Sign	Rev #	4/9
REQUIREMENT: Management Review, QSM 4.1.3			
SUBJECT: Some Basic Information			
QOP 011	DOCUMENT REQ'D: Forms A-003, A-006, A-007, A-034		

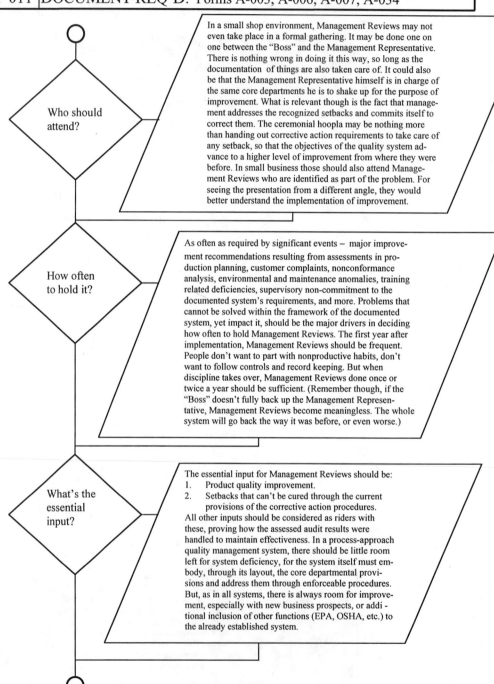

Who should attend?

In a small shop environment, Management Reviews may not even take place in a formal gathering. It may be done one on one between the "Boss" and the Management Representative. There is nothing wrong in doing it this way, so long as the documentation of things are also taken care of. It could also be that the Management Representative himself is in charge of the same core departments he is to shake up for the purpose of improvement. What is relevant though is the fact that management addresses the recognized setbacks and commits itself to correct them. The ceremonial hoopla may be nothing more than handing out corrective action requirements to take care of any setback, so that the objectives of the quality system advance to a higher level of improvement from where they were before. In small business those should also attend Management Reviews who are identified as part of the problem. For seeing the presentation from a different angle, they would better understand the implementation of improvement.

How often to hold it?

As often as required by significant events – major improvement recommendations resulting from assessments in production planning, customer complaints, nonconformance analysis, environmental and maintenance anomalies, training related deficiencies, supervisory non-commitment to the documented system's requirements, and more. Problems that cannot be solved within the framework of the documented system, yet impact it, should be the major drivers in deciding how often to hold Management Reviews. The first year after implementation, Management Reviews should be frequent. People don't want to part with nonproductive habits, don't want to follow controls and record keeping. But when discipline takes over, Management Reviews done once or twice a year should be sufficient. (Remember though, if the "Boss" doesn't fully back up the Management Representative, Management Reviews become meaningless. The whole system will go back the way it was before, or even worse.)

What's the essential input?

The essential input for Management Reviews should be:
1. Product quality improvement.
2. Setbacks that can't be cured through the current provisions of the corrective action procedures.

All other inputs should be considered as riders with these, proving how the assessed audit results were handled to maintain effectiveness. In a process-approach quality management system, there should be little room left for system deficiency, for the system itself must embody, through its layout, the core departmental provisions and address them through enforceable procedures. But, as in all systems, there is always room for improvement, especially with new business prospects, or additional inclusion of other functions (EPA, OSHA, etc.) to the already established system.

TRAINING METRICS	Iss dt.	Rev dt.	Pg.
QUALITY OPERATING PROCEDURE	Sign	Rev #	5/9
REQUIREMENT: Management Review, QSM 4.1.3			
SUBJECT: Some Basic Information			
QOP 011 |DOCUMENT REQ'D: Forms A-003, A-006, A-007, A-034			

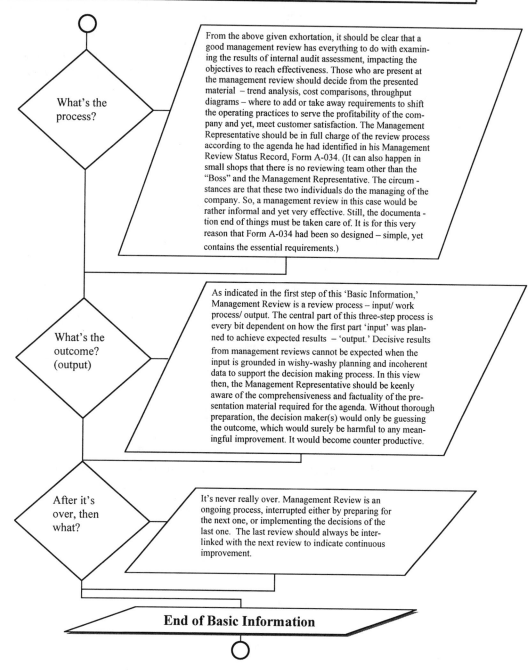

What's the process?

From the above given exhortation, it should be clear that a good management review has everything to do with examining the results of internal audit assessment, impacting the objectives to reach effectiveness. Those who are present at the management review should decide from the presented material – trend analysis, cost comparisons, throughput diagrams – where to add or take away requirements to shift the operating practices to serve the profitability of the company and yet, meet customer satisfaction. The Management Representative should be in full charge of the review process according to the agenda he had identified in his Management Review Status Record, Form A-034. (It can also happen in small shops that there is no reviewing team other than the "Boss" and the Management Representative. The circumstances are that these two individuals do the managing of the company. So, a management review in this case would be rather informal and yet very effective. Still, the documentation end of things must be taken care of. It is for this very reason that Form A-034 had been so designed – simple, yet contains the essential requirements.)

What's the outcome? (output)

As indicated in the first step of this 'Basic Information,' Management Review is a review process – input/ work process/ output. The central part of this three-step process is every bit dependent on how the first part 'input' was planned to achieve expected results – 'output.' Decisive results from management reviews cannot be expected when the input is grounded in wishy-washy planning and incoherent data to support the decision making process. In this view then, the Management Representative should be keenly aware of the comprehensiveness and factuality of the presentation material required for the agenda. Without thorough preparation, the decision maker(s) would only be guessing the outcome, which would surely be harmful to any meaningful improvement. It would become counter productive.

After it's over, then what?

It's never really over. Management Review is an ongoing process, interrupted either by preparing for the next one, or implementing the decisions of the last one. The last review should always be interlinked with the next review to indicate continuous improvement.

End of Basic Information

TRAINING METRICS	Iss dt.	Rev dt.	Pg.
QUALITY OPERATING PROCEDURE	Sign	Rev #	6/9
REQUIREMENT: Management Review, QSM 4.1.3			
SUBJECT: Preparation, Presentation, Conclusion, and Follow-up			
QOP 011 |DOCUMENT REQ'D: Forms A-003, A-006, A-007, A-034			

1.0   PURPOSE

Define the Company's management review requirements in order to assess, as often as necessary, the suitability and effectiveness of the quality management system.

2.0   APPLICATION

The provisions of this procedure shall apply to the preparation, presentation, conclusion, and follow-up of management reviews.

3.0   PROCEDURE

3.01

General Requirement

Management Reviews shall be based on the results of effective assessment of the internal audits of the Quality Management System. Effectiveness of performance shall be measured against targeted limits (threshold) established in line with the internal (organizational) and external (customer satisfaction) business interest (risk limits) of the Company. Reason: Without being able to demonstrate compliance versus noncompliance (ratios) against realistically established threshold limits, the goals to accomplish the set objectives of the quality system would miss the reason for additional improvements beyond the routine improvements accomplished through the corrective action provisions of QOP 009.

The Management Representative shall be responsible for establishing reasonable threshold limits (maximum allowable non-compliance figures) in all the areas where this can be objectively implemented in order to know when to trigger (plan-do-check-act) corrective measures to remove the causes that negatively impacted the goals to achieve the business objectives of the Company. (See paragraph 3.02, step 2 on pg 241.) When the corrective measures, identified through assessment, cannot be implemented through the provisions of QOP 009, Sections Two and Three, the Management Representative shall set the agenda for Management Review.

Management shall not be called to session to review issues for which the answers to solve them are given in the documented procedures, for that would be dereliction (shirking one's duty). Conversely, when the lack of enforcement on the part of the delegated authority (supervision) impacts compliance with the provisions of the documented system, management shall be called to session to implement corrective action.

TRAINING METRICS	Iss dt.	Rev dt.	Pg.
QUALITY OPERATING PROCEDURE	Sign	Rev #	7/9
REQUIREMENT: Management Review, QSM 4.1.3			
SUBJECT: Preparation, Presentation, Conclusion, and Follow-up			
QOP 011 |DOCUMENT REQ'D: Forms A-003, A-006, A-007, A-034			

Preparation for Management Review

3.02

The Management Representative does:

1. Review the statistical results of the internal audit assessment. Determine the levels of non-compliance. Determine whether they have been cured through the corrective action provisions of QOP 009, Sections Two and Three. If they have not been cured, set the agenda for Management Review to resolve them. Determine new objectives, recommendations, and solutions

Trend analysis of recurring problems may reveal major flaws in training methods, or how enforcement is carried out, or even the requirement to change procedures. Trend analysis should be a major tool in determining system C/A.

Your NCMRs are a major source for determining the dispersion of problems. Pick the high numbers for C/A.

2. Gather the supporting data to justify Management Review – supporting data to demonstrate to management both the positive and the negative areas of the internal audit assessment. Show the effectiveness versus the lack of effectiveness in respect to overall performance objectives. Prepare proposal for how to resolve the lack of effectiveness. Describe this on form A-003 to support your discussion (persuasion) points during Management Review. Issue it to all.

Before calling management into session take a good look whether lack of training is not the core of the problem!

3. Select the participants to the Management Review. Ensure that those against whom corrective action may be issued are present. Management wants to hear their input for solutions

4. Compile the supporting data in order of presentation. Make enough copies of these documents so that each participant can have a copy ahead of time

5. Draw up the agenda for the Management Review under sections 1 and 2 of form A-034. Complete the "Attendees, Place and Date" sections. Distribute form A-034 to all participants

TRAINING METRICS	Iss dt.	Rev dt.	Pg.
QUALITY OPERATING PROCEDURE	Sign	Rev #	8/9
REQUIREMENT: Management Review, QSM 4.1.3			
SUBJECT: Preparation, Presentation, Conclusion, and Follow-up			
QOP 011 DOCUMENT REQ'D: Forms A-003, A-006, A-007, A-034			

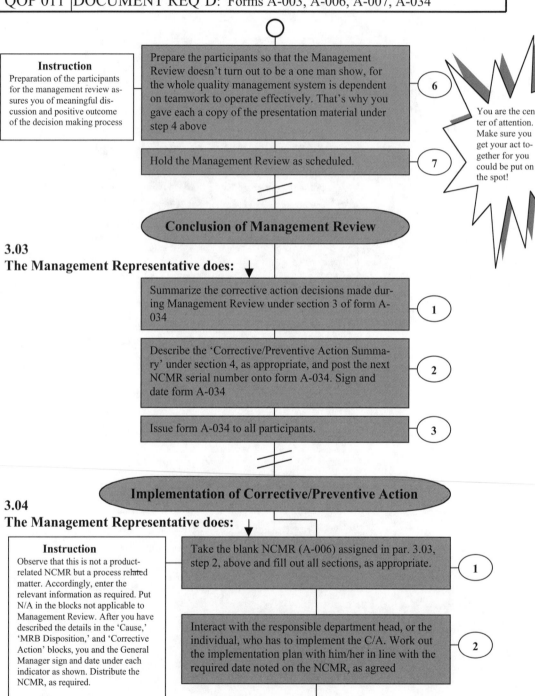

Instruction

Preparation of the participants for the management review assures you of meaningful discussion and positive outcome of the decision making process

Prepare the participants so that the Management Review doesn't turn out to be a one man show, for the whole quality management system is dependent on teamwork to operate effectively. That's why you gave each a copy of the presentation material under step 4 above

6

You are the center of attention. Make sure you get your act together for you could be put on the spot!

Hold the Management Review as scheduled.

7

Conclusion of Management Review

3.03
The Management Representative does:

Summarize the corrective action decisions made during Management Review under section 3 of form A-034

1

Describe the 'Corrective/Preventive Action Summary' under section 4, as appropriate, and post the next NCMR serial number onto form A-034. Sign and date form A-034

2

Issue form A-034 to all participants.

3

Implementation of Corrective/Preventive Action

3.04
The Management Representative does:

Instruction

Observe that this is not a product-related NCMR but a process related matter. Accordingly, enter the relevant information as required. Put N/A in the blocks not applicable to Management Review. After you have described the details in the 'Cause,' 'MRB Disposition,' and 'Corrective Action' blocks, you and the General Manager sign and date under each indicator as shown. Distribute the NCMR, as required.

Take the blank NCMR (A-006) assigned in par. 3.03, step 2, above and fill out all sections, as appropriate.

1

Interact with the responsible department head, or the individual, who has to implement the C/A. Work out the implementation plan with him/her in line with the required date noted on the NCMR, as agreed

2

TRAINING METRICS	Iss dt.	Rev dt.	Pg.
QUALITY OPERATING PROCEDURE	Sign	Rev #	9/9
REQUIREMENT: Management Review, QSM 4.1.3			
SUBJECT: Preparation, Presentation, Conclusion, and Follow-up			
QOP 011 DOCUMENT REQ'D: Forms A-003, A-006, A-007, A-034			

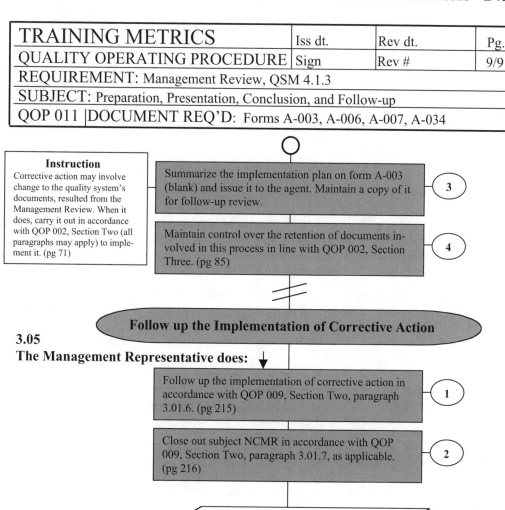

Instruction

Corrective action may involve change to the quality system's documents, resulted from the Management Review. When it does, carry it out in accordance with QOP 002, Section Two (all paragraphs may apply) to implement it. (pg 71)

Summarize the implementation plan on form A-003 (blank) and issue it to the agent. Maintain a copy of it for follow-up review. **3**

Maintain control over the retention of documents involved in this process in line with QOP 002, Section Three. (pg 85) **4**

Follow up the Implementation of Corrective Action

3.05

The Management Representative does:

Follow up the implementation of corrective action in accordance with QOP 009, Section Two, paragraph 3.01.6. (pg 215) **1**

Close out subject NCMR in accordance with QOP 009, Section Two, paragraph 3.01.7, as applicable. (pg 216) **2**

End of QOP 011

OVERVIEW TO QOP 012
(PERFORMANCE STANDARD)

QOP 012 is designed to give you a quick review of linkage as to how a process-approach system integrates assignments to procedures in handling two types of product categories, – conforming and non-conforming products.

Both must be simultaneously dealt with, right through a product's realization processes. If we don't, the conforming product eventually will be shipped, but the non-conforming product will die on the vine in many cases. Sometimes it will become a quickly fixed product, without corrective action. We don't prevent problems this way. We repeat them. This is not the *continual improvement* opportunity ISO 9001/2000 has in mind. Continuous improvement in a process-approach quality management system primarily takes place by implementing corrective action through process control, hence preventing problems from flowing down, process to process. It is for a very good reason then, why we must also control non-conforming products as we do conforming products.

Section One and Two of this QOP communicate the put-through mechanism, first for the conforming products, second for the non-conforming products. These two procedures impact directly or indirectly every other Quality Operating Procedure in the quality system, simply because they effectively handle the two product categories mentioned above, throughout all the realization processes. They are performance standards, for every identified process has a performance requirement linked to it, which requires work result documentation and end result measurement to evaluate achievement in both category types.

Section Three is a completely different procedure, dealing with developmental projects to qualify processing methods, and from it, formalizing the project related work instructions for an ongoing production effort, often required with new customer purchase orders.

DIAGRAMMATIC PROCEDURE – QOP 012

PERFORMANCE STANDARD, PROCESSING CONTROL
(Put Through)

CONTENTS   Page

CONTENTS (Continued)

MINFOR INCORPORATED	Iss dt.		Rev dt		Pg.
QUALITY OPERATING PROCEDURE	Sign AG		Rev #		3/22
REQUIREMENT: QUALITY SYSTEM MANUAL SEC: 4.9					
SUBJECT: Performance Standard, Processing Control (put through)					
QOP 012	DOCUMENT REQ'D: Form A-008 (Job Traveler)				

REVISION HISTORY

Rev Date	Rev No	Description	Approval

TRAINING METRICS	Iss dt.	Rev dt.	Pg.
QUALITY OPERATING PROCEDURE	Sign	Rev #	4/22
REQUIREMENT: Performance Standard, Processing Control, QSM 4.9			
SUBJECT: Some Basic Information			
QOP 012 |DOCUMENT REQ'D: Form A-008 (Job Traveler), A-001			

SECTION ONE

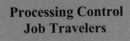

**Processing Control
Job Travelers**

Some Basic Information

What's a Job Traveler?

A Job Traveler is a Process Router, indicating the sequences of the processing steps as defined by Engineering and Quality. It also contains the specific work instruction under each processing step as to what is required to be done. It gives references, when required, to other controlling documents such as Operation Sheets or quality procedures, etc. A Job Traveler is the single most important trace document for a specified product. Once a processing step has been completed, it clearly indicates what quantities were accepted or rejected. A Job Traveler controls the product's inspection and test status indication at each defined operation (number). This document, when all the specified operations have been completed and the product has been final inspected and released, is kept by Quality in the Inspection Product Folder as a permanent record. Without it, there is no reliable processing accountability, no history to indicate in-process control of a specified product – the evidence to DEMONSTRATE compliance.

Who issues them?

Job Travelers originate from Engineering. After all the processing sequences and the related work instruction requirements have been incorporated, Engineering releases them to Production Planning. Job Travelers can also be temporarily issued when certain new orders require the determination of capability and processing development. What is important to remember is that Job Travelers are controlled documents. Once they have been finally released, changes to them require document control. See QOP 002, Section Two, par. 3.03.1. (pg 73)

Who maintains them?

The Job Travelers are maintained by Engineering and are part of the specified Production Folders, although they may be released separately depending on urgency of setup preparation. The preparation and release of processing related technical work instructions, which will be contained in the Production Folders, decisively depend on the complexity of the product itself. There are times when a Job Traveler will be the sole document in a Production Folder. And that's fine as long as it clearly defines what to do. The retention of Job Travelers is defined under QOP 002, Section Three.

TRAINING METRICS	Iss dt.	Rev dt.		Pg.
QUALITY OPERATING PROCEDURE	Sign	Rev #		5/22
REQUIREMENT: Performance Standard, Processing Control, QSM 4.9				
SUBJECT: Some Basic Information				
QOP 012 |DOCUMENT REQ'D: Form A-008 (Job Traveler), A-001				

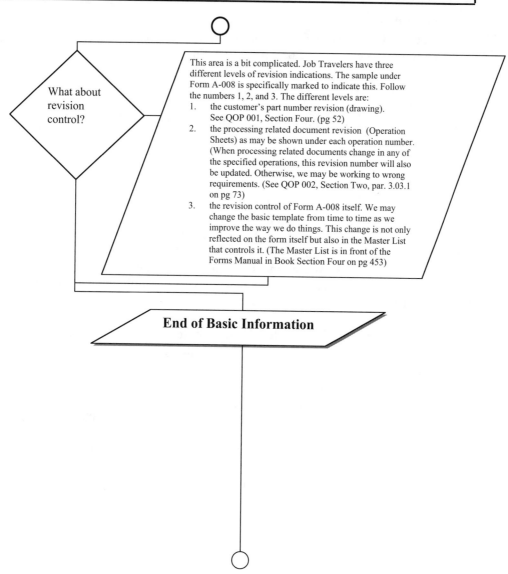

What about revision control?

This area is a bit complicated. Job Travelers have three different levels of revision indications. The sample under Form A-008 is specifically marked to indicate this. Follow the numbers 1, 2, and 3. The different levels are:

1. the customer's part number revision (drawing). See QOP 001, Section Four. (pg 52)
2. the processing related document revision (Operation Sheets) as may be shown under each operation number. (When processing related documents change in any of the specified operations, this revision number will also be updated. Otherwise, we may be working to wrong requirements. (See QOP 002, Section Two, par. 3.03.1 on pg 73)
3. the revision control of Form A-008 itself. We may change the basic template from time to time as we improve the way we do things. This change is not only reflected on the form itself but also in the Master List that controls it. (The Master List is in front of the Forms Manual in Book Section Four on pg 453)

End of Basic Information

TRAINING METRICS	Iss dt.	Rev dt.	Pg.
QUALITY OPERATING PROCEDURE	Sign	Rev #	6/22
REQUIREMENT: Performance Standard, Processing Control, QSM 4.9			
SUBJECT: **Processing Evolution from Beginning to End (Put Through)**			
QOP 012 |DOCUMENT REQ'D: Form A-008 (Job Traveler), A-001			

1.0   PURPOSE

Standardize work assignments in order to demonstrate the process-approach put through system in an actual work environment.

2.0   APPLICATION

Subject procedure applies to contractually defined products, going through the phases of planning, processing, verification, release and delivery to the customer.

3.0   PROCEDURE

General

A Job Traveler is that controlled document which defines all the required process steps that regulate the movement of a product from the beginning of a process to the end of a process in a way that process owners can determine from it, step by step, the identification of a product and its related documents, the equipment and tooling, the work instruction on what to do and what not to do in order to process that product to meet the scheduled work process requirement and ultimately, the customer's requirement, without unnecessary interruptions.

In order for a Job Traveler to become a useful, valid, and an approved process control document for all those who are required to follow it, the issuing authority (Process Engineer) must interact with Contracts, Production Planning, Purchasing, Manufacturing, and Quality to integrate the processing cycle evolution of the product, equipment, people, and methods. This would have to be done in a manner such that not one of the members of the operation's team would have to wait to fulfill its responsibilities to carry out his/her work assignment in the product's realization processes due to poor planning in the beginning phases of the layout processes. After all, the Job Traveler is the road map of the processing cycle, which requires continuous input and follow-up by those in charge of it, so that when it is released to the traffic (the work realization processes), the people (process owners) would not have to be rerouted to take alternate paths to reach their targets (production goals), because somebody failed to do his part to make the road map (Job Traveler) an accurately reflected schedule-execute-verify-control driven directive (Plan-Do-Check-Act).

This process-approach quality management system planned the Job Traveler (including Split Travelers) to be the single most important document to control the product's entire routing requirements in a manner so as to ensure that accurate directions are given, in sequential order, to the process owners. This would enable

TRAINING METRICS	Iss dt.	Rev dt.	Pg.
QUALITY OPERATING PROCEDURE	Sign	Rev #	7/22
REQUIREMENT: Performance Standard, Processing Control, QSM 4.9			
SUBJECT: **Processing Evolution from Beginning to End (Put Through)**			
QOP 012 |DOCUMENT REQ'D: Form A-008 (Job Traveler), A-001			

them to carry out their work assignments so that the product could be processed through the cycle the first time in a controlled manner, ensuring that quality is continuously built into the product. Of course, to make this a reality, everybody would have to be entrusted to take charge of his work assignment in order to help each other to work as a team. This brings us then to integrating the work assignments within the process-approach system so the above can be realized, ensuring along the way that each process owner verifies his/her work results and documents it, as required, to prevent transferring mistakes, regardless who made them, from process to process.

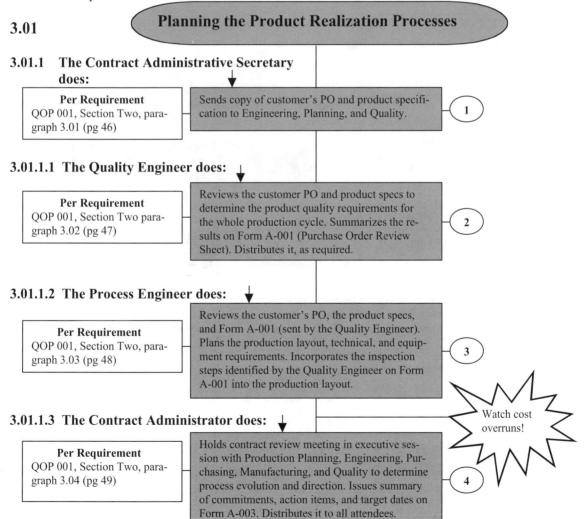

3.01

Planning the Product Realization Processes

3.01.1 The Contract Administrative Secretary does:

Per Requirement
QOP 001, Section Two, paragraph 3.01 (pg 46)

Sends copy of customer's PO and product specification to Engineering, Planning, and Quality. (1)

3.01.1.1 The Quality Engineer does:

Per Requirement
QOP 001, Section Two paragraph 3.02 (pg 47)

Reviews the customer PO and product specs to determine the product quality requirements for the whole production cycle. Summarizes the results on Form A-001 (Purchase Order Review Sheet). Distributes it, as required. (2)

3.01.1.2 The Process Engineer does:

Per Requirement
QOP 001, Section Two, paragraph 3.03 (pg 48)

Reviews the customer's PO, the product specs, and Form A-001 (sent by the Quality Engineer). Plans the production layout, technical, and equipment requirements. Incorporates the inspection steps identified by the Quality Engineer on Form A-001 into the production layout. (3)

3.01.1.3 The Contract Administrator does:

Watch cost overruns!

Per Requirement
QOP 001, Section Two, paragraph 3.04 (pg 49)

Holds contract review meeting in executive session with Production Planning, Engineering, Purchasing, Manufacturing, and Quality to determine process evolution and direction. Issues summary of commitments, action items, and target dates on Form A-003. Distributes it to all attendees. (4)

TRAINING METRICS	Iss dt.	Rev dt.	Pg.
QUALITY OPERATING PROCEDURE	Sign	Rev #	8/22
REQUIREMENT: Performance Standard, Processing Control, QSM 4.9			
SUBJECT: **Processing Evolution from Beginning to End (Put Through)**			
QOP 012 |DOCUMENT REQ'D: Form A-008 (Job Traveler), A-001			

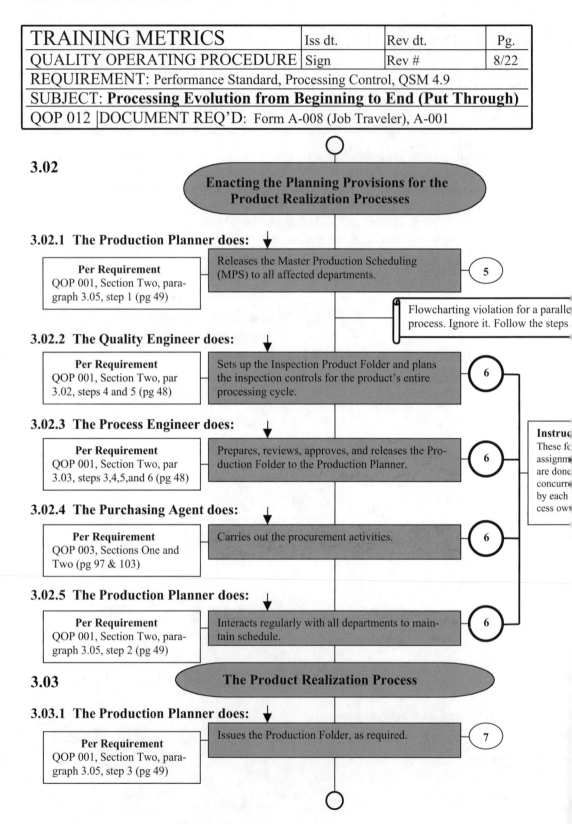

3.02

Enacting the Planning Provisions for the
Product Realization Processes

3.02.1 The Production Planner does:

Per Requirement
QOP 001, Section Two, paragraph 3.05, step 1 (pg 49)

Releases the Master Production Scheduling
(MPS) to all affected departments.

5

Flowcharting violation for a paralle
process. Ignore it. Follow the steps

3.02.2 The Quality Engineer does:

Per Requirement
QOP 001, Section Two, par
3.02, steps 4 and 5 (pg 48)

Sets up the Inspection Product Folder and plans
the inspection controls for the product's entire
processing cycle.

6

3.02.3 The Process Engineer does:

Per Requirement
QOP 001, Section Two, par
3.03, steps 3,4,5,and 6 (pg 48)

Prepares, reviews, approves, and releases the Production Folder to the Production Planner.

6

Instruc
These fo
assignm
are done
concurre
by each
cess ow

3.02.4 The Purchasing Agent does:

Per Requirement
QOP 003, Sections One and
Two (pg 97 & 103)

Carries out the procurement activities.

6

3.02.5 The Production Planner does:

Per Requirement
QOP 001, Section Two, paragraph 3.05, step 2 (pg 49)

Interacts regularly with all departments to maintain schedule.

6

3.03   The Product Realization Process

3.03.1 The Production Planner does:

Per Requirement
QOP 001, Section Two, paragraph 3.05, step 3 (pg 49)

Issues the Production Folder, as required.

7

TRAINING METRICS	Iss dt.	Rev dt.	Pg.
QUALITY OPERATING PROCEDURE	Sign	Rev #	9/22
REQUIREMENT: Performance Standard, Processing Control, QSM 4.9			
SUBJECT: **Processing Evolution from Beginning to End (Put Through)**			
QOP 012 |DOCUMENT REQ'D: Form A-008 (Job Traveler), A-001			

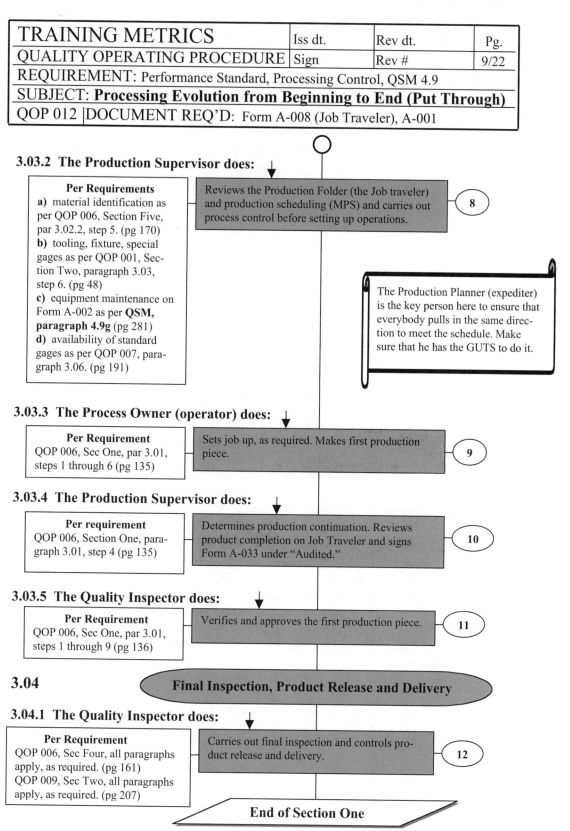

3.03.2 The Production Supervisor does:

Per Requirements
a) material identification as per QOP 006, Section Five, par 3.02.2, step 5. (pg 170)
b) tooling, fixture, special gages as per QOP 001, Section Two, paragraph 3.03, step 6. (pg 48)
c) equipment maintenance on Form A-002 as per **QSM, paragraph 4.9g** (pg 281)
d) availability of standard gages as per QOP 007, paragraph 3.06. (pg 191)

Reviews the Production Folder (the Job traveler) and production scheduling (MPS) and carries out process control before setting up operations.

8

The Production Planner (expediter) is the key person here to ensure that everybody pulls in the same direction to meet the schedule. Make sure that he has the GUTS to do it.

3.03.3 The Process Owner (operator) does:

Per Requirement
QOP 006, Sec One, par 3.01, steps 1 through 6 (pg 135)

Sets job up, as required. Makes first production piece.

9

3.03.4 The Production Supervisor does:

Per requirement
QOP 006, Section One, paragraph 3.01, step 4 (pg 135)

Determines production continuation. Reviews product completion on Job Traveler and signs Form A-033 under "Audited."

10

3.03.5 The Quality Inspector does:

Per Requirement
QOP 006, Sec One, par 3.01, steps 1 through 9 (pg 136)

Verifies and approves the first production piece.

11

3.04

Final Inspection, Product Release and Delivery

3.04.1 The Quality Inspector does:

Per Requirement
QOP 006, Sec Four, all paragraphs apply, as required. (pg 161)
QOP 009, Sec Two, all paragraphs apply, as required. (pg 207)

Carries out final inspection and controls product release and delivery.

12

End of Section One

TRAINING METRICS	Iss dt.	Rev dt.	Pg.
QUALITY OPERATING PROCEDURE	Sign	Rev #	10/22
REQUIREMENT: Performance Standard, Processing Control, QSM 4.13			
SUBJECT: **Handling Non-conforming Products (Put Through)**			
QOP 012 |DOCUMENT REQ'D: Form A-006 (NCMR for MRB action)			

SECTION TWO

1.0   **PURPOSE**

Integrate the handling and control of non-conformities as part of process control during the product's overall processing cycle.

2.0   **APPLICATION**

This procedure provides specific instructions on how to handle and control non-conformities identified at any point in the processing cycle of the product, including customer complaints and product returns.

3.0   **PROCEDURE**

General

Since a non-conforming state could contribute to process deficiencies anywhere in the processing cycle of a product, timely resolution of the constraint, created by it, must be implemented to ensure effective process continuation. A nonconforming state does not only mean hardware in a process-approach system, but it also includes errors in documents, in processes, and in the methods. These may eventually affect production and product quality. To stop the transfer of deficiencies creeping from process to process, we have implemented the controls to isolate them and cure them when and where they occur. Not every nonconformity leads to hardware deficiency and when it doesn't, we have implemented other methods in this process-approach system to cure them. Carefulness must be exercised therefore between the two distinct varieties – hardware vs. non-hardware –

for the cost difference in handling is substantial. Applying the NCMR (MRB) process is primarily hardware related. It should also be used when no other documented procedure exists to handle corrective action. The four main categories in which we may discover non-conformities that we must control are:

1. **hardware related (handled by the application of the NCMR (MRB) process);**
2. **production related documents (handled by the application of Form A-009);**
3. **customer related documents (handled through QOP 001, Section Four);**
4. **documents making up the Quality System manual and the Quality Operating Procedures (handled by draft revisions).**

TRAINING METRICS	Iss dt.	Rev dt.	Pg.
QUALITY OPERATING PROCEDURE	Sign	Rev #	11/22
REQUIREMENT: Performance Standard, Processing Control, QSM 4.13			
SUBJECT: **Handling Non-conforming Products (Put Through)**			
QOP 012 |DOCUMENT REQ'D: Form A-006 (NCMR for MRB action)			

3.01

Hardware Related Non-conforming Products Identified During Production

3.01.1 The Process Owners (operators) do:

Per Requirement
QOP 006, Section One, paragraph 3.01, step 6. (pg 135)

Notify supervision when you have verified that a product is non-conforming.

1

3.01.2 The Supervisor does:

Per Requirement
QOP 006, Section One, paragraph 3.01, steps 1 through 5. (pg 135)

Review that the process owner properly verified the product non-conformance. Take the necessary action to continue production. Process NCMR (for MRB) through the Quality Inspector, as required

2

3.01.3 The Quality Inspector does:

Per Requirement
a) QOP 006, Section One, paragraph 3.01, step 8. (pg 137)
b) QOP 009, Section Two, paragraph 3.01.1, steps 1 through 3. (pg 207)

a)   verify the product non-conformance and document it on Form A-018.
b)   process non-conforming material report Form A-006.

3

Instruction
Further processing of NCMR is picked up under paragraph 3.02.2 below.

3.02

Hardware Related Non-conforming Product Identified At Any Point During the Processing Cycle of a Product

3.02.1 The Quality Inspector does:

Per Requirement
QOP 009, Section Two, paragraph 3.01.1, steps 1 through 3. (pg 207)

a)   verify the non-conforming product and document the results onto Form A-018. Tag it with Form A-012. Segregate it, as practical.
b)   Obtain a blank form of A-006 and fill it out.

1

TRAINING METRICS	Iss dt.	Rev dt.	Pg.
QUALITY OPERATING PROCEDURE	Sign	Rev #	12/22
REQUIREMENT: Performance Standard, Processing Control, QSM 4.13			
SUBJECT: **Handling Non-conforming Products (Put Through)**			
QOP 012 │DOCUMENT REQ'D: Form A-006 (NCMR for MRB action)			

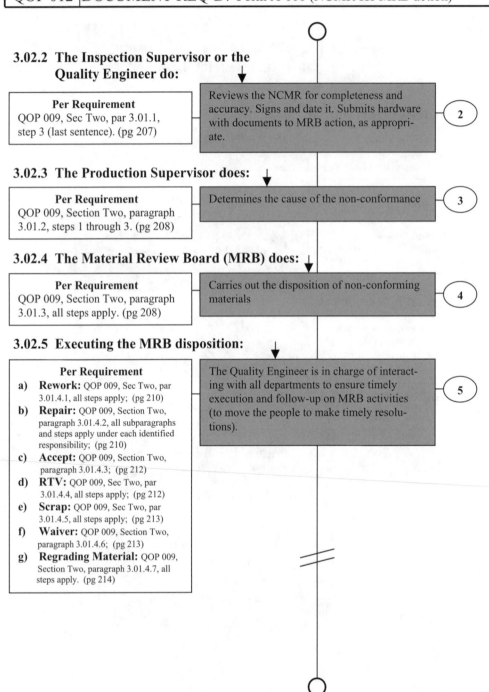

3.02.2 The Inspection Supervisor or the
**   Quality Engineer do:**

> **Per Requirement**
> QOP 009, Sec Two, par 3.01.1,
> step 3 (last sentence). (pg 207)

> Reviews the NCMR for completeness and accuracy. Signs and date it. Submits hardware with documents to MRB action, as appropriate.   ② 2

3.02.3 The Production Supervisor does:

> **Per Requirement**
> QOP 009, Section Two, paragraph 3.01.2, steps 1 through 3. (pg 208)

> Determines the cause of the non-conformance   ③ 3

3.02.4 The Material Review Board (MRB) does:

> **Per Requirement**
> QOP 009, Section Two, paragraph 3.01.3, all steps apply. (pg 208)

> Carries out the disposition of non-conforming materials   ④ 4

3.02.5 Executing the MRB disposition:

> **Per Requirement**
> a) **Rework:** QOP 009, Sec Two, par 3.01.4.1, all steps apply; (pg 210)
> b) **Repair:** QOP 009, Section Two, paragraph 3.01.4.2, all subparagraphs and steps apply under each identified responsibility; (pg 210)
> c) **Accept:** QOP 009, Section Two, paragraph 3.01.4.3; (pg 212)
> d) **RTV:** QOP 009, Sec Two, par 3.01.4.4, all steps apply; (pg 212)
> e) **Scrap:** QOP 009, Sec Two, par 3.01.4.5, all steps apply; (pg 213)
> f) **Waiver:** QOP 009, Section Two, paragraph 3.01.4.6; (pg 213)
> g) **Regrading Material:** QOP 009, Section Two, paragraph 3.01.4.7, all steps apply. (pg 214)

> The Quality Engineer is in charge of interacting with all departments to ensure timely execution and follow-up on MRB activities (to move the people to make timely resolutions).   ⑤ 5

TRAINING METRICS	Iss dt.	Rev dt.	Pg.
QUALITY OPERATING PROCEDURE	Sign	Rev #	13/22
REQUIREMENT: Performance Standard, Processing Control, QSM 4.13			
SUBJECT: **Handling Non-conforming Products (Put Through)**			
QOP 012 DOCUMENT REQ'D: Form A-006 (NCMR for MRB action)			

○

3.03

Handling Corrective Action

3.03.1 The Management Representative does: ↓

Per Requirement
QOP 009, Section Two, paragraph 3.01.5, all steps apply. (pg 214)

Carries out corrective/preventive action planning and determines the implementation process together with the responsible department head or individual. Documents the planning results. Implements training requirement and its recording.

(1)

3.03.2 The Management Representative does: ↓

Per Requirement
QOP 009, Section Two, paragraph 3.01.6, all steps apply. (pg 215)

Follows up the corrective/preventive action's effective implementation.

(2)

3.03.3 The Management Representative does: ↓

Per Requirement
QOP 009, Sec Two, par 3.01.7 and 3.01.8, all steps apply. (pg 216)

Closes out the NCMR (MRB) process and answers any outstanding SCAR. Maintains document retention.

(3)

3.04

Production Related Document-Change Requirement, Identified By Process Owners, Anywhere in the Product's Life Cycle

3.04.1 The Process Owners do: ↓

Per Requirement
QOP 002, Sec Two, par 3.0, under "General Requirement." (pg 72)

Any process owner who discovers errors, mistakes, or misleading instructions in any document which they are required to use during the process of carrying out their work assignments must report the problem to their immediate supervisor.

(1)

○

TRAINING METRICS	Iss dt.	Rev dt.	Pg.
QUALITY OPERATING PROCEDURE	Sign	Rev #	14/22
REQUIREMENT: Performance Standard, Processing Control, QSM 4.13			
SUBJECT: **Handling Non-conforming Products (Put Through)**			
QOP 012 │ DOCUMENT REQ'D: Form A-006 (NCMR for MRB action)			

3.04.2 The Supervisor does:

Per Requirement QOP 002, Sec Two, par 3.0, under "General Requirement". (pg 72)	Determine the nature of the document anomaly. If product quality is impacted, stop production. If not, continue production. In either case call the Process and Quality Engineers to the site and point out to them the document problem.	2

3.04.3 The Process and Quality Engineers do:

Per Requirement (a) QOP 002, Section Two, paragraph 3.03.1, all steps apply. (pg 73)	a) carry out the "Immediate and Provisional" change requirements when the process and/or product is impacted;	
Per Requirement (b) QOP 002, Section Two, paragraph 3.03.2.1, all steps apply. (pg 74)	b) carry out the "Standard Document Changes" when the process and/or product is not immediately impacted;	3
Per Requirement (c) QOP 002, Section Two, paragraph 3.03.2.2, all steps apply. (pg 75)	c) carry out "NCMR (MRB) affected document changes";	
Per Requirement (d) QOP 002, Section Two, paragraph 3.03.2.3. (pg 75)	d) carry out "Software Changes";	

3.05

Document-Change Requirements, Affecting the Quality System's Manual and the Quality Operating Procedures

3.05.1 The Management Representative does:

Per Requirement (a) QOP 002, Section Two, paragraph 4.0 and 4.01. (pg 76)	a) review the type of change requirements needed.	1

TRAINING METRICS	Iss dt.	Rev dt.	Pg.
QUALITY OPERATING PROCEDURE	Sign	Rev #	15/22
REQUIREMENT: Performance Standard, Processing Control, QSM 4.13			
SUBJECT: **Handling Non-conforming Products (Put Through)**			
QOP 012 |DOCUMENT REQ'D: Form A-006 (NCMR for MRB action)			

Per Requirement   (b)
QOP 002, Section Two, paragraph 4.01.1, all steps apply. (pg 76)

b)   carry out the **indirect** changes to the procedures.

2

Per Requirement   (c)
1.   QOP 002, Section Two, paragraph 4.02. (pg 78)
2.   QOP 002, Section Two, paragraph 4.02.1, specific requirement, all steps apply. (pg 78)

c)   carry out the **direct** changes to the procedures.

3

Per Requirement   (d)
1.   QOP 002, Section Two, paragraph 5.0. (pg 79)
2.   QOP 002, Section Two, paragraph 5.01, specific requirement, all steps apply. (pg 79)

d)   carry out the **revision indication changes** in the Quality System Manual.

4

Per Requirement   (e)
QOP 002, Section Two, paragraph 5.02, all steps apply. (pg 80)

e)   carry out the **revision indication changes** in the Quality Operating Procedures.

5

Per Requirement   (f)
QOP 002. Section Two, paragraph 5.03, all steps apply. (pg 82)

f)   carry out the **revision indication changes** to the Quality System Forms.

6

3.06

Handling Non-conforming Products Resulting From Document Changes

3.06.1 The Quality Engineer does:

Per Requirement
QOP 002, Section Two, paragraph 6.0 and 6.01. (pg 83)

Handle the non-conforming material resulted from imposing document changes.

TRAINING METRICS	Iss dt.	Rev dt.	Pg.
QUALITY OPERATING PROCEDURE	Sign	Rev #	16/22
REQUIREMENT: Performance Standard, Processing Control, QSM 4.13			
SUBJECT: **Handling Non-conforming Products (Put Through)**			
QOP 012 |DOCUMENT REQ'D: Form A-006 (NCMR for MRB action)			

3.07 — **Handling Customer-Related Document Changes**

General Information

Customer-related document changes involve two separately initiated efforts. The first one is initiated by the customer through the provisions of Purchase Order Amendments. The second one is initiated internally, due to deficiencies discovered in the customer's documents. The customer-initiated changes involve extensive review process, due to cost factors, and therefore they are handled much like a quotation process (RFQ). It is covered under QOP 001, Section Four, pg 52. The internally initiated changes involve the application of a "Waiver Request" through the provisions of QOP 009, Section Two, paragraph 3.01.4.2.1, step 1, under "Repair." (pg 210)

3.07.1 — **Customer Initiated Contract (PO) Amendments**

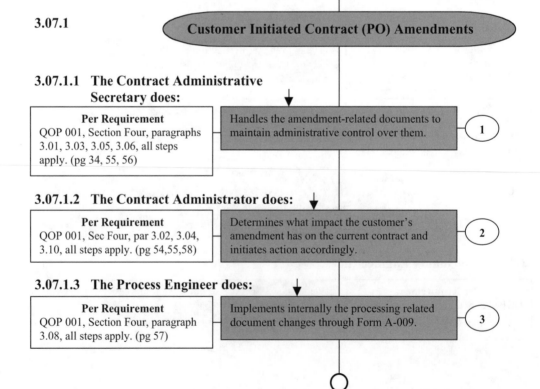

3.07.1.1 The Contract Administrative Secretary does:

Per Requirement
QOP 001, Section Four, paragraphs 3.01, 3.03, 3.05, 3.06, all steps apply. (pg 34, 55, 56)

Handles the amendment-related documents to maintain administrative control over them.   1

3.07.1.2 The Contract Administrator does:

Per Requirement
QOP 001, Sec Four, par 3.02, 3.04, 3.10, all steps apply. (pg 54,55,58)

Determines what impact the customer's amendment has on the current contract and initiates action accordingly.   2

3.07.1.3 The Process Engineer does:

Per Requirement
QOP 001, Section Four, paragraph 3.08, all steps apply. (pg 57)

Implements internally the processing related document changes through Form A-009.   3

TRAINING METRICS	Iss dt.	Rev dt.	Pg.
QUALITY OPERATING PROCEDURE	Sign	Rev #	17/22
REQUIREMENT: Performance Standard, Processing Control, QSM 4.13			
SUBJECT: **Handling Non-conforming Products (Put Through)**			
QOP 012 DOCUMENT REQ'D: Form A-006 (NCMR for MRB action)			

3.07.1.4 The Process and Quality Engineers do: ↓

Per Requirement	Work together to implement the customer's amendment	4
QOP 001, Section Four, paragraphs 3.09 and 3.11, all steps apply. (pg 58)		

3.08

Customer Complaints, Product Returns, and Customer Evaluation (Satisfaction) Reporting

3.08.1

Customer Complaints

3.08.1.1 The Complaint Receiver does: ↓

Per Requirement	Directs the complaint reporter to the Quality Manager.	1
QOP 009, Sec Three, paragraph 3.01.1 "General Requirement." (pg 218)		

3.08.1.2 The Quality Manager or his/her Designate does: ↓

Per Requirement	Investigates and interacts to resolve the complaint.	2
QOP 009, Section Three, paragraph 3.01.1.1, all steps apply. (pg 218)		

3.08.2

Product Return

3.08.2.1 The Receiver of the Product does: ↓

Per Requirement	Receives the customer returned product and initiates action.	1
QOP 009, Section Three, paragraph 3.02.1, all steps apply. (pg 219)		

TRAINING METRICS	Iss dt.	Rev dt.	Pg.
QUALITY OPERATING PROCEDURE	Sign	Rev #	18/22

REQUIREMENT: Performance Standard, Processing Control, QSM 4.13

SUBJECT: Handling Non-conforming Products (Put Through)

QOP 012 | DOCUMENT REQ'D: Form A-006 (NCMR for MRB action)

3.08.2.2   The Quality Engineer does:

| **Per Requirement** QOP 009, Section Three, paragraph 3.02.2 and 3.03, all steps apply. (pg 219 and 220) | Investigates background, subjects the returned product to verification, cause determination, and corrective action. | 2 |

3.08.3   **Customer Satisfaction Reporting**

3.08.3.1 All Personnel do:

| **Per Requirement** QOP 009, Sec Three, paragraph 3.04 "General Requirement". (pg 222) | Understand the nature of customer satisfaction reporting. | 1 |

3.08.3.2   The Final Inspector does:

| **Per Requirement** QOP 009, Section Three, paragraph 3.04.1.1, all steps apply. (pg 222) | Initiates the customer evaluation and satisfaction report,, Form A-038. | 2 |

3.08.3.3   The Management Representative does:

| **Per Requirement** QOP 009, Section Three, paragraph 3.04.1.2, all steps apply. (pg 223) | Handles the replied customer evaluation and satisfaction report. | 3 |

End of Section Two

TRAINING METRICS	Iss dt.	Rev dt.	Pg.
QUALITY OPERATING PROCEDURE	Sign	Rev #	19/22

REQUIREMENT: Performance Standard, Processing Control, QSM 4.9

SUBJECT: **Developmental Process Control (Put Through)**

QOP 012 |DOCUMENT REQ'D: Forms A-008, A-001, A-018

SECTION THREE

1.0 PURPOSE

Maintain control over those activities that involve the product's processing qualification when there are no documented procedures exist on how to do it, as with
new customer Purchase Orders.

2.0 APPLICATION

This procedure applies to the planning and implementation of qualification processes on new jobs where the capability as to how to do it has not been previously
documented and demonstrated.

3.0 PROCEDURE

General Information

Processing qualification of new products is often regarded by many customers as
nothing more than setting up a new job in an otherwise similar production environment. While the equipment, tools, and operator's skill may be the same, the
processing steps and work methods involved could be completely different. This
type of product cannot be regarded as identical to those products for which the
processing capability has already been demonstrated and documented. Under
uncontrolled conditions, the entire order could become scrap. A process-approach
quality management system must provide the procedural mechanism under which
the processing qualification can be orderly carried out and the product released to
the customer according to his/her PO requirement. **Since we have already provided controlling procedures for First Article Production Approval (ESA)
under QOP 006, Section Two, we will be applying certain provisions of the
same (text) procedures to implement basic requirements for the processing
qualification of new jobs. The applicable steps are marked with a star "☆."**
(see pg 368)

3.01

Planning

3.01.1 The Process Engineer does:

Per Requirement	
1. QOP 006, Sec Two, par 2.0, steps 1,2,4, and 5. (pg 368) **2.** QOP 006, Section Two, paragraph 3.0, step 2. (pg 368)	Reviews the customer's PO and product specs in order to determine the planning and processing layout for developing the procedures on how to implement the qualification steps on new jobs, which procedures when formalized would become the standard work instruction for process owners.

1

TRAINING METRICS	Iss dt.	Rev dt.	Pg.
QUALITY OPERATING PROCEDURE	Sign	Rev #	20/22
REQUIREMENT: Performance Standard, Processing Control, QSM 4.9			
SUBJECT: **Developmental Process Control (Put Through)**			
QOP 012 |DOCUMENT REQ'D: Forms A-008, A-001, A-018			

3.01.2 The Quality Engineer does:

Per Requirement
1. QOP 006, Sec Two, paragraph 2.0, steps 1,2,3, 4, and 5. (pg 368)
2. QOP 006, Sec Two, paragraph 3.0, steps 1, 3, and 4. (pg 368)

Reviews the customer's PO and product specification in order to determine the planning of product quality verification requirements for the processing cycle.

2

3.02

Issuance of Developmental Procedures

3.02.1 The Quality Engineer does:

Per Requirement
QOP 006, Sec Two, par 4.0, steps 1,2,3,4, and 5. (pg 368 & 369)

Issues the Purchase Order Review Sheet, Form A-001, in order to identify the product quality verification requirements for the whole processing cycle.

1

3.02.2   The Process Engineer does:

Per Requirement
QOP 006, Sec Two, par 5.0, steps 1,2,3,4,5, and 6. (pg 369 & 370)

Determines, documents, reviews, approves, and issues the developmental Production Folder.

2

3.03

Process Control

3.03.1   The Quality Engineer does:

Per Requirement
QOP 006, Sec Two, paragraph 6.0, steps 1,2,3,4,5,6, and 7. (pg 370) (Apply these requirements as suitable for developmental processing effort.)

Maintains process control, as required.

TRAINING METRICS	Iss dt.	Rev dt.	Pg.
QUALITY OPERATING PROCEDURE	Sign	Rev #	21/22
REQUIREMENT: Performance Standard, Processing Control, QSM 4.9			
SUBJECT: **Developmental Process Control (Put Through)**			
QOP 012	DOCUMENT REQ'D: Forms A-008, A-001, A-018		

3.03.2 The Quality Engineer does:

Handles developmental non-conforming materials as follows:

(1) If the contract (PO) has material (parts) allowance for developmental purposes, no accountability for the non-conformance is required. Correct the nonconformance as part of processing development. Note that this type of allowance many times is not stated in the contract. What may be stated, however, is that you are allowed a certain percentage of parts for attrition (scrap). Note also that physical scrapping may still be controlled under certain contract provisions **(as in the case of AS9000).**

Per Requirement
QOP 006, Section Two, paragraph 7.0, steps 1 through 8, as required. (pg 370 & 371)

(2) If there is no allowance for developmental purposes given and the material (parts) were supplied by the customer, handle the non-conforming product as stated under:

(3) If the material (parts) is company purchased, any fallout due to processing development should not be controlled as non-conforming material. Document it on Form A-006 (NCMR) as part of development and destroy the product, as required.

3.04   **Closure of Developmental Procedures**

3.04.1 The Process Engineer does:

Per Requirement
QOP 006, Section Two, paragraph 5.0, step 7. (pg 370)

After the developmental processes are over, you are required to formalize all procedures to become a standard operating work instruction issued in Production Folders.

TRAINING METRICS	Iss dt.	Rev dt.	Pg.
QUALITY OPERATING PROCEDURE	Sign	Rev #	22/22

REQUIREMENT: Performance Standard, Processing Control, QSM 4.9

SUBJECT: **Developmental Process Control (Put Through)**

QOP 012 |DOCUMENT REQ'D: Forms A-008, A-001, A-018

3.04.2   The Process and Quality Engineers do:

> **Per Requirement**
> QOP 006, Section Two, paragraph 11.0, all provisions apply as required. (pg 373)

Maintain retention of the Developmental Procedures as stated under:

3.05

Release of Product Completed Under Development

3.05.1   The Final Inspector does:

> **Per Requirement**
> **QOP 006, Section Four, all provisions apply as required.** (Make up a standard Inspection Product Folder at this time in order to retain the final inspection records separately from the "Developmental Records." Interact with the Quality Engineer to work out document handling details.) (pg 161)

Carry out final inspection on products made under Developmental Process the same way as products made under standard operating conditions. The customer's product specification (drawing) is the controlling document. If the developmental product meets all the specification requirements, the product can't be rejected. For background information, rely on documents contained in the First Article Inspection Product Folder now marked as "Development." Carry out the final inspection, release, and delivery requirements as stated under:

End of Section Three

End of QOP 012

BOOK SECTION TWO

OVERVIEW TO THE QUALITY SYSTEM MANUAL

Seven years ago as I was writing the first edition of this Process-Approach Quality System Manual for manufacturing shops, I was searching for a "conveyor" (the flow process mechanism) which would identify the quality requirements suitable for shop operations. I wanted the quality requirements to represent the true nature of how a manufactured product was put through the various stages from planning through product completion and delivery to the customer. What I was looking for was a process approach that would tie the various departmental activities into a structured dependency so that everybody would be focused in one direction to bring about managing their work processes in line with commitments to product quality in a product's flow cycle. I was looking for something that did not exist. I gave up the search and settled down instead to find only the quality requirements. As I was mulling over the 1987 edition of ISO 9000, I recognized that it had already identified for me the core departmental requirements that I needed, less the linkage to a process-approach system. I decided that I would follow the layout mechanism of the Standard and develop to it, separately, the process-approach procedures. In this second edition, I have retained the quality requirements of the same Standard (ISO 9001/1994) for controlling product quality in shop operations because the need exists for them, and because they represent reality.

Change in layout in the upcoming year 2000 Standard will not remove this reality. The Standard will shift the requirements around and add more to them so that nobody can recognize them anymore as a manufacturing quality standard. The upcoming changes in ISO 9001/2000 will not affect this process-approach system, because I have already incorporated them as good business practice in manufacturing. Appendix One at the end of the book explains the alignment, while the operating procedures have implemented them. Appendix Five, on the other hand, identifies the changes and demonstrates how the implementation occurred. The encircled numbers that appear next to the clause titles in this quality manual are corresponding to the clause number requirements of ISO 9001/2000.

MINFOR INCORPORATED

QUALITY SYSTEM MANUAL

The provisions of SAE AS9000 have been incorporated as complementary requirement when imposed by contract

(xxx)   **Encircled numbers represent alignment compatibility with ISO 9001/2000**

Controlled Document

MINFOR INCORPORATED	Issued	Rev Date	Pg No
	1/1/ 99	4/30/2000	2/23
QUALITY SYSTEM MANUAL	Signed AG	Rev No 2	

CONTENTS   PAGE

1.0 SCOPE (1, 4.2.2)

This process-approach quality system manual outlines the general and specific requirements for organizational commitment to product quality, customer satisfaction, and continuous improvement. These commitments are summarized in this manual and communicated in written procedures to the members of the organization in a manner that the contracted product is systematically controlled from process to process in the product's life cycle and accordingly manufactured within regulatory and statutory requirements. Management's goals are to achieve the prevention of non-conformities in document maintenance, in process attainment, and in product realization. Management shall require from process owners demonstration of work results through documentation to enable the organization to measure satisfactory compliance to the stated product specifications and related work instructions in order to achieve its objectives.

1.1 MISSION (5.1)

Achieve the organizational commitments in product quality, customer satisfaction, and continuous improvement to the requirements of this quality system manual and in association with the requirements of ISO 9001/2000.

1.2 OBJECTIVES (5.4.1)

Plan interrelated processes and communicate them to affected parties. Carry out assignments to meet the planned requirements. Document the work results and measure performance, as required. Implement corrective/preventive action, as required. Review overall effectiveness to consistently achieve product quality, customer satisfaction, and continuous improvement. Set new objectives in line with business goals.

2.0 EXCLUSION (1.2, 4.2.2)

This process-approach quality system shall exclude design and service related activities for they are not part of the organization's business commitments.

3.0 DEFINITION

Process-approach management

Plan, integrate and control the interrelated processes in a product's life-cycle in a systematic manner for maintaining operational cohesion and effectiveness in preventing the transfer of non-conformities from process to process, that would otherwise impact the continuing activities and consequently the end product.

3.1 REFERENCES

ISO 9002/1994 – Model for Quality Assurance in Production, Installation, and Service
ISO 9001/2000 – Quality Management System – Requirements

ANSI Z1.4/1993 – Sampling Statistics
ANSI/NCSL Z540-1 – Calibration System
SAE AS9000 – Aerospace Standard

3.2   REVISION HISTORY   (5.4.2b, 4.2.3c)

SECTION	REV DATE	REASON FOR REVISION	MGMT REP
All	2/10/99	Added AS9000	AG
All	4/30/00	Added ISO 9001/2000	AG

3.3 ANNUAL REVIEW $\quad$ 5.3e

DATE	MGMT REP	RESULTS OF ANNUAL REVIEW
NOTES:		

4 QUALITY SYSTEM REQUIREMENTS

4.1 MANAGEMENT RESPONSIBILITY

4.1.1 Quality policy ⟨ 5.1, 5.3, 5.4.1, 5.5.1 ⟩

The organizational quality policy has been planned with commitments regarding resources and quality objectives to meet product quality, customer satisfaction, and continual improvement within regulatory and statutory requirements. This policy is being communicated through process instructions at all levels within a logical framework. The quality objectives are measurable to ensure conformity to requirements and provide the basis for reviewing continuing suitability.

4.1.2 Organization (Manufacturing containment circle for continual improvement)

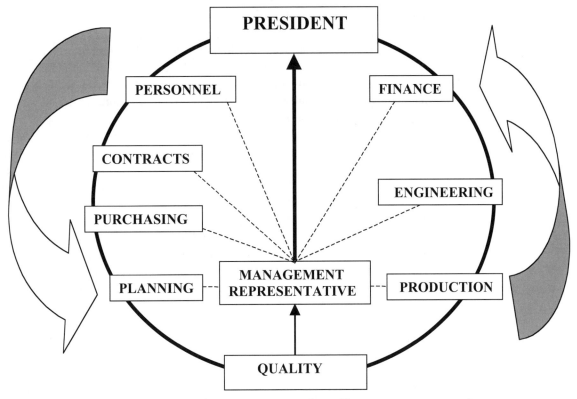

This organization chart represents the process-approach quality management containment circle. The organizational activities are cross-functionally interlocked in a closed-loop structure in order to achieve continual improvement regarding quality objectives in relation to internal/external customer satisfaction.

ISO 9001:2000 for SMALL BUSINESSES 273

4.1.2.1   Responsibility and authority   ⟨ 5.5.1, 5.5.2, 6.1 ⟩

The President shall:

Oversee that the organizational responsibilities are carried out in accordance with the company's objectives stated in the documented procedures of the Quality Management System. Provide the necessary resources to maintain the Quality Management System. Preside over Management Reviews to ensure the effective and efficient operation of the Quality Management System.

The Director of Personnel shall:

Administer employee relations in accordance with federal, state, and local regulations. Interact with all department managers to ensure that these regulations have been effectively communicated to all employees, implemented and maintained in line with organizational commitment. Ensure that all new employees receive adequate training regarding the company's quality management system as directed by the Management Representative.

The Director of Finance shall:

Plan and maintain cost accounting within budgeted limits. Ensure that cost incurred as a result of work accomplishment is continuously monitored in all areas in order to realize the planned objectives of the company. Interact with all department managers to periodically review unfavorable conditions. Interact with the Management Representative to implement corrective action.

The Director of Contracts shall:

Plan and maintain customer relations regarding sales, marketing, and procurement. Administer the quality management policies regarding contract activities in line with documented procedures. Interact with Planning, Purchasing, Engineering, Manufacturing, and Quality in a timely fashion to ensure that cost control, product quality, and delivery schedules are maintained, as planned.

The Director of Planning shall:

Plan and coordinate scheduling and dispatching activities in line with delivery requirements. Maintain timely liaison with all departments in order to remove bottlenecks hindering planning, production, and delivery schedules. Observe product quality requirements according to the documented procedures.

The Director of Engineering shall:

Plan and direct all engineering activities to meet the customers' product quality and delivery requirements. Carry out all processing related planning and work effort in line with cost control, product quality, and scheduling requirements. Observe product quality requirements in all engineering activities as stated in the documented procedures. Interact with customers to resolve bottlenecks impacting product related engineering problems. Communicate with all departments to coordinate engineering activities to ensure effective and efficient production.

The Director of Manufacturing shall:

Plan and carry out manufacturing activities according to cost control, product quality, and delivery requirements. Oversee effective equipment maintenance in line with production requirements. Interact with all departments to remove bottlenecks hindering production operations. Ensure the maintenance of all federal, state, and local regulations, as required. Provide a safe and suitable work environment.

The Quality Assurance Manager shall:

Plan and direct the quality department in accordance with the stipulated requirements of the documented Quality Management System. Interact with all departments to remove bottlenecks impacting product quality enforcement, including the resolution of non-conforming products and customer complaints. Coordinate all inspection work effort in line with cost control, scheduling and processing requirements. Maintain close relationship with the Management Representative regarding all matters impacting the effectiveness of the quality system.

The Management Representative shall:

(5.5.2)

Establish, implement, and maintain the processes of the Quality Management System. Plan and coordinate the training of employees regarding the maintenance and awareness of customer requirements of the Quality Management System. Plan and coordinate the internal quality audit effort. Plan and coordinate Management Reviews on a regular basis. Plan, implement, and follow up the corrective action system in line with the requirements to maintain continual improvement.

The Process Owners shall:

Observe and carry out the product related work assignments as directed by supervision. Verify and document work results as required by the applicable work instructions. Maintain a clean and safe work environment surrounding work centers. Maintain product-related documents in the applicable production folders, as required.

4.1.2.2 Resources

(6.1 6.2.1)

All department managers shall have unlimited access to the president of the company in order to determine and allocate resource requirements in a timely manner to ensure the effective implementation and continuous improvement of the process-approach quality management system in meeting internal and external customer satisfaction in order to:

(6.2.2)

1. identify the need, evaluate competency for the assigned personnel and accordingly maintain and document training requirements, assuring the adequacy and awareness in understanding the process and quality requirements;

(6.3)

2. provide and maintain adequate and suitable work environment within the infrastructure, ensuring equipment and service to meet product conformity;

(6.4)

3. manage the work environment in order to maintain product conformity, customer satisfaction, and continuous improvement.

4.1.2.3 Management representative 〔5.5.2〕

The management representative shall have the authority, regardless of other responsibilities, to establish, implement, and maintain the processes of the quality management system as stated above under "Responsibility and Authority."

4.1.2.4 SAE AS9000

AS9000 requirements have been incorporated throughout the product's processing cycle by the application of Quality Operating Procedures.

4.1.3 Management review 〔5.6.1, 5.6.2, 5.6.3〕

Management review shall be planned in order to evaluate the suitability, adequacy, and effectiveness of the quality management system. The review input shall be based on assessed results of internal audits carried out to the specified quality objective procedures. The measured audit results, when reviewed and evaluated by upper management, will demonstrate whether any corrective action is required to improve the quality management system to achieve organizational objectives. Management reviews shall be conducted and documented as stated in QOP 011 (pg 235 or 428).

4.2 QUALITY SYSTEM

4.2.1 General 〔4.1, 4.2.1, 4.2.2, 5.1, 5.4.1, 5.5.3〕

This Quality System Manual outlines the requirements of ISO 9001/2000 as to how the product quality objectives are identified, documented, and maintained in line with organizational commitments, including permissible exclusions. The Manual references the QOPs, which outline the specific process instructions relating to specified tasks, sequences and interactions. In certain cases, the Manual itself states the applicable process instructions. Work results are recorded on specified forms, identified, controlled, and retained according to documented procedures. Taken together, the Quality System Manual, the Quality Operating Procedures, and the Forms comprise the documented system through which the criteria and methods are implemented and maintained to ensure effective operation and control. They are controlled documents. Through these efforts, the requirements for product quality, customer satisfaction, and continuous improvement have been communicated at appropriate levels and functions within the organization.

4.2.2 Quality system procedures 〔4.2.1, 4.2.4〕

Identified process instructions shall serve as procedures to ensure the effective and consistent operation of the process-approach quality management system. These process instructions have been systematically documented in twelve Quality Operating Procedures (QOP) and issued and controlled to enforce the process-approach quality system. Documentation of work results, when required, is defined, maintained, and monitored to ensure compliance. Distribution of the quality system's procedures is controlled on form A-010 (pg 465). **(AS9000, paragraph 4.2.2c)**

4.2.3   Quality planning $\qquad$ 4.1, 5.4.1, 5.4.2, 5.5.3, 7.1, 8.1, 8.5.1

In order to effectively manage the quality system's objectives, the requirements for planning have been made an inherent responsibility on the part of those issuing process and work instructions to process owners that execute work assignments to achieve product quality, customer satisfaction, and continuous improvement:

a) Quality planning forms an inherent part of every Quality Operating Procedure, beginning with QOP 001 (Contract Review, pg 46 or 301). The Quality Engineer issues Form A-001 (pg 455) to summarize product quality requirements for the entire processing cycle of a contracted product;

b) Resources needed in the acquisition of new processes, equipment, personnel and product support are under the responsibility of each department manager in collaboration with the president of the company;

   (1) AS9000 the design, manufacture, and use of tooling so that variable measurements can be taken, particularly for key characteristics;
   This provision in incorporated under QOP 001, Section Two, paragraph 3.02, step 2, pg 47 or 302, as a standard requirement.

c) Compatibility of processes, procedures, and assignments is maintained through the integrated documented system in all areas of a product's processing cycle.

d) Updating of quality requirements in documents, inspection and test techniques, including equipment, is consistently carried out with the technical and quality requirements of a contracted product (See QOP 001 on pg 38 or 295, all sections, and Form A-001. Also see QOP 002, Section Two on pg 71 or 317)

e) The determination of measurement capability is consistently carried out under the provisions of QOP 001 and accordingly acquired in time for the needed application;

f) Product verification requirements are stipulated in the applicable Quality Operating Procedures as well as in the Job Travelers and Operation Sheets throughout a product's processing cycle;

   (1) AS9000 the identification of in-process verification points when adequate verification of conformance cannot be performed at a later stage of realization;
   This provision is enforced under 4.9 of this QSM. When it is in-house required for in-process verification and recording, Form A-033 (pg 492) or A-018 (pg 474) is used (See QOP 006, Section One);

g) Standards of acceptability, including workmanship requirements, are defined, product by product, and process by process in the applicable Job Travelers, Operation Sheets, and Product Specifications (dwg). All process owners are required to verify and document the acceptance or rejection of product quality results as stated in the work instructions.

h) The identification, maintenance, and retention of quality records are made a requirement in every Quality Operating Procedure according to the need of process realization. More specifically, the retention of documents is detailed in QOP 002, Section Three (pg 85 or 328).

i) **AS9000** – the identification and selection of subcontractors is incorporated under QOP 003, Section Two (pg 103 or 341). Flow-down requirement is incorporated under par 4.6.2 of this quality system manual.

j) **AS9000** – key characteristics are identified, executed, verified and documented via the provisions defined in QOP 006, Section One and Three.

k) Internal communication, to dispense relevant information to personnel, is manifested in all quality objective procedures. Specific action items are issued on Form A-003 (pg 457), as required.

$\boxed{5.5.3}$

4.3   CONTRACT REVIEW

4.3.1   General

Written procedures shall define contract reviews and the coordination of those activities. (See QOP 001, Section One, Two, Three, and Four, pg 38 or 295)

4.3.2   Review $\boxed{5.2, 7.1, 7.2.1, 7.2.2, 7.2.3}$

The contract review process is communicated as stated below:

a) Determining the technical, quality, and processing capability to manufacture a product to meet the customer, regulatory and statutory requirements shall be carried out by Contracts as defined in the RFQ provisions in Section One of QOP 001. Both specified and unstated requirements shall be determined. Verbal agreements shall be handled in accordance with QOP 001, Sec Three.

b) After the receipt of a contract (PO), its implementation shall be carried out in accordance with the provisions of QOP 001, Section Two.

4.3.3   Amendment to contract $\boxed{7.2.2}$

Customer requested changes, known as amendment to contract, are reviewed before implementation in order to determine the impact of implementation and the associated cost. Implementation of subject changes is carried out in accordance with the provisions stated in QOP 001, Section Four (pg 52 or 307).

4.3.4   Records $\boxed{4.2.4, 7.2.2}$

Records of contract reviews are maintained as stated in QOP 001, Section One, Two, Three, and Four. Also, as stated in QOP 002, Section Three (pg 85 or 328).

4.4   NOT ENGAGED IN DESIGN ACTIVITY (AS9000 does not apply)

4.5 DOCUMENT AND DATA CONTROL

4.5.1 General — 4.2.1, 4.2.2, 4.2.3, 4.2.4, 5.4.2, 7.2.2, 7.2.3

Written procedures have been issued and implemented as to how internal and external documents and data are communicated and controlled. All the provisions stated in QOP 002, Section One, Two, and Three shall apply, as required.

4.5.2 Document and data approval and issue — 4.2.3

The approval, issue, and control of the quality system's documents are carried out in accordance with the provisions of QOP 002, Section One (pg 64 or 315).

4.5.3 Document and data changes — 4.2.3, 5.4.2b, 7.2.2

The specific requirements needed to control document and data changes, including the removal of obsolete documents from circulation, are handled in accordance with the stated provisions of QOP 002, Section Two (pg 71 or 317).

4.5.3.1 AS9000 Document Change Incorporation: – amendments to customer documents are controlled through the provisions of QOP 001, Section Four. The internal implementation of those amendments are carried out in accordance with the provisions of QOP 002, Section Two, paragraph 3.03 (pg 73 or 317), as required.

4.5.4 Use of mark and registration certificate (required only with registration)

The use of mark and registration certificate is strictly limited to the scope of registration granted. The organization may refer to the registration in publications, advertising and promotions. It shall not be permitted to display the Mark or the Registration Certificate on any product or advertising for a product on any packaging or containers, or infer in any manner product endorsement.

4.6 PURCHASING

4.6.1 General — 7.4.1, 7.4.2, 7.4.3, 8.4d

Written procedures as issued in QOP 003, Section One and Two, clearly state how product related purchases are documented, reviewed, approved, and controlled, including customer designated sources. **(AS9000 "NOTE" under 4.6.1)**

4.6.2 Evaluation of subcontractors (Suppliers) — 4.1, 7.4.1

Subcontractors are selected according to their capability to furnish products to flow-down requirements of the contract. **(AS9000 par. 4.6.5)** Provisions of QOP 003, Section One (pg 97 or 338) and Two (pg 103 or 341), govern all requirements of the evaluation and control processes. Including:

a) periodic evaluation by survey (self-survey is permitted);

b) requirements for product quality, verification, and documentation to determine approval or disapproval; **(AS9000 "NOTE")**

c) maintenance of "Supplier Product Folders" containing the recorded evidence of subcontractors' inspection data;

d) special process approval by customers, as required; **(AS9000)**

(8.4d) e) subcontractor (supplier) product quality assessment and rating as per QOP 003, Section Two, par 3.03 (pg 108 or 343).

4.6.3   Purchasing data   **(7.4.2)**

Product related purchase orders shall contain clear definition and ordering data:

a) precise identification of the product ordered;

b) product and technical specifications: quality provisions: approval of documents, processes, product, equipment and personnel, as required;

c) identification of the subcontractor's quality system.

Provisions of QOP 003, Section One (pg 97 or 338), governs the review and release approval of product related purchase orders.

4.6.4   Verification of purchased product   **(7.4.3)**

AS9000 – verification methods are enforced through QOP 006, Section Five.

4.6.4.1   Supplier verification at subcontractor's premises

Source inspection at the subcontractor's facility is carried out in accordance with the provisions of QOP 006, Section Seven (pg 176 or 388).

4.6.4.2   Customer verification of subcontracted product

The requirements to submit products to customer source inspection at our facility or at the subcontractor's premises are defined in QOP 006, Section Six (pg 172 or 386).

4.6.4.3   AS 9000 – Right of Entry

Right of entry provision is defined in the purchase orders issued to subcontractors, as required. (See QOP 003, Section One)

4.6.4.4   AS9000 – Delegation of Supplier Verification to Subcontractors

Delegation of product inspection activities to subcontractors, including the delegated requirements, is defined in QOP 003, Section One, paragraph 3.01.

4.6.5   AS9000 – Requirements Flow-down

Not verifiable and key characteristics are required to be flown down in the purchase order as specified in QOP 003, Section One, paragraph 3.01.

4.7   CONTROL OF CUSTOMER-SUPPLIED PRODUCT   (7.5.4)

Written procedures have been established and implemented to control customer supplied products as stated in QOP 004 (pg 110 or 345). To enforce the handling of non-conforming products supplied by customers, the provisions of QOP 009 are imposed as a requirement.

4.8   PRODUCT IDENTIFICATION AND TRACEABILITY (7.5.3, 7.5.4)

Contractually identified products and the related documentation, including traceability and record retention are controlled in accordance with the provisions of QOP 005 (pg 119 or 351). The methods of identification, applied to the product, which may be required by contract and/or internal control, are also imposed via the same QOP.

4.9   PROCESS CONTROL (6.3, 6.4, 7.1, 7.2.3, 7.5.1, 7.5.2, 8.1, 8.2.3, 8.4)

The below listed controls form only part of the product quality planning and measurement requirements. Additional controls have been incorporated, as required, in the relevant quality operating procedures to ensure problem prevention, control of non-conformance from process to process, and timely corrective action implementation, regarding customer complaints and product returns.

a) In a Process Approach Quality Management System, the planning, identification, and documentation of process controls are implemented and part of every procedure that comprise the product realization processes, beginning with the RFQ phase and ending with the delivery of a product;

(6.4)   b) Processing equipment and a suitable working environment is provided by management (see paragraph 4.1.2.2, item 2, in this QSM);

(8.1)   c) Work results are recorded on form A-033 (pg 492) by process owners to demonstrate compliance;

(8.1 8.4c)   d) Supervisory surveillance to measure product quality compliance during process realizations are stated in QOP 006, Section One, paragraph 3.01 (pg 135 or 364); **(AS9000 (1))**

(7.5.5)   e) Process and equipment approval is ensured when the first production unit is verified, accepted, and documented through the provisions of QOP 006, Sections One and Two, QOP 012, Section Three, QOP 003, Section Two;

f) Process Engineering specifies workmanship standards in the Job Travelers and Operation Sheets as part of processing requirement, when applicable (See QOP 001, Section Two, pg 48 or 303);

g) Equipment maintenance is continuously monitored and recorded on Form A-002 (pg 456);

h) AS9000 – accountability for all manufactured products is maintained in the Job Traveler (A-008), Inspection Report (A-018), Non-conforming Material Report (A-006);

i) AS9000 – evidence of processing and inspection completion is documented in the Job Travelers (A-008) and Inspection Reports (A-018);

j) AS9000 – the removal of foreign objects is made a processing related requirement through clearly defined instructions in the Job Traveler;

(7.5.5) When a customer specifies in the contract (PO) the requirement of special process provisions and the process is performed outside the premises, the Quality Engineer is responsible for identifying and flowing down that requirement: (1) in the Purchase Order Review Sheet (A-001): (2) to determine approval requirement and obtain approval from the customer: (3) to carry out survey at the subcontractor (mail-in survey is acceptable): (4) to invoke Certificate of Conformance at time of approving the internally issued PO.

(7.5.5) When a customer specifies in the contract (PO) the requirement of special process provisions and the process is performed in-house, the Quality Engineer is responsible for identifying and flowing down that requirement: (1) in the Purchase Order Review Sheet (A-001): (2) to ensure qualification maintenance of the process, the equipment, and the process owner: (3) to maintain certification records of the approved process, equipment, and process owner: (4) to obtain customer approval before process commencement.

4.9.1 AS9000 – Process Specification Requirement:

The customer approval provision is made a requirement as stated just above.

4.9.2 AS9000 – Tooling:

Maintenance and control of production tooling is a standard requirement and enforced through the provisions of first production piece approval as stated in QOP 006, Section One.

4.10 INSPECTION AND TESTING

7.1, 7.2.3, 7.4.3, 7.5.1, 8.1, 8.2.1, 8.4, 8.2.3, 8.2.4

4.10.1 General

This process-approach quality management system enforces planning for inspection and testing to maintain product quality as a process-linked responsibility for all process owners. The documented procedures, detailed in QOP 006, cover the monitoring and measurement requirements for a product's entire lifecycle. QOP 006 identifies the process owners' responsibilities, the process assignments and product quality substantiation with documentation of work results, process by process, as applicable. Inspection and testing is an inter-linked process with other organizational commitments (NCMR process) in the quality system to maintain uninterrupted continuity in a specified product's process realization cycle to ensure customer satisfaction through the analysis of data and the resulting corrective action, when required.

4.10.1.1 AS9000 – Subcontracting Inspection Activities:

This provision is incorporated under paragraph 4.6.4.4 of this manual.

4.10.2 Receiving inspection and testing

7.4.3, 8.2.4

4.10.2.1 Receiving inspection and testing requirement, including documentation, are covered under the provisions of QOP 006, Section Five (pg 167 or 383).

4.10.2.2 The amount of inspection and test requirement is dependent on Source Inspection and the subcontractor's previously demonstrated capability.

4.10.2.3 In case of urgent production requirement, an unverified product may be released for processing in accordance with the provisions of QOP 006, Section Five, paragraph 3.02.2 (see Special Instruction under step 2).

4.10.2.4 **AS9000 – When certification test reports are used as a means of product acceptance, procedures shall document the types and frequencies of analysis to validate certifications.**

This provision is part of QOP 005, paragraph 3.01.2, step 4, and also invoked during Receiving Inspection under the provisions of QOP 006, Section Five, paragraph 3.02.2, step 2c, as required (see pg 170 or 384).

4.10.3 In-process inspection and testing

5.3, 7.4.3, 7.5.1, 8.2.3, 8.2.4

In-process inspection and testing requirements are controlled under the provisions of QOP 006, Section One and Three. Record maintenance and supervisory surveillance over process owners' performance results to ensure compliance with the requirements of documented procedures are part of these provisions.

4.10.4 Final inspection and testing

7.4.3, 8.2.4

Final inspection and testing is carried out according to the provisions stated in QOP 006, Section Four (pg 161 or 380). All the requirements to maintain the end item quality provisions, including release approval, have been incorporated.

4.10.4.1 Customer Source Inspection

Customer source inspection when imposed by purchase order is carried out in accordance with QOP 006, Section Six (pg 172 or 386).

4.10.4.2 Source Inspection at the Subcontractor

The Company's source inspection at the subcontractor is carried out in accordance with QOP 006, Section Seven (pg 176 or 388).

4.10.5   Inspection and test records $4.2.4$

Under the documented provisions of this Quality Management System, inspection and test records are organized, filed, and maintained. These requirements have been incorporated under QOP 006 in all sections. Specific information on record retention is stated under QOP 002, Section Three.

4.10.5.1 AS9000 – First Production Article

First production article inspection, verification, and documentation have been implemented under the provisions of QOP 006, Section One. First Article Qualification requirements for "Engineering Source Approval" is covered under the provisions stated in QOP 006, Section Two.

4.11   CONTROL OF INSPECTION, MEASURING AND TEST EQUIPMENT

4.11.1   General $4.2.4, 7.6$

Record maintenance is the only acceptable process to demonstrate control of inspection, measuring and test equipment, including the confirmation of test software, and comparative references under this QMS. The control of measuring and test equipment is carried out in accordance with ISO 10012-1 and/or ANSI/NCSL Z540-1. Electronic data collection is authorized to maintain records of calibration. All the controlling provisions are incorporated under QOP 007, except software change control, which is defined under QOP 002, Sec Two, par 3.03.2.3.

4.11.1.1 AS9000 "Definition"
The type of inspection, measuring and test equipment covered under this clarification statement is incorporated under QOP 007 (pg 180 or 390) as a general gage calibration requirement.

4.11.2   Control procedure 7.6

QOP 007 incorporates the control procedure to regulate calibration of the inspection, measuring and test equipment, including traceability of test masters and the control over personal, customer provided, or subcontractor applied equipment.

4.12   INSPECTION AND TEST STATUS $7.5.1, 7.5.3$

Inspection and test status on products and/or product related documents, indicating the acceptance or rejection of the product, is carried out under the provisions of QOP 008 (pg 192 or 397). The issuance of passwords and inspection stamps is documented and maintained on Form A-032 under the same QOP provisions. The releasing authority may substitute signature and date instead of the authorized media, as required. Production personnel are authorized to apply signature and date when releasing products on the authority of Job Travelers.

4.12.1   AS9000 Acceptance Authority Media – the controlling provisions are stated under clause paragraph 4.12 above.

4.13   CONTROL OF NON-CONFORMING PRODUCT

4.13.1   General ⟨ **4.2.4, 5.3, 7.2.3, 8.1, 8.2.4, 8.3, 8.4, 8.5.1** ⟩

In a process-approach quality management system, non-conforming product is more than just hardware. Any error in documents or processes that impacts product quality, or the methods and processes that may affect or create a bottleneck to achieve product quality is also called non-conforming product. Therefore, managing non-conforming products in all the areas of core departmental functions is as important as managing conforming products. Documented procedures, therefore, have been so integrated that a total quality management of the non-conforming products becomes cohesively a de facto operation within the quality management system to provide for continual improvement to enable the organization to attain its management objectives. The controlling provisions to handle the various types of non-conformance are detailed in QOP 009 (pg 199 or 403).

AS9000 Notes 1 and 2, regarding notification of non-conformance and non-conforming customer returns, have been incorporated in QOP 009 as a routinely required quality management objective.

⟨ **8.5, 8.5.2, 8.5.3** ⟩

4.13.2   Review and disposition of non-conforming product
The responsibility and authority to evaluate and disposition non-conforming products is defined in detail in QOP 009, Sections One, Two, and Three. Documented procedures give direction on how to control the various types of dispositions (rework, repair, accept, regrade, scrap, etc., see pg 210 or 412).

4.13.2.1   AS9000 Material Review Authority:
Notwithstanding the requirements of 4.13.2, the supplier shall not use dispositions of use-as-is or repair, unless specifically authorized by the customer, if (1) the product is produced to customer design, or (2) the nonconformity results in a departure from the contract requirements.
This provision is incorporated in QOP 009, Section Two, paragraph 3.01.4.2 and 3.01.4.3 (pg 210 or 412).

4.13.2.2   AS9000 Regrading Material:
Product dispositioned for regrade requires a change in product identification to preclude the product's original use. Adequate test reports and certifications shall reflect the regrading. This provision is incorporated in QOP 009, Section Two, paragraph 3.01.4.7, step 2 (pg 214 or 415).

4.13.2.3 AS9000 Scrap Material:

Product dispositioned for scrap shall be conspicuously and permanently marked until physically rendered unsuitable for use in completed products. This provision is incorporated in QOP 009, Section Two, paragraph 3.01.4.5.

4.13.2.4 AS9000 Notification:

The supplier's system shall provide for timely reporting of nonconformances that may affect product already delivered. This provision is incorporated in QOP 009, Section One, paragraph 3.01b (pg 205 or 408).

4.13.3 Customer Complaints, Product Return, and Customer Satisfaction

(8.2.1, 8.3, 8.4, 8.5)

The handling and controlling of customer complaints, product returns, and customer satisfaction reporting is assigned to the Quality Department and Management Representative. They shall control the processes in accordance with QOP 009, Section Three.

4.14 CORRECTIVE AND PREVENTIVE ACTION

4.14.1 General

(4.2.4, 8.3, 8.4, 8.5.1, 8.5.2, 8.5.3)

This process-approach quality management system implements corrective and preventive action as a major force in maintaining continuous improvement in all the core departmental responsibilities. Therefore, the implementation process is procedurally integrated with all other procedures of the quality management system in order to ensure effectiveness of the operating system. Enforcement details are defined in QOP 009, Section Two, paragraph 3.01.5 (pg 214 or 415).

4.14.2 Corrective action

(8.4, 8.5.2)

a) Customer complaints, reports, and product returns are handled in accordance with QOP 009, Section Three (pg 217 or 417);

b) Cause determination of non-conformities in product, process, or the quality system is dependent on the type of non-conformance reported as required under the provisions of QOP 009, Section One. Documentation of the cause(s) of non-conformity is handled through the application of Form A-006 in accordance with QOP 009, Section Two, paragraph 3.01.2 (pg 208 or 410).

c) The MRB authority is responsible for determining corrective/preventive action as part of the disposition process stated under QOP 009, Section Two, paragraph 3.01.3 and document it on Form A-006 (pg 208 or 411);

d) Follow-up of corrective/preventive action implementation is handled by the Management Representative in accordance with QOP 009, Section Two, paragraph 3.01.6 (pg 215 or 415).

4.14.3 Preventive action (8.4, 8.5.3)

Preventive action requires the same method of investigation and cause determination as corrective action does. See the requirements above in clause paragraph 4.14.2.

4.15 HANDLING, STORAGE, PACKAGING, PRESERVATION, AND DELIVERY

4.15.1 General (7.5.5)

In a process-approach quality management system, the documented procedures dealing with defining processing related activities must put emphasis on preventive quality such as product handling and preservation requirements. By providing adequate instructions for these efforts along with other subjective requirements (workmanship), the final product will be made ready for packaging and delivery to customers or to storage at the end of final inspection, without additional and separate work instruction procedures and process loading requirements. QOP 001, Section Two, paragraph 3.03, steps 2 and 3 (pg 48 or 302) incorporate this requirement for the Process Engineer to do at the time when he/she develops the product's processing requirement procedures. At Final Inspection, the Quality Inspector is required to verify and document the results of these activities prior to releasing the product for delivery to the customer or to storage. (See QOP 006, Section Four, paragraphs 3.02 and 3.03, pg 163 or 381)

4.15.2 Handling (7.5.5)

Throughout the entire processing cycle from the receipt of products through production and delivery, the Job Traveler and/or Operation Sheet is providing instructions for product handling.

4.15.3 Storage (7.5.5)

Contract related raw materials or the finished goods are stored either in temporary holding areas or in other suitable places to enforce control over them. The raw materials are tagged with Form A-027. The finished goods are tagged with Form A-013. These tags indicate inspection approval to receive or release the product for further processing in accordance with instructions stated in the Job Traveler.

Product preservation shall be a planned requirement as stated in QOP 001, Section Two, paragraph 3.03, step 2 (pg 48 or 302).

Non-conforming or unidentified products are not permitted to be stored together with conforming products, as appropriate.

Semi-finished or in-process related products may be held or stored anywhere, provided that the Job Traveler and/or the process related documents are together with the product. Indication of the last inspection status is mandatory.

Shelf-life controlled products (chemical compounds, surfactants, etc.) are controlled according to EPA regulations as defined in the EPA standards.

4.15.4 Packaging (7.5.5)

Packaging of products for delivery to subcontractors (Suppliers) and for shipment to customers is maintained to best commercial practice or as otherwise stated in the contract (PO) and/or Form A-001 (Purchase Order Review Sheet).

4.15.5 Preservation (7.5.5)

Preservation methods are defined as required in the Job Traveler and/or Operation Sheet in accordance with QOP 001, Section Two, paragraph 3.03, step 2, as required.

Mixing of different products or different part numbers in accomplishing the preservation process is not allowed unless the segregation process is also defined in the Job Traveler and/or Operation Sheet.

All foreign objects (chips, shredding, dirt, oil, etc.) shall be eliminated from the finished products according to the work instructions stated in the Job Traveler and/or Operation Sheet, as required.**(AS9000 paragraph 4.9j above)**

4.15.6 Delivery (7.5.1, 7.5.5)

Product delivery is carried out in accordance with QOP 006, Section Four, paragraph 3.03. When it is a contractual requirement, the protection of the quality of the product during delivery to destination shall be complied with.

4.16 CONTROL OF QUALITY RECORDS (4.2.4)

Control of the quality documents and records from contract activities through the delivery of the product is maintained in accordance with documented procedures as stated in QOP 002, Sections One, Two, and <u>Three</u> (pg 60 or 311). Additionally, each QOP has specific instruction as to the control of process related records.

4.16.1 AS9000 Record Availability:
Records shall be readily available for review by the customer or regulatory agencies. This provision in incorporated under QOP 002, Section Three, paragraph 3.0 (pg 328).

4.17   **INTERNAL QUALITY AUDITS**   4.2.4, 8.1, 8.2.2, 8.2.3, 8.4

Internal quality audits are planned, carried out, and documented in accordance with the procedures stated in QOP 010. Accordingly, when required by assessment, corrective action and follow-up are implemented with those responsible in order to maintain the effective operation of the Quality Management System. In addition, process control surveillance is also implemented in accordance with QOP 006, Section Three (pg 151 or 374), to effect timely isolation and corrective action of process related non-conformance. Through the analysis of data and communication of the audit results, upper management determines the state of the quality system and initiates additional improvement, as required. Corrective/preventive action is implemented in accordance with the provisions of QOP 010, paragraph 3.03.3 (pg 231 or 426).

4.18   **TRAINING**   6.2.1, 6.2.2

This process-approach quality management system considers the training of all employees to be the most important tool in understanding (awareness) and maintaining product quality objectives in all areas of work activities. Accordingly, management has developed and implemented documented procedures to maintain effective training and the evaluation of it in line with individual needs. For each Quality Operating Procedure, there is a corresponding Training Metrics instituted for training purposes and is administered by the Management Representative. Records of training activities are maintained on Form A-016 (pg 471), and the evaluation of training results is maintained on form A-039 (pg 498).

4.19   **SERVICING -- Minfor Incorporated is not engaged in servicing related activities. (AS9000 par. 4.19.1 does not apply)**

4.20   **STATISTICAL TECHNIQUES**

4.20.1   **Identification of need**   7.5.1, 8.1, 8.2.3, 8.2.4, 8.4

Standard control charts (uncontrolled forms) on which to record variable or attribute data are used in the production areas, as required, to determine process variations. Sampling plans (Mil-Std-105 and/or ANSI Z1.4-1993) are applied in accordance with the contract provisions. Other statistical tools such as Pareto Chart, Cause-and-Effect Diagrams, Tree Diagrams, Failure Mode Effect Analysis (FMEA), etc., are also used as required by customers, or as determined by the Management Representative.

4.20.2   Procedure

The Quality Engineer determines during Purchase Order review the contractual need for Statistical Process Control Charts and identifies this requirement on Form A-001 (Purchase Order Review Sheet). The Process Engineer picks up this information from Form A-001 (see QOP 001, Section Two, paragraph 3.02) and enters the requirement in the Job Traveler. The process owner follows the noted information in the Job Traveler and picks up the control charts from the Quality Department. He/she enters the measurement data on the control chart and monitors the work result data between the upper and lower level tolerance boundaries and adjusts the equipment when the trend becomes unfavorable. The control chart becomes part of the Inspection Product Folder and may be submitted to the customer with other inspection data at time of product delivery.

The Management Representative may apply any statistical tools as part of his/her responsibility in determining root causes of problems in the process of implementing corrective actions to maintain continual improvements.

Inspection personnel may use sampling statistics regularly when permitted by the quality provisions of the contract (PO). **(AS9000, par. 4.20.3 apply, as required)**

4.20.3   AS9000 Sampling Inspection:

When the supplier uses sampling inspection as a means of product acceptance, the plan shall be statistically valid and appropriate for use. The plan shall preclude the acceptance of known defectives in the lot. This requirement is incorporated as a standard operating procedure.

End of Document

BOOK SECTION THREE

QUALITY OPERATING PROCEDURES
PROCEDURES MANUAL

OVERVIEW TO THE PROCEDURES MANUAL

The Quality Operating Procedures (QOPs) are stand-alone documents under their individual number, embodying one or more sections. These sections are known as **quality objective procedures.** Each QOP is designed to include specific process instructions and performance procedures relating to the type of requirement it needs to control. Some of the QOPs contain several interrelated requirements. These requirements are divided into dedicated sections within the QOP. Example: QOP 006 has seven sections. Each section is defining a specific type of inspection requirement, **the quality objective requirement**. QOP 009 has three sections. Each section defines how to control non-conforming products and interrelated matters. QOP 002 has also three sections, and each section defines how to control the system's documents. Other QOPs define how to control a single requirement, and for procedural reasons, they couldn't be combined with any other QOP. 'Stand-alone' means that each QOP has its own control mechanism (page plate) to take care of its own revision, page number, approval, and other internal requirements.

The control mechanism is uniformly applied to all the twelve operating procedures. Collectively, these Quality Operating Procedures are also controlled through a Master List in order to maintain the 'Procedures Manual.' The 'Master List' is located two pages behind this page. These procedures collectively control all the process and quality requirements needed to maintain an effective process-approach system in a manufacturing shop environment. From the beginning to the end each procedure maintains process flow in line as to how assignments are carried out in a real work environment. Only experience is needed to carry out each identified assignment in order to effectively fulfill each requirement. The method of how each assignment is carried out is dependent on the experience level of the process owner. That is why these procedures are universally suitable in a manufacturing work environment with only a few employees. For each procedure there is an equivalent Training Metrics provided to better understand product quality requirements. This explains the fact that in a small shop environment several processes may be performed by one individual. While it is performed, he/she bears the job title and the responsibility for that process assignment.

Note that page numbers have been added next to the references in order to facilitate easy navigation between documents.

CONTROLLED DOCUMENTS

QUALITY OPERATING PROCEDURES

CONTENTS   **Page**

MINFOR INCORPORATED		Iss dt. 1/1/99	Rev dt. 9/3/99		Pg.
QUALITY OPERATING PROCEDURE		Sign AG	Rev # 3		1
REQUIREMENT: QUALITY SYSTEM MANUAL SEC: 4.01 – 4.20					
SUBJECT: MASTER LIST					
QOP (all) DOCUMENT REQ'D: QOP 001 TROUGH QOP 012					

QOP	TITLE	REV	DATE	STATUS
001	Contract Review	0	1/1/99	New
	Sec 1 Control of RFQ			
	Sec 2 Purchase Order Review			
	Sec 3 Verbal Purchase Orders			
	Sec 4 Amendment to Contract			
002	Document and Data Control	0	1/1/99	New
	Sec 1 Approval and Issue			
	Sec 2 Document and Data Changes			
	Sec 3 Handling and Retention			
003	Control of Purchases (Internal)	0	1/1/99	New
	Sec 1 Approval and Issue of Purchase Orders			
	Sec 2 Evaluation of Subcontractors			
004	Control of Customer Supplied Product	0	1/1/99	New
005	Product Identification and Traceability	1	9/2/99	Revised
006	Inspection and Test Control	1	9/3/99	Revised
	Sec 1 First Piece Inspection			
	Sec 2 First Article Inspection			
	Sec 3 In-Process Inspection			
	Sec 4 Final Inspection			
	Sec 5 Receiving Inspection			
	Sec 6 Customer Source Inspection			
	Sec 7 Source Inspection at Subcontractors			
007	Control of Inspection, Measuring and Test Equipment	0	1/1/99	New
008	Inspection and Test Status	0	1/1/99	New
009	Control of Non-conforming Products	0	1/1/99	New
	Sec 1 Reporting Process			
	Sec 2 Disposition and Follow-up Process			
	Sec 3 Customer Complaints and Product Return			
010	Internal Quality Audits	0	1/1/99	New
011	Management Review	0	1/1/99	New
012	*Performance Metrics, Processing Control*	*0*	*7/16/99*	*New*
	Sec 1 Processing Control (Put-Through)			
	Sec 2 Handling Non-conformance (Put-Through)			
	Sec 3 Developmental Process Control			

MINFOR INCORPORATED	Iss dt. 1/1/99	Rev dt. 9/3//99	Pg.
QUALITY OPERATING PROCEDURE	Sign AG	Rev # 3	2
REQUIREMENT: QUALITY SYSTEM MANUAL SEC: 4.1 – 4.20			
SUBJECT: MASTER LIST			
QOP (all) |DOCUMENT REQ'D: QOP 001 through QOP 012			

REVISION HISTORY

Rev Date	Rev No	Description	Approval
7/16/99	1	Added QOP 012	AG
9/2/99	2	QOP 005 has been completely rewritten to define part, batch, and lot identification marking control	AG
9/3/99	3	In QOP 006, under Section Five, par. 3.02.2, item 4 has been added;	AG

OVERVIEW TO QOP 001
(CONTRACT REVIEW)

Fulfilling customer satisfaction starts at the earliest phase when we establish a relationship with our customers. Most of us don't realize this fact until later when we find out that the customer's product specification often is riddled with mistakes. The customer doesn't pay for downtime that you incur as a result of these mistakes. A process-approach quality management system starts right here, ensuring that you work with the customer up front, at the quoting phase of the job, so that you don't quote a job with all its mistakes already in it. Section One of this QOP deals with the quotation process, the RFQ.

Section Two of this QOP deals with controlling the Purchase Order Review, giving you a second opportunity to make sure that you take care of those mistakes that either you or the customer may have made between the time the RFQ was reviewed and the Purchase Order issued. Although many times Quality is not part of the job quoting process, this time it takes a serious review of the issued Purchase Order. At this time Quality determines what is really needed to make sure that product quality verification is planned for during the entire build cycle of the product and that everybody becomes aware of it up front so that they also can plan for it. As at this time Quality summarizes these requirements and issues them on Form A-001. Now preventive quality starts rolling into the processing documents issued by Engineering.

Section Three controls how verbal agreements should be handled in order that we avoid confrontation with the customer later over what was said versus what should have been said. Again, preventive quality plays a major role here to avoid problems later.

Section Four takes care of those issues that have everything to do with customer initiated document changes. It tells you how to control and implement them in order that customer satisfaction is maintained for both the organization and the buyer.

MINFOR INCORPORATED	Iss dt. 1/1/99	Rev dt. 1/1/99	Pg.
QUALITY OPERATING PROCEDURE	Sign AG	Rev # 0	1/15
REQUIREMENT: QUALITY SYSTEM MANUAL SEC: 4.3			
SUBJECT: CONTRACT REVIEW			
QOP 001 DOCUMENT REQ'D: QOP 001 Sec: 1,2,3, and 4			

QUALITY OPERATING PROCEDURE   001

CONTRACT REVIEW

Form A-003 Rev 0 1999

MINFOR INCORPORATED	Iss dt. 1/1/99	Rev dt. 1/1/99	Pg.
QUALITY OPERATING PROCEDURE	Sign AG	Rev # 0	2/15
REQUIREMENT: QUALITY SYSTEM MANUAL SEC: 4.3			
SUBJECT: CONTRACT REVIEW			
QOP 001 | DOCUMENT REQ'D: QOP 001 Sec: 1,2,3, and 4			

QUALITY OPERATING PROCEDURE   001

CONTRACT REVIEW

Contents   **Page**

Form A-003 Rev 0 1999

MINFOR INCORPORATED	Iss dt. 1/1/99	Rev dt. 1/1/99	Pg.
QUALITY OPERATING PROCEDURE	Sign AG	Rev # 0	3/15
REQUIREMENT: QUALITY SYSTEM MANUAL SEC: 4.3			
SUBJECT: CONTRACT REVIEW			
QOP 001 DOCUMENT REQ'D: This page			

REVISION HISTORY

Rev Date	Rev No	Description	Approval

MINFOR INCORPORATED	Iss dt. 1/1/99	Rev dt. 1/1/99		Pg.
QUALITY OPERATING PROCEDURE	Sign AG	Rev # 0		4/15
REQUIREMENT: QUALITY SYSTEM MANUAL SEC: 4.3				
SUBJECT: REQUEST FOR QUOTE				
QOP 001 DOCUMENT REQ'D: RFQ, Technical Data Package, Forms A-004, A-005				

SECTION ONE

1.0   PURPOSE

Evaluate "Request For Quote" (RFQ) in order to determine cost and capability.

2.0   APPLICATION

This procedure shall apply to accepted RFQs. Turned down RFQs shall not be controlled.

3.0   PROCEDURE

3.01   The Administrative Secretary shall: With the receipt of an RFQ do:
1. Log in on form A-005 (RFQ, PO, and Amendment Log) the company's name, the RFQ number (P/N) and date.

   Note 1. you will be handling three types of document entries on this same form, (RFQ, PO, and Amendment documents). Therefore, record the data for each on a separate sheet of the same form (A-005) to avoid confusion. Make sure you add the appropriate title to each form.
2. Make up the RFQ folder. Title it. Place in it the RFQ documents and form A-004.
3. Turn this folder over to the Contract Administrator.

3.02   The Contract Administrator shall: Review the RFQ content information and determine whether the quote request is compatible with the company's interest. If not, decline the RFQ and return the folder to the secretary.

   Note 2. Review the product specification of the RFQ. Look for errors, incomplete data, and requirements. Work out any discrepancies now, as the customer will not pay for downtime later.

3.03   The Administrative Secretary shall: Log out the RFQ on form A-005 (pg 460) by indicating the "turned down" status. Do this in line with the original entry. Then, notify the customer and file the RFQ folder, as required.

3.04   The Contract Administrator shall: On acceptance of an RFQ:
1. Determine the quote process. (Involve Engineering, Planning, Purchasing, Manufacturing, and Quality, as required.)

MINFOR INCORPORATED	Iss dt. 1/1/99	Rev dt. 1/1/99		Pg.
QUALITY OPERATING PROCEDURE	Sign AG	Rev # 0		5/15
REQUIREMENT: QUALITY SYSTEM MANUAL SEC: 4.3				
SUBJECT: REQUEST FOR QUOTE				
QOP 001 DOCUMENT REQ'D: RFQ, Technical Data Package, Forms A-004, A-005				

2. Do the quote on form A-004 (RFQ Worksheet).
3. Review it. Make adjustments as needed.
4. Place all documents back into the RFQ folder and return it to the secretary.

3.05   The Administrative Secretary shall:

1. Formalize the RFQ response on company letterhead and have it signed by the Contract Administrator.
2. Send this letter to the customer.
3. Log out on form A-005 this transaction in line with initial entry.
4. Place all those documents that relate to this transaction in the RFQ folder and file it for future record.

MINFOR INCORPORATED	Iss dt. 1/1/99	Rev dt. 1/1/99	Pg.
QUALITY OPERATING PROCEDURE	Sign AG	Rev # 0	6/15
REQUIREMENT: QUALITY SYSTEM MANUAL SEC: 4.3			
SUBJECT: Purchase Order Review and Processing Preparation			
QOP 001│DOCUMENT REQ'D: PO, Tender, Forms A-001, A-003, A-005, A-011, A-031			

SECTION TWO

1.0 PURPOSE

Ensure that the Purchase Order content information agrees with that of the tender and to determine and document quality requirements imposed in the PO.

2.0 APPLICATION

The purchase order requirements shall apply to all activities affecting documents and processes, during the entire processing cycle of the identified product.

3.0 PROCEDURE

3.01 The Administrative Secretary shall: On receipt of PO:

1. Log in the receipt of the PO on form A-005.
2. Review the content information and determine its accuracy against the tender. (The "tender" is the formal response sent to the customer earlier and is located in the RFQ folder.)
   a) If the content information of the PO agrees with that of the tender, accept it by stamping on it "Accepted." Then, sign and date it.
   b) If it doesn't agree with the tender, have the Contract Administrator resolve the differences with the customer. When that's done, accept the PO as in "a" above. Then, do the following:
3. Make up the PO folder. Put the customer's name and PO number on it.
4. Fill out form A-011 (Control of Customer Documents). Record on this form the customer sent document numbers and their revision levels. When that's done, attach it to the inside of the front cover of the PO folder.
5. Make copies of the PO and its attachment(s) now, as required.
   a) Send a copy to Engineering, Planning, and Quality.
   b) Send other copies as required. Then:
6. Fill out form A-031 (Issue and Traceability of Drawings). Record on this form the part number and revision level of the documents that you have distributed (new) and retrieved (obsolete). Dispose of the obsolete documents, as required. Stamp "Obsolete" on those retained.
   Note 3. From this day on update form A-011 and A-031 whenever the customer makes changes to the listed documents. Then, follow up with the distribution and retrieval process as before.

MINFOR INCORPORATED	Iss dt. 1/1/99	Rev dt. 1/1/99	Pg.
QUALITY OPERATING PROCEDURE	Sign AG	Rev # 0	7/15
REQUIREMENT: QUALITY SYSTEM MANUAL SEC: 4.3			
SUBJECT: Purchase Order Review and Processing Preparation			
QOP 001|DOCUMENT REQ'D: PO, Tender, Forms A-001, A-003, A-005, A-011, A-031			

7. Place the "Accepted" PO, attachment(s), and A-031 in the folder and file it. Put the tender back in the RFQ file.

8. Log out the PO activity on form A-005 by recording the "Acceptance" status.

Note 4. Repeat all the above steps for each customer PO. Multiple POs from the same customer require the same individual log-in, handling, and log-out.

3.02   The Quality Engineer shall:

1. Review the PO, Contracts provided, for the imposed quality requirements.

2. Transcribe the requirement(s) onto form A-001 (Purchase Order Review Sheet). Determine adequacy, especially for variable measurement of key characteristics. **(AS9000, paragraph 4.2.3b(1))**

Note 5. When no quality requirements are given, impose requirements on the form that are necessary to control the product during its entire processing cycle. (Don't forget identification marking, workmanship, and shipping requirements.)

3. Make two copies of this form. Send one copy to Engineering; the other to Contracts (file copy).

Note 6. Reissue of this form will not be required again on a repeated PO, unless the customer has amended the PO, impacting the product and/or quality.

4. Make up the Inspection Product Folder now and place all the documents inside. Maintain this folder on file during the entire processing cycle of the product and retain it as long as the internal and/or the customer's PO require it;

5. Plan and carry out your quality engineering assignments in order to meet all the controlling requirements for an effective product quality realization for the contracted product.

3.03   The Process Engineer shall:

1. Review the PO and related technical documents received from Contracts. Review also form A-001, received from Quality.

2. Plan the processing layout to meet the PO requirements. The quality requirements recorded on form A-001 are part of the processing layout, as stated. So are workmanship, handling, cleaning and preservation, and safety. If this order is new and you don't have the process capability developed for it, determine the need for process qualification and do it as defined under QOP 012, Section Three (Developmental Process, pg 449).

3. Prepare, review, and approve all the required documents needed for product realization. (Job Traveler, Operation Sheets, Software Program, etc.)

4. Maintain master documents in Engineering Folder(s).

5. Prepare Production Folder(s) as required. Ensure current revision levels in all the included documents.

MINFOR INCORPORATED	Iss dt. 1/1/99	Rev dt. 1/1/99		Pg.
QUALITY OPERATING PROCEDURE	Sign AG	Rev # 0		8/15
REQUIREMENT: QUALITY SYSTEM MANUAL SEC: 4.3				
SUBJECT: Purchase Order Review and Processing Preparation				
QOP 001|DOCUMENT REQ'D: PO, Tender, Forms A-001, A-003, A-005, A-011, A-031				

Note 7. Keep track of PO amendments. Don't release any Production Folder if it requires the implementation of PO amendments. Refer to Section Four of this QOP for details as to how to do it. (See pg 307)

6. Follow up your planning regarding the readiness of special tooling, gages, and setup related equipment (fixtures, machine, etc.). Do this prior to releasing a Production Folder. Release the Production Folder(s) to Production Planning, as required, in line with production scheduling requirement. (See QOP 012, Sec One, par 3.01.1.3, pg 439)

3.04   The Contract Administrator shall:

1. Ensure central planning to prepare enactment of the customer's PO require-ments. Schedule meeting to set agenda. List action item(s) with commitment dates on form A-003 and issue it to the affected parties. Comply with QOP 012, Sec One, par 3.01.1.3, pg 339.

2. Coordinate the PO flow-down requirements with Production Planning, Engineering, Quality, and others. <u>Keep the customer informed on progress.</u>

Note 8. Do your project meetings regularly. Issue action item(s) on form A-003 with commitment dates routinely. Combine other business in your agenda, as required. Follow up on critical issues and <u>keep the customer informed.</u>

3.05   The Production Planner shall:

1. Release the Master Production Scheduling (MPS), as required;
2. Interact regularly with all departments in order to maintain schedule;
3. Issue the Production Folder in line with scheduling requirement.

MINFOR INCORPORATED	Iss dt. 1/1/99	Rev dt. 1/1/99	Pg.
QUALITY OPERATING PROCEDURE	Sign AG	Rev # 0	9/15
REQUIREMENT: QUALITY SYSTEM MANUAL SEC: 4.3			
SUBJECT: VERBAL PURCHASE ORDER			
QOP 001|DOCUMENT REQ'D: PO, Forms A-001, A-003, A-005, A-011, A-031			

SECTION THREE

1.0 PURPOSE

Initiate processing and quality planning, based on customer's verbal commitment, pending issuance of formal purchase order.

2.0 APPLICATION

When a verbal commitment from a customer has been formalized and he/she has issued a Purchase Order, the requirements thereof shall apply to all activities affecting processes and documents during the entire production cycle of the contracted product.

3.0 PROCEDURE

General: After a verbal agreement has been concluded with a customer, the Contract Administrator shall request from the customer the issuance of the formal Purchase Order in which the verbal commitments have been accurately reflected. He/she shall keep record of his negotiation on form A-003 because verbal agreements don't have RFQ records on file as precondition to a contract. After the receipt of the Purchase Order, the following shall be carried out:

3.01 The Administrative Secretary shall: On receipt of a PO:
1. Log in the receipt of it on form A-005. Then:
2. Turn the whole PO package over to the Contract Administrator.

3.02 The Contract Administrator shall:
1. Review the PO content information and determine whether it is in agreement with the verbally declared statements made between himself and the customer. (Refer to the recorded information previously made on form A-003)
   a) If it is in agreement, write on the front page "Accepted." Sign and date it. Then, turn the PO package over to the secretary.
   b) If it is not in agreement, work out the differences with the customer and ask him to resubmit the PO, containing the corrections.
   c) Repeat the review process until satisfied. Then, do as stated in "a" above.

MINFOR INCORPORATED	Iss dt. 1/1/99	Rev dt. 1/1/99		Pg.
QUALITY OPERATING PROCEDURE	Sign AG	Rev # 0		10/15
REQUIREMENT: QUALITY SYSTEM MANUAL SEC: 4.3				
SUBJECT: VERBAL PURCHASE ORDER				
QOP 001|DOCUMENT REQ'D: PO, Forms A-001, A-003, A-005, A-011, A-031				

3.03 The Administrative Secretary shall:

1. Make up the PO folder now. Put the customer's name and PO number on it.
2. Fill out form A-011 (Control of Customer Documents). Record on this form the customer sent document numbers and their revision levels. When that's done, attach it to the inside of the front cover of the folder.
3. Make copies of the PO and its attachment(s) now, as required.
   a) Send copy to Engineering, Planning, and Quality.
   b) Send other copies as required.
4. Fill out Form A-031 (Issue and Traceability of Drawings). Record on this form the part number and revision level of the documents that you have distributed (new) and retrieved (obsolete). Dispose the obsolete documents, as required. Stamp "Obsolete" on those retained.

Note 9. From this day on update form A-011 and A-031 whenever the customer makes changes to the listed documents. Then, follow up with the distribution and retrieval process as before.

5. Place the "Accepted" PO, attachment(s) and A-031 in the folder and file it.
6. Log out the PO activity on form A-005 by recording the "Acceptance" status.

Note 10. Repeat all the above steps for each customer PO. Multiple POs from the same customer require the same individual log-in and handling.

3.04 The Quality Engineer shall:

1. Review the PO, Contracts provided, for the imposed quality requirements.
2. Transcribe the requirement(s) onto form A-001 (Purchase Order Review Sheet). Determine adequacy. Add other requirements as needed.

Note 11. If there are no quality requirements stipulated in the customer's PO, record those requirements on the form that are necessary to control the product during its entire processing cycle.(Don't forget identification marking, workmanship, and shipping requirements.)

3. Make two copies of this form. Send one copy to Engineering; the other to Contracts (file copy).

Note 12. Reissue of this form will not be required again on a repeated PO, unless the customer has amended the PO, impacting the product.

4. Make up the Inspection Product Folder and place all the documents inside. Maintain this folder on file during the entire processing cycle of the product and retain it as long as the internal and/or the customer's PO require it.

3.05 The Process Engineer shall:

1. Review the PO and related technical documents received from Contracts. Review also form A-001, received from Quality.

MINFOR INCORPORATED	Iss dt. 1/1/99	Rev dt. 1/1/99	Pg.
QUALITY OPERATING PROCEDURE	Sign AG	Rev # 0	11/15
REQUIREMENT: QUALITY SYSTEM MANUAL SEC: 4.3			
SUBJECT: VERBAL PURCHASE ORDER			
QOP 001|DOCUMENT REQ'D: PO, Forms A-001, A-003, A-005, A-011, A-031			

2. Plan the processing layout to meet the PO requirements. The quality requirements recorded on form A-001 are part of the processing layout, as stated. So are workmanship, handling, cleaning and preservation, and safety.

3. Prepare and approve all the required documents needed for product realization. (Job Traveler, Operation Sheets, Software Program, etc.) Follow the provisions of QOP 012 (Performance Standard) in the preparation of Job Travelers.

4. Keep master documents in Engineering Folder(s).

5. Prepare Production Folder(s) as required. Ensure current revision levels in all the included documents.

Note 13. Keep track of PO amendments. Don't release any Production Folder if it requires the implementation of PO amendments. Refer to Section Four of this QOP for details as to how to do it.

6. Follow up your planning regarding the readiness of special tooling, gages, and setup related equipment (fixtures). Do this prior to releasing a Production Folder. See QOP 012, Sec One, par 3.02.3, pg 440, in order to comply with the performance standard.

7. Release the Production Folder(s) in line with Production Planning requirement.

3.06   The Contract Administrator shall:

1. Ensure that central planning prepares enactment of the customer's PO requirements. Schedule meeting to set agenda. List action item(s) on form A-003 and issue it to the participants.

2. Coordinate with Production Planning, Engineering, Quality, and others the PO flow-down requirements. Observe "Note 8" under Section Two, par 3.04 of this QOP (pg 303).

MINFOR INCORPORATED	Iss dt. 1/1/99	Rev dt. 1/1/99	Pg.
QUALITY OPERATING PROCEDURE	Sign AG	Rev # 0	12/15
REQUIREMENT: QUALITY SYSTEM MANUAL SEC: 4.3.3			
SUBJECT: AMENDMENT TO CONTRACT AND PROCEDURES			
QOP 001│DOCUMENT REQ'D: PO,RFQ, Forms A-001,A-003,A-004,A-005,A-006, A-009, A-011, A-031			

SECTION FOUR

1.0 PURPOSE

Establish documented procedures for implementing contract amendments and the resulting internal document changes.

2.0 APPLICATION

Contract amendments and the resulting changes shall apply to all the affected product(s) and document(s). Changes to internal documents shall be implemented by the use of form A-009 as stated in QOP 002, Sec Two, par 3.03.1, pg 318, as required.**(AS9000, par 4.5.3.1)**

3.0 PROCEDURE

3.01   The Administrative Secretary shall:
1. Log in the receipt of the PO amendment on form A-005.
2. Review amendment for critical action items. Send out "Alerts" (stop orders), as required to Production Planning, Engineering, and Quality on form A-003 (or E-mail).
3. Prepare the Amendment Folder and attach it to the existing PO and RFQ folders. (Verbally based POs don't have RFQs)
4. Forward this package to the Contract Administrator.

Note 14. Do the above 4 steps routinely for each follow-on amendment(s) for each customer PO.

3.02   The Contract Administrator shall:
1. Review the amendment against the customer's PO and determine the cost impact to processing and product quality.

Note 15. In some cases a new PO may have already been issued, embodying the amendment requirement. Remember that any pre-negotiated amendment still must go through the implementation process.
2. If there is no impact determined, record this information on the front page of the amendment document. Sign and date it. Return the whole package to the secretary now.

MINFOR INCORPORATED	Iss dt. 1/1/99	Rev dt. 1/1/99	Pg.
QUALITY OPERATING PROCEDURE	Sign AG	Rev # 0	13/15
REQUIREMENT: QUALITY SYSTEM MANUAL SEC: 4.3.3			
SUBJECT: AMENDMENT TO CONTRACT AND PROCEDURES			
QOP 001|DOCUMENT REQ'D: PO,RFQ, Forms A-001,A-003,A-004,A-005,A-006, A-009, A-011, A-031			

3.03 The Administrative Secretary shall:
1. Stamp on the front page of the amendment document "Accepted." Sign and date it.
2. Implement the "paper change."
3. Send "information" copy of the amendment to Engineering, Quality and others, as required.
4. File the Amendment Folder. File the RFQ and PO Folders.

Note 16. An amendment may be filed with its PO, as internally determined.

5. Log out this activity on form A-005 (in line where it was logged in).

3.04   The Contract Administrator shall: When the amendment impacts processing and/or product quality, do the following:
1. Figure out the cost.
2. Update form A-004 (RFQ Worksheet) from the RFQ folder.

Note 17. Determining cost impact may require Engineering and Quality input.

3. Write on front of the Amendment document: "Impact to processing and product quality." Sign and date it. Clip form A-004 onto the amendment document. Pause! Then:
4. Plan enactment of the customer's amendment. Schedule meeting to set agenda. Note action item(s) on form A-003 and issue it to the affected parties when the meeting is over. (See QOP 012, Sec One, par 3.01.1.3, pg 439)
5. Return all the folders to the secretary now.

3.05   The Administrative Secretary shall:
1. Formalize cost on company letterhead (copy the added cost from form A-004) and have it signed by the Contract Administrator.
2. Send this letter to the Customer and wait for the reply. Put the folders on hold.

3.06   The Administrative Secretary shall: After the receipt of cost approval:
1. Stamp on the front of the amendment document "Accepted." Sign and date it.
2. Update form A-011 and A-031 regarding revisions to documents, if any.
3. Remove from circulation the obsolete documents and dispose of them, as required. Stamp "Obsolete" on those retained.

Note 18. Retain obsolete documents for background information, as required.

4. Make two copies of the amendment document and its attachment(s) now. Send one copy to Engineering and the other to Quality.
5. Put form A-004 back to the RFQ folder. Then, file the RFQ, PO, and Amendment folders. Now, log out the amendment activity on form A-005.

MINFOR INCORPORATED	Iss dt. 1/1/99	Rev dt. 1/1/99		Pg.
QUALITY OPERATING PROCEDURE	Sign AG	Rev # 0		14/15
REQUIREMENT: QUALITY SYSTEM MANUAL SEC: 4.3.3				
SUBJECT: AMENDMENT TO CONTRACT AND PROCEDURES				
QOP 001|DOCUMENT REQ'D: PO,RFQ, Forms A-001,A-003,A-004,A-005,A-006, A-009, A-011, A-031				

3.07 The Quality Engineer shall:

1. Review the amendment document and attachment(s) received from Contracts.
2. Determine Quality's action item(s).
3. If there is impact, reissue or revise form A-001, as required.
4. Make two copies of form A-001 now. Send one copy to Engineering and the other copy to Contracts (file copy).

Note 19. If there is no impact, the first issue of form A-001 stays in effect.

5. Put your documents on hold and contact Engineering. Set time and agenda. Work out details to implement the PO Amendment.

3.08 The Process Engineer shall:

1. Review the amendment document and attachment(s) received from Contracts, and also review form A-001 received from Quality.
2. Determine impact to processing and documents. Draw up plans accordingly.
3. Work together with the Quality Engineer as previously agreed (under 3.07).

3.09 The Process and Quality Engineers shall:

1. Draw up action items according to amendment requirement(s).
2. Work out the details during the project meeting scheduled by the Contract Administrator. (Noted under 3.04, item 4 above)

3.10 The Contract Administrator shall:

1. Hold project meeting as scheduled.
2. Discuss the amendment, impacting the product and processing, with Production Planning, Engineering, Quality, and others, as required.
3. Give action items to each department according to requirement(s). Formalize the action items on form A-003 with commitment dates and hand these out to the participants.
4. Set up the next meeting to review progress on previously made commitments.

Note 20. Do your project meetings regularly. Issue the action item list routinely, requiring commitment date. Combine other business in your agenda, as required. Follow up on critical issues and keep the customer informed.

3.11 The Process Engineer and Quality Engineers shall:

1. Implement all the action items affecting PO amendments. Interact in all areas affecting product, documents, and processing.
2. Carry out all internal document changes by the application of form A-009 in accordance with QOP 002, Sec Two, par 3.03.1 (pg 318) and Three, par 3.03c (pg 330). **Comply with AS9000 par. 4.5.3.1, as required.**

MINFOR INCORPORATED	Iss dt. 1/1/99	Rev dt. 1/1/99	Pg.
QUALITY OPERATING PROCEDURE	Sign AG	Rev # 0	15/15
REQUIREMENT: QUALITY SYSTEM MANUAL SEC: 4.3.3			
SUBJECT: AMENDMENT TO CONTRACT AND PROCEDURES			
QOP 001|DOCUMENT REQ'D: PO,RFQ, Forms A-001,A-003,A-004,A-005,A-006, A-009, A-011, A-031			

3. Carry out the handling of non-conforming products by the application of form A-006 in accordance with QOP 009, Sec One (pg 408) and Two (pg 410).
4. Observe performance requirements as stated in QOP 012, Sec Two, par 3.04, 3.05, and 3.06 (pg 445, 446).

4.00 Contracts

Implementation of customer initiated **verbal** changes into the production processes shall be denied. Amendments shall be formally submitted by the customers.

OVERVIEW TO QOP 002
(CONTROL OF DOCUMENT AND DATA)

The control of documents and data is the prerequisite required to manage an effective quality system. Without it, you have no system. This QOP 002 has three sections in order to ensure that we manage the documents and data in line with a process-approach quality management system. We give you instructions on how to design and control the documents, comprising the quality system's procedures. We give you instructions on how to carry out revisions, how to identify documents and how to retain them. In fact, this process-approach system is exactly constructed, implemented, and maintained the way it tells you how you should do yours. What is really good here is that whatever you need to know regarding document and data control is in one procedure. Other QOPs often refer you here to do your thing according to the stated requirements laid out for you.

There are so many steps involved here that I could easily write pages to tell you about them, but I won't. Rather, I'll do it where it counts – inside the procedure. This QOP is one of the longest procedures, along with QOP 006 (inspection control, pg 357) and QOP 009 (non-conformance control, pg 403), to make sure that the documented system is effectively laid out, inter-linked, coordinated, and maintained for all the quality system's documents in the process-approach system.

MINFOR INCORPORATED	Iss dt. 1/1/99	Rev dt. 1/1/99		Pg
QUALITY OPERATING PROCEDURE	Sign	Rev #	0	1/23
REQUIREMENT: QUALITY SYSTEM MANUAL SEC: 4.5				
SUBJECT: CONTROL OF DOCUMENT AND DATA				
QOP 002 DOCUMENT REQ'D: This QOP 002, Sec: 1, 2, and 3				

QUALITY OPERATING PROCEDURE   002

CONTROL OF DOCUMENTS AND DATA

Contents   **Page**

Form A-003 Rev 0 1999

MINFOR INCORPORATED	Iss dt. 1/1/99	Rev dt. 1/1/99		Pg
QUALITY OPERATING PROCEDURE	Sign	Rev #	0	2/23
REQUIREMENT: QUALITY SYSTEM MANUAL SEC: 4.5				
SUBJECT: CONTROL OF DOCUMENT AND DATA				
QOP 002 DOCUMENT REQ'D: This QOP 002, Sec: 1, 2, and 3				

QUALITY OPERATING PROCEDURE   002

CONTROL OF DOCUMENTS AND DATA

Contents   **Page**

Form A-003 Rev 0 1999

MINFOR INCORPORATED	Iss dt. 1/1/99	Rev dt. 1/1/99		Pg.
QUALITY OPERATING PROCEDURE	Sign	Rev #	0	3/23
REQUIREMENT: QUALITY SYSTEM MANUAL SEC: 4.5				
SUBJECT: CONTROL OF DOCUMENT AND DATA				
QOP 002 | DOCUMENT REQ'D: This QOP 002, Sec: 1, 2, and 3				

REVISION HISTORY

Rev Date	Rev No	Description	Approval

MINFOR INCORPORATED	Iss dt. 1/1/99	Rev dt. 1/1/99	Pg.
QUALITY OPERATING PROCEDURE	Sign	Rev # 0	4/23
REQUIREMENT: QUALITY SYSTEM MANUAL SEC: 4.5.2			
SUBJECT: DOCUMENT AND DATA APPROVAL AND ISSUE			
QOP 002 DOCUMENT REQ'D: This QOP 002, sec. 1, 2, and 3			

SECTION ONE

1.0   PURPOSE

To control the planning, approval and issuance of the quality system's documents.

2.0   APPLICATION

Planning, approval and issuance of the quality system's documents and data shall apply to all the identified documents within the Quality System Manual, the Quality Operating Procedures and the Forms. **(SAE AS9000 par 4.5.3.1)**

3.0   PROCEDURE

General:

The Management Representative shall be responsible to establish, implement, and maintain the Quality Management System's documented procedures and assign responsibilities under delegated authority. He/she shall enforce the following:

3.01   Establish, Implement, and Maintain the Quality System Manual (QSM)

1. The QSM shall follow the numerical sequencing of ISO 9002/1994. Sections that don't apply to the business shall be omitted and explained as to why. Additional requirements shall be sequenced in the subparagraphs. The order in sequencing the numerical and/or alphabetical line items shall be maintained and the lower tier documents referenced, as applicable.
2. The text of the QSM shall be written in policy statements and in the imperative ("shall") in order to express the mandatory nature of requirements. Process instructions in the manual shall be added only if there is no separate procedure available.
3. Each page shall have a header (page plate), indicating the applicable data – issue date, approval, revision status and date, page number, and other linkage information.
4. The QSM shall be reviewed and approved prior to release. Distribution shall be controlled by the application of Form A-010.
5. Amendments to the QSM shall be recorded on its "Revision History" sheet in accordance with Section Two of this QOP. Amendments shall be highlighted. Obsolete manuals shall be recalled and disposed, and revised manuals promptly issued.
6. Copies of the approved QSM shall be internally issued as needed and given to the customers as requested. All the controlled copies shall be maintained.

316   ISO 9001: for SMALL BUSINESSES

MINFOR INCORPORATED	Iss dt. 1/1/99	Rev dt. 1/1/99	Pg.
QUALITY OPERATING PROCEDURE	Sign	Rev # 0	5/23
REQUIREMENT: QUALITY SYSTEM MANUAL SEC: 4.5.2			
SUBJECT: DOCUMENT AND DATA APPROVAL AND ISSUE			
QOP 002 |DOCUMENT REQ'D: This QOP 002, sec. 1, 2, and 3			

3.02   Establish, Implement, and Maintain the Quality Operating Procedures

1. Each page in the Quality Operating Procedures (QOPs) shall have a header (page plate) within which the applicable data: issue date, approval, revision status and date, page number and other linkage information shall be indicated.
2. The QOPs shall be constructed as work processes are sequenced. Paragraph sequencing shall be maintained in order of logical continuity.
3. Similar tasks shall be combined under one procedure. They shall be given title and a separate section within the procedure. (Example: QOP 006 on pg 357)
4. Linkage between the QSM and the QOPs shall be maintained by reference.
5. The QOPs and associated forms shall be reviewed and approved prior to release. They shall be combined into "Procedures Manual," then issued. Distribution shall be controlled on form A-010 (pg 465). The QOP's are proprietary documents. Customers may not receive them.
6. Change to a QOP shall be recorded and controlled on its "Revision History" page. Each page of the impacted QOP shall be updated. With each change the Master List that controls the procedures collectively shall be updated. Changes in the text shall be uniformly *highlighted*.
7. Obsolete sections from the QOPs shall be removed and disposed, and replaced by the changed sections which shall be reviewed and approved prior to release.

3.03   Establish, Implement, and Maintain the system's Forms

1. Each form shall be suitably designed and titled to represent the applicable record keeping requirement of the work assignment. Signature and date lines shall be provided for those responsible to acknowledge and/or review the work results.
2. Each form shall have its own number and shall indicate its revision status and date.
3. The form(s) shall be referenced in the procedure that describes the requirement for recording work results. Note that a form is part of the respective procedure but it has been split off from the procedure for ease of handling and control. Approval of a procedure shall approve the inter-linked form(s).
4. There shall be a Master List designed summarizing all the controlled forms. It shall have its own Revision Sheet to record changes made to the form(s), (pg 453, 454).
5. The original forms shall be kept in sequential order as noted in the Master List.
6. After the forms have been approved and issued, they shall be speedily distributed. Obsolete forms shall be removed from circulation and disposed of as appropriate.

3.04   Assignment: The Quality Management System's documented procedures shall be controlled and maintained under delegated authority assigned by the Management Representative to the Quality Manager. (Often, both are one and the same.)

MINFOR INCORPORATED	Iss dt. 1/1/99	Rev dt. 1/1/99	Pg.
QUALITY OPERATING PROCEDURE	Sign	Rev # 0	6/23
REQUIREMENT: QUALITY SYSTEM MANUAL SEC: 4.5.3			
SUBJECT: DOCUMENT AND DATA CHANGES			
QOP 002 |DOCUMENT REQ'D: QOP 002, sec. 1, 2 and 3; Forms A-009, A-010			

SECTION TWO

1.0   PURPOSE

To maintain control over document and data changes.

2.0   APPLICATION

The control over document and data changes shall apply to all the identified documents of the quality management system.

3.0   PROCEDURE

General Requirement

Errors, mistakes, and misleading instructions in documents are adversely affecting performance, processing efficiency, and product quality. Therefore, this process-approach system provides the tools to remove them when and where they occur. Process owners in all departments are required to report them to their immediate supervisors as soon as they discover them. The supervisors in turn are required to determine what immediate impact they have on the ongoing process, foremost on product quality, and take the necessary steps to eliminate them. QOP 012 (the Performance Standard), Section Two, paragraph 3.04 (pg 445) gives the required guidance as to how to go about curing these document deficiencies.

3.01   Customer Controlled Documents

Changes to customer controlled documents and data shall be approved and released only by the customer(s) through the issuance of amendment(s) against the already placed Purchase Order(s). Internal implementation of Purchase Order amendments shall be carried out as defined in QOP 001, Section Four (pg 307) and in this procedure.
Note 1. Implementation of customer initiated **verbal** changes into the processing operations shall be prohibited without procedural documentation as stated above.

3.02   Internally Controlled Processing Related Documents

Changes to processing related documents and data shall be carried out by the application of form A-009 (pg 464) as stated below under paragraph 3.03.1 and 3.03.2 "Implementation." These changes may or may not impact the quality system's procedures. (See paragraph 4.0 below.) When they do, they are known as <u>indirect</u> changes.

3.03   Types of Processing Related Changes

MINFOR INCORPORATED	Iss dt. 1/1/99	Rev dt. 1/1/99	Pg.
QUALITY OPERATING PROCEDURE	Sign	Rev # 0	7/23
REQUIREMENT: QUALITY SYSTEM MANUAL SEC: 4.5.3			
SUBJECT: DOCUMENT AND DATA CHANGES			
QOP 002 |DOCUMENT REQ'D: QOP 002, sec. 1, 2 and 3; Forms A-009, A-010			

3.03.1 Immediate and Provisional Changes (Needed to allow uninterrupted processing.)

Implementation: (Obtain form A-009 from page 464 in order to understand the process.)

1. The Process Engineer fills out the upper portion of form A-009, then fills out items 1, 2, and 3 and submits it to the supervisor for review. The supervisor reviews the change. Signs and dates "Reviewed By." Now, the Process Engineer forwards this document to the Quality Engineer;

2. The Quality Engineer reviews the proposed change(s). Upon concurrence, he/she signs and dates "Accepted By." If the proposed change is unsatisfactory, he/she sends it back to the Process Engineer for correction. Processes in item "1" shall be repeated;

3. After the correction has been made, reviewed, and accepted, the Quality and the Process Engineers execute the immediate change(s) in the processing related document(s) on the processing floor.

   The following steps shall be carried out:
   - cross out the current data in the applicable document(s);
   - write above it the new data as described in A-009;
   - both, sign and date the provisional change in the document(s).

Note 2. This change will be valid until the next production lot is scheduled, or as otherwise determined.

4. The Process Engineer now changes the master document(s), including revision level(s). Submits the revised document(s) for review and approval to supervision. After approval, he/she makes as many copies of the revised master document(s) as needed for distribution. Returns the masters to the Engineering file. Then:
   - attaches one copy of the master document(s) to A-009 and turns these over to the Quality Engineer;
   - distributes the rest of the copies to all the affected departments;
   - removes the obsolete document(s) at the same time and disposes them, as required;
   - then, updates the Production Folder(s);
   - maintains both the Engineering and Production Folders, as required.

Note 3. Obsolete document(s) shall be stamped "Obsolete" and retained for background information, as required.

5. The Quality Engineer now takes the documents sent by the Process Engineer. He/she determines the impact caused by this revision to the quality system's documents. Then, completes items 4 and 5 on form A-009. If there is no impact, he/she checks off "N/A" in item 8. Signs and dates the document and files A-009 together with the revised document(s) in the "Document Change Folder." On the other hand, if there is impact, he/she lists the impacted documents under item 6 on A-009 and does the following:

MINFOR INCORPORATED	Iss dt. 1/1/99	Rev dt. 1/1/99		Pg.
QUALITY OPERATING PROCEDURE	Sign	Rev # 0		8/23
REQUIREMENT: QUALITY SYSTEM MANUAL SEC: 4.5.3				
SUBJECT: DOCUMENT AND DATA CHANGES				
QOP 002 |DOCUMENT REQ'D: QOP 002, sec. 1, 2 and 3; Forms A-009, A-010				

✿ makes copies of the impacted master document(s) and highlights (with a "yellow" marker) in the copies the text impacted by the change imposed by A-009. Then, returns the master(s) to original file; Then:

✿ turns over to the Management Representative all the documents (A-009, the document(s) revised by Engineering, and the highlighted copies of the quality documents).

6. The Management Representative obtains the processing related document package from the Quality Engineer and carries out his/her responsibility as stated in paragraph 4.01 below.

3.03.2   Standard and NCMR Processing Related Document Changes

3.03.2.1 Standard Document Changes

General

Standard document changes are routine changes without having immediate impact on the current processing requirements. They shall be implemented in accordance with production scheduling, established by the customer's Purchase Order delivery requirement.

Implementation:

1. All the provisions stated above in paragraph 3.03.1, except item 3, shall be carried out when incorporating standard changes to the processing related document(s);
2. Implementation takes effect with the next processing cycle, or as determined by the customer's Purchase Order.

3.03.2.2 NCMR Affected Changes

General

When non-conforming products are dispositioned, changes to processing and/or quality documents may be imposed through corrective action, depending on the cause of the problem. The corrective action may require either an immediate or standard document change.

Implementation:

1. When the corrective action requires an immediate change to the processing docu - ments, all the provisions stated in paragraph 3.03.1 above shall be carried out;
2. When the corrective action requires a standard change to documents, all the provisions stated in paragraph 3.03.1, except item 3, shall be carried out;

MINFOR INCORPORATED	Iss dt. 1/1/99	Rev dt. 1/1/99	Pg.
QUALITY OPERATING PROCEDURE	Sign	Rev # 0	9/23
REQUIREMENT: QUALITY SYSTEM MANUAL SEC: 4.5.3			
SUBJECT: DOCUMENT AND DATA CHANGES			
QOP 002 |DOCUMENT REQ'D: QOP 002, sec. 1, 2 and 3; Forms A-009, A-010			

3. Implementation takes effect as determined by the corrective action date.

3.03.2.3 Software Changes

Implementation
Software programs are controlled as documents. Changes to these documents shall be accomplished in accordance with the requirements stated under paragraphs 3.03.1 and 3.03.2 above and as stated in QOP 001, Section Four (Amendment to Contract, pg 307).

Exception applies:
Developmental projects shall be exempted from the provisions of paragraphs 3.03.1 and 3.03.2 until product development has been accomplished. (See QOP 012, Section Three on page 449)

Note 4. Document changes should promote improvement to the operating system's effectiveness and efficiency. Cost and customer satisfaction should be evaluated.

4.0 Change Control of the Quality System's Documents

General
The processing related documents (Job Travelers, Operation Sheets, Software Programs, and other setup-related documents) have their changes accomplished through the application of form A-009. The quality system documents, however, have their changes carried out directly within their own structural system when not impacted by change requirements in form A-009. Processing related document changes may or may not trigger off changes to the quality system's documents. When they do, they will indirectly impact the quality system's procedures. Therefore, when direct changes are required to the quality system's documents, form A-009 shall not be processed. Rather the direct change method shall be carried out through the process of draft changes in the copies of the impacted master document(s). (See paragraph 4.02 below on page 322)

Note 5. Changes made in the text of the quality system's documents after original issue shall be highlighted. (*italicized* or "**bold**")

Note 6. Typing and syntax errors in the masters can be corrected without formal change control only if they don't impact product quality and/or processing. The Management Representative shall control these types of corrections, including the uniform distribution of the revised documents along with the retraction of obsolete documents.

MINFOR INCORPORATED	Iss dt. 1/1/99	Rev dt. 1/1/99	Pg.
QUALITY OPERATING PROCEDURE	Sign	Rev # 0	10/23
REQUIREMENT: QUALITY SYSTEM MANUAL SEC: 4.5.3			
SUBJECT: DOCUMENT AND DATA CHANGES			
QOP 002 |DOCUMENT REQ'D: QOP 002, sec. 1, 2 and 3; Forms A-009, A-010			

Note 7. The quality system's forms are part of the quality operating procedures by reference. They have not been attached to the quality operating procedures for reasons of practical handling and control. When making changes to the procedures, ensure that any change due to flow-down requirement has been incorporated.

4.01   Indirect Change(s) to the Quality System's Documents

4.01.1 Procedure to Implement Indirect Changes

The Management Representative shall:

1. Review the documents submitted by the Quality Engineer and research the validity of the change. Based on research results, he/she evaluates the overall impact of this change to the quality operating system, including performance, customer satisfaction, and flow-down requirements. Then, makes decision whether to put it on hold and determine other measures, reject it for the time being, or implement it. He/she records this decision with justification onto form A-009 under "Notes." Then:
2. In case of "hold", he/she finds alternate solution(s), which would not require revision to the manual or to the operating procedures (QOPs), and implements it.
3. In case of "rejection," he/she signs and dates A-009 and files it together with the attachment(s) in the "Document Change Folder" and waits for recurring symptoms to reopen it.
4. In case of "implementation," he/she takes the already highlighted documents, crosses out the current text and writes above it the required new text. Then:
5. Takes care of the flow-down requirements, if required. In this effort, he/she takes the impacted master document(s) and makes copies. Marks the copies by crossing out the current text and writing above it the required new text. These will become the draft-change copies. Hold the masters and drafts at hand.
6. Evaluate the impact this change has on performance, customer satisfaction, and implementation. Plan accordingly, and set up training for the affected personnel before formalizing and implementing the change(s). After training:
7. Takes all the impacted master documents and updates them according to the draft changes (remove old data, insert new data) and highlights the new data in the master(s) - (*italicize* or "**bold**"). Then, completes A-009, item 7, and signs and dates it. (Keep A-009 together with the draft-change documents as they will be needed under paragraph 5.0 below.)
8. Now, carries out the revision indication changes in all the affected documents as stated under paragraph 5.0 below before distribution.

MINFOR INCORPORATED	Iss dt. 1/1/99	Rev dt. 1/1/99		Pg.
QUALITY OPERATING PROCEDURE	Sign	Rev #	0	11/23
REQUIREMENT: QUALITY SYSTEM MANUAL SEC: 4.5.3				
SUBJECT: DOCUMENT AND DATA CHANGES				
QOP 002 |DOCUMENT REQ'D: QOP 002, sec. 1, 2 and 3; Forms A-009, A-010				

9. After the revision indication changes, prepares to distribute the revised documents. Does this in the following manner:

   a) If the quality manual and/or the quality operating procedures are impacted, mak the in-shop distribution of the revised documents and retraction of the obsolet documents at the same time according to the document owners listed on form A-010 (pg 465).

   For the manual:

   ❀ Replace the obsolete manuals with the revised manuals. Dispose of the obsolet manuals, as required. (Don't forget to recall and replace the quality manual issued to customers, as required (controlled vs. uncontrolled)).

   For the operating procedures:

   ❀ Replace the obsolete quality operating procedures and the master list and it revision history page in the "Procedures Manual" for each owner. Dispose of the collected obsolete quality operating procedures, the master lists and their revi sion history pages, as required.

   b) When the forms are impacted, remove the obsolete forms and replace them wit the revised forms from all the shop files, as required. Do the same in the master's file. Remember to place the forms' "Master List" and its "Revision History" page in front of the pile. Dispose of obsolete forms, as required.

4.02 Direct Changes to the Quality System's Documents

General

Direct changes to the quality system's documents may become necessary as a result o system audits, performance evaluations, statistical analysis in MRB activities, policy change due to customer requirements, changes in goals and objectives for future business, implementation of sector specific standards, updating changes due to revision in the applicable standards, management reviews, and others. The driving forces behind implementing changes to the quality system's documents should be targeted to achieve improvements to the overall operating system in line with business objectives. The Management Representative is responsible for implementing changes to the quality system's documents.

4.02.1 Procedure to Implement Direct Changes

The Management Representative shall:

MINFOR INCORPORATED	Iss dt. 1/1/99	Rev dt. 1/1/99	Pg.
QUALITY OPERATING PROCEDURE	Sign	Rev # 0	12/23
REQUIREMENT: QUALITY SYSTEM MANUAL SEC: 4.5.3			
SUBJECT: DOCUMENT AND DATA CHANGES			
QOP 002	DOCUMENT REQ'D: QOP 002, sec. 1, 2 and 3; Forms A-009, A-010		

1. Consolidate supporting materials to justify the change. Determine requirement. Identify which document is affected and make a copy of the master(s). When done, return the master(s) to original file.

2. Now, take the copies and determine where they belong. Is it the quality manual, or a quality operating procedure, or a form, or all three? Conduct a thorough evalua - tion regarding flow-down requirements. Start this process with the quality manual, then the operating procedures, and the forms. Observe references in the chain of events as stated in the manual, carried over to the operating procedures and from there to the forms. Establish linkage within the procedural system. Keep a note where things are in order to be able to pinpoint linkage to flow-down requirement.

3. After flow-down has been identified, make a copy of each page of the master document affected by the flow-down impact. Now, take all the copies and mark the change(s). Cross out the text needed to be changed and write above it the required new text.

4. Evaluate the impact this change has on performance, customer satisfaction, and implementation. Plan accordingly, and set up training for the affected personnel before formalizing and implementing the change(s).

5. While training is going on, take the master documents impacted by the draft changes. Delete in them the impacted data and type in the required new data as it appears in the draft-change document(s). Highlight the new data, as required - (*italicize* or make it "**bold**"). Now, go to paragraph 5.0 below and carry out the revision indication changes in the affected documents before distribution.

6. By the time training is over, prepare to distribute the revised documents. Do this in the following manner:

   a) If the quality manual and/or the quality operating procedures are impacted, make the in-shop distribution of the revised documents and retraction of the obsolete documents at the same time according to the document owners listed on form A-010 (pg 465).

   For the manual:
   ✿ Replace the obsolete manuals with the revised manuals. Dispose of the obsolete manuals, as required. (Don't forget to recall and replace the quality manuals issued to customers, as required (controlled vs. uncontrolled)).
   For the operating procedures:
   ✿ Replace the obsolete quality operating procedures and the master list and its revision history page in the "Procedures Manual" for each owner. Dispose of the collected obsolete quality operating procedures, the master lists and their revision history pages, as required.

MINFOR INCORPORATED	Iss dt. 1/1/99	Rev dt. 1/1/99		Pg.
QUALITY OPERATING PROCEDURE	Sign	Rev # 0		13/23
REQUIREMENT: QUALITY SYSTEM MANUAL SEC: 4.5.3				
SUBJECT: DOCUMENT AND DATA CHANGES				
QOP 002 |DOCUMENT REQ'D: QOP 002, sec. 1, 2 and 3; Forms A-009, A-010				

b) When the forms are impacted, remove the obsolete forms and replace them with the revised forms from all the shop files, as required. Do the same in the master's file. Remember to place the forms' "Master List" and its "Revision History" page in front of the pile. Dispose of obsolete forms, as required.

Note 8. Form A-010 has been designed to reassign documents in line with personnel changes. Update this form when the turnover takes place.

5.0 Revision Indication Control of the Quality System's Documents

General

The quality system's documents are grouped in three separately controlled entities: the quality manual, the quality operating procedures, and the forms. Each entity has built-in mechanism to take care of revision indication changes. The quality manual has its own "Revision History" page, located at the beginning pages in the manual (pg 271). The quality operating procedures have two different "Revision History" pages, one to control the revision of <u>each</u> quality operating procedure and the second one to control all of them together. This second one is attached to the "Master List" (pg 294) which controls all the procedures as a "Procedures Manual." The forms are controlled through a different "Master List" and its related "Revision History" page. (All are located in front of the respective group.)

Note 9. <u>Revision indication changes must be carried out whether the quality system's documents have been impacted by indirect changes (through the application of form A-009), or by direct changes (through the application of draft changes).</u>

Note 10. The Management Representative will have to determine which entity out of the three has to be updated for the revision indication changes.

5.01 Revision Indication Change to the <u>Quality System Manual</u>

The <u>text</u> changes in the master manual have now been updated from the draft changes. Before the distribution of the revised manual can take place, the revision indication in the manual's "Revision History" page and then on every page of the manual must be done. This shall be accomplished as stated hereunder:

The Management Representative shall:

1. Make a copy of the "Revision History" page of the master manual and start the change process in the page header. Cross out the current revision date and number and write above the effective date and number. Next, go to the recording section and

MINFOR INCORPORATED	Iss dt. 1/1/99	Rev dt. 1/1/99	Pg.
QUALITY OPERATING PROCEDURE	Sign	Rev # 0	14/23
REQUIREMENT: QUALITY SYSTEM MANUAL SEC: 4.5.3			
SUBJECT: DOCUMENT AND DATA CHANGES			
QOP 002 |DOCUMENT REQ'D: QOP 002, sec. 1, 2 and 3; Forms A-009, A-010			

write in the required information according to each column's description. This will be the draft-change copy.

2. When this is done, take the master manual and formally change the revision date and number in the header on every page according to the draft-change. Update at this time also the manual's "Revision History" page. Adjust page numbers, as required. (Don't forget to realign the "Contents" page numbers, as required.) Then:

3. Make as many copies of the master manual as needed in line with the "Document Issuance Control," form A-010, which lists the manuals' numbers and owners. Organize the pages of each manual in sequential order and bind them into individual manuals. Number each manual as they are listed on form A-010. Then, do the distribution of the manuals as stated in paragraph 4.02.1, item 6a, above, on page 323, for the directly impacted changes, or paragraph 4.01.1, item 9a, above, on page 322, for the indirectly impacted changes.

4. When finished, put all the draft-change copies in sequential order and file them as one interrelated package in the "Document Change Folder." Then, file the masters in their original folder(s). Maintain file control.

5.02 Revision Indication Change to the <u>Quality Operating Procedures</u>

The <u>text</u> changes in the master pages of the impacted quality operating procedure have now been updated from the draft changes. Before the distribution of the revised quality operating procedure can take place, the revision and change indication in the "Revision History" page must be done. Then, on every page of the procedure, the revision indication and date must be updated. In addition to this, the "Master List" and its "Revision History" page must also be updated. This shall be accomplished as stated hereunder:

The Management Representative shall:
1. Make a copy of the "Revision History" page from the master of the affected Quality Operating Procedure and start the change process in the page header. Cross out the current revision date and number and write above the effective date and number.

2. If other data requires change, do the same thing. Next, go to the recording section and write in the required information under each column's description.
    This page will be the draft copy.

3. Now, take the master procedure and formally change the revision date and number in the header on every page according to the draft changes done in "1" above. Update at this time also the procedure's "Revision History" page. Adjust page numbers, as required. (Don't forget to realign the "Contents" page numbers, as required.) Then:

MINFOR INCORPORATED	Iss dt. 1/1/99	Rev dt. 1/1/99	Pg.
QUALITY OPERATING PROCEDURE	Sign	Rev # 0	15/23
REQUIREMENT: QUALITY SYSTEM MANUAL SEC: 4.5.3			
SUBJECT: DOCUMENT AND DATA CHANGES			
QOP 002 |DOCUMENT REQ'D: QOP 002, sec. 1, 2 and 3; Forms A-009, A-010			

4. Make a copy from the master of the "Master List" and its "Revision History" page. Start with the "Revision History" page first. Cross out the current revision date and number in the header and write above the effective date and number. Then, go to the recording section and write in the required information under each column's description.

5. Now, take the "Master List" copy and cross out the current revision date and number in the header and write above the same date and number as in its already marked "Revision History" copy. Then, identify the procedure's number and title in the listing section and cross out there the current "revision, date, and status" indications and write above the effective revision and date as it appears in the already marked "Revision History" copy of the affected operating procedure (done under step 1, above). Indicate "Revised" under the "status" column. These two pages, the "Revision History" copy under step 4 and the "Master List" copy under this step, will be the draft copies.

6. Now, take the masters of the "Master List" and its "Revision History" page and formally change them according to the marked data in the draft-change copies.

7. Then, take all the masters – the formally revised Quality Operating Procedure, the Master List and its Revision History page – and make as many copies as needed in line with the "Document Issuance Control" form A-010, which list the manuals' numbers and owners. Organize the pages of each operating procedure in sequential order. (Punch the holes if required.) Then, do the distribution (insertion into the "Procedures Manual") of the Quality Operating Procedure, the Master List and its Revision History page as stated in paragraph 4.02.1, item 6a, above, on page 323, for the directly impacted changes, or paragraph 4.01.1, item 9a, above, on page 322, for the indirectly impacted changes.

8. When finished, put all the draft-change copies in sequential order and file them in the "Document Change Folder" as one interrelated package. Then, file the masters in their original folder(s). Maintain file control.

5.03   **Revision Indication Change to the Quality System's Forms**

The Management Representative shall:

1. Make a copy from the master of the impacted form and cross out the current revision line. Write above it the new data. (The revision line is subscripted at the right bottom of each form.) Then:

2. Make a copy of the forms' "Master List" and its "Revision History" page. Start with the "Revision History" copy. Cross out the current revision date and number in the header section. Write above the new date and number. Now, go to the recording

MINFOR INCORPORATED	Iss dt. 1/1/99	Rev dt. 1/1/99	Pg.
QUALITY OPERATING PROCEDURE	Sign	Rev # 0	16/23
REQUIREMENT: QUALITY SYSTEM MANUAL SEC: 4.5.3			
SUBJECT: DOCUMENT AND DATA CHANGES			
QOP 002	DOCUMENT REQ'D: QOP 002, sec. 1, 2 and 3; Forms A-009, A-010		

section and write in under the "rev. dt., rev #, description, approval" columns the updating information.

3. Now, go to the "Master List" copy. Look up the form's number in the listing and cross out its current "rev., dt., and status" indications. Write above it the new requirement from the already marked form (the new subscripted change). Write "Revised" under the "status" column. These will be the draft-change copies. Now:

4. Formalize the changes in the masters from the draft copies. When finished, make enough copies of them to cover the floor's and the central file's requirements. Do the distribution as stated under paragraph 4.02.1, item 6b, above, on page 324, for the directly impacted changes, or paragraph 4.01.1, item 9b, above, on page 322, for the indirectly impacted changes.

5. File all draft-change copies as one interrelated package in the "Document Change Folder." Then, file the masters in their original folder(s). Keep the "Master List" and its "Revision History" page in front of the forms' pile. Maintain file control.

6.0   Handling Non-conforming Product(s) Resulting from Document Changes

General

Document change(s) may cause a contracted product to become non-conforming at any stage in the processing cycle. It could also impact the already delivered product. Non-conforming product(s), resulting from "Amendment to Contract," may trigger off trace-ability requirement as well. Processing related non-conforming product(s) may require notification to the customer, via the "Waiver" process, in order to determine disposition, may also impact the already delivered product(s). All phases in controlling nonconforming product(s), resulting from document changes, should be carried out as stated in Quality Operating Procedure 009, Section One (pg 408) and Two (pg 410).

6.01   Implementation
The Quality Engineer shall:

1. Identify and tag the non-conforming product through the application of form A-012 (Rejected Material Ticket) and separate it, as practical.

2. Process completely form A-006 (Non-conforming Material Report) in accordance with QOP 009, Section One and Two.

3. Follow up all the requirements stated in QOP 009, Sec Two (pg 410) as applicable, until the non-conformance has been eliminated and the product disposed of.

Note 11.   When document changes are imposed as a result of Corrective Action through the disposition process of an NCMR, the non-conforming product(s) should be handled on the same NCMR. (see paragraph 3.03.2.2, above, on page 319)

MINFOR INCORPORATED	Iss dt. 1/1/99	Rev dt. 1/1/99		Pg.
QUALITY OPERATING PROCEDURE	Sign	Rev #	0	17/23
REQUIREMENT: QUALITY SYSTEM MANUAL SEC: 4.16				
SUBJECT: HANDLING AND RETENTION OF QUALITY RECORDS				
QOP 002	DOCUMENT REQ'D: All records of the Quality System			

SECTION THREE

1.0 PURPOSE

Summarize the retention, maintenance, and responsibility in managing the Quality System's records.

2.0 APPLICATION

The hereunder written procedure shall apply to all the identified and documented records within the Quality System.

3.0 PROCEDURE

General

The quality system's effectiveness is demonstrated through the control of its records which serve as evidence to prove product conformance or non-conformance to the specified requirements within the established and implemented system. (see Summary of Forms, shown on the Master List, page 453)

Since record keeping is evidenced through the application of the various forms, the various departments within the organization shall be responsible for managing them in accordance with procedural requirements as described in the applicable QOPs.

From time to time, subject records shall be made readily available for review by the customer or regulatory agencies. **(see QSM and SAE AS9000 paragraph 4.16.1)**

To prove the effective operation of the quality system, subject records shall be periodically audited in accordance with documented procedures (QOP 010, pg 421). Obsolete documents shall be retrieved from all affected areas. Their retention shall be determined according to need and those that are retained shall be stamped "Obsolete."

The quality system's records demonstrate the documentation of work results, recorded by those process owners who bear the responsibility for their accuracy and completeness each step of the way in the product's realization processes. The record retention responsibility, therefore, shall be assigned to those process owners who carry out the last step in the documentation, review, approval, or final handling of the designated record(s). According to work assignment responsibilities, as defined in the quality system's procedures, the retention of records shall be done as defined below.

MINFOR INCORPORATED	Iss dt. 1/1/99	Rev dt. 1/1/99		Pg.
QUALITY OPERATING PROCEDURE	Sign	Rev #	0	18/23
REQUIREMENT: QUALITY SYSTEM MANUAL SEC: 4.16				
SUBJECT: HANDLING AND RETENTION OF QUALITY RECORDS				
QOP 002	DOCUMENT REQ'D: All records of the Quality System			

Implementation:

3.01   **The Management Representative shall:**

a) Maintain and retain **Form A-034** – Management Review Status Record -- under separate file, identified as "Management Review Records";

b) Maintain and retain **Form A-021** – Audit Schedule – under separate file, identified as "Audit Records";

c) Maintain and retain **Form A-022** – Audit Plan – identified under "Audit Records";

d) Maintain and retain **Form A-010** – Document Issuance Control – under separate file, identified as "QSM and QOP Distribution Control";

e) Maintain and retain **Form A-016** – Training Log – under separate file (binder), identified as "Training Log";

f) Maintain and retain **Form A-006** – Non-conforming Material Report – under separate folder, identified as "Non-conforming Material Records" after corrective action review, acceptance, and NCMR close out (on A-007);

g) Maintain and retain **Form A-009** – Amendment to Procedures – in the "Document Change Folder" after he/she has completed all the indirect related changes to the quality system's documents, affected by the application of this form;

h) Maintain and retain **the masters** of the Quality System Manual, the Quality Operating Procedures and the Forms in their respective folders, identified under their titles. Update these documents as required. Organize all change-related documents (drafts) and file them in the "Document Change Folder(s)."

i) Maintain **Form A-039** – Training Evaluation Sheet – in the Training Log.

3.02   **The Contract Administrative Secretary shall:**

a) Maintain and retain **Form A-004** – RFQ Worksheet – under separate folder, identified by the prospective customer's name. The "RFQ Folder" shall contain all the documents that relate to the specified transaction;

b) Maintain and retain **Form A-005** – RFQ, PO, and Amendment Register under separate folder, identified by the same title. When the form is full, the next one shall be consecutively numbered. To avoid mixing the RFQ, PO, and Amendment entries on one form, it is expected that each title will have its own entry carried out separately on duplicates of the same form;

c) Maintain and retain **Form A-011** – Control of Customer Documents – attached to the backside of the front cover of the respective customer's "Purchase Order Folder";

MINFOR INCORPORATED	Iss dt. 1/1/99	Rev dt. 1/1/99	Pg.
QUALITY OPERATING PROCEDURE	Sign	Rev # 0	19/23
REQUIREMENT: QUALITY SYSTEM MANUAL SEC: 4.16			
SUBJECT: HANDLING AND RETENTION OF QUALITY RECORDS			
QOP 002	DOCUMENT REQ'D: All records of the Quality System		

d) Maintain and retain **Form A-031** – Issue and Traceability of Customer Drawing - in the respective customer's "Purchase Order Folder";

e) Maintain and retain copy of **Form A-001** – Purchase Order Review Sheet – in the respective customer's "Purchase Order Folder." (This form was issued by Quality.) When reissued with PO Amendments, it shall be maintained in the "PO Amendment Folder." The PO and Amendment Folders may be combined, depending on the methods of internal handling;

f) Maintain and retain **Form A-015** – Record of Received Materials – in the respective customer's "Purchase Order Folder." (This form will be sent by Quality when the received material has been released. Data from this form will be used to record the identity and status of the received product in order to facilitate Production Planning.);

g) Maintain and retain copy of **Form A-028** – Packing Slip (Shipper) – in the respective customer's "Purchase Order Folder." (Original is shipped with the product to the customer, after the release of the product by Quality.)

3.03   **The Process Engineer shall:**

a) Maintain and retain the master of **Form A-008** – Job Traveler – in the respective customer's "Engineering Folder." When the Job Traveler is amended, the master shall be updated. The obsolete Job Traveler shall be stamped "Obsolete" and retained in the "Document Change Folder" under the provisions of change control triggered by the application of Form A-009;

b) Maintain and retain copy of **Form A-001** – Purchase Order Review Sheet – and **Form A-015** – Record of Received Materials – (issued by Quality) in the respective customer's Engineering Folder. Whenever these forms are amended due to customer's PO Amendment, or to internal requirements, they shall be attached to the first issued copy and maintained in the Engineering Folder. The Process Engineer shall use the provisions recorded on these forms to implement in the Job Traveler the stipulated inspection and product identification requirement(s) to maintain product quality during the product's processing realization;

c) Maintain and retain **Form A-009** in the "Document Change Folder."

3.04   **The Purchasing Agent shall:**

a) Maintain and retain copy of **Form A-017** – Purchase Order (Internal) – under the respective subcontractor's "Purchase Order Folder." Amendments issued shall be attached to the primary PO. Any product related correspondence with

MINFOR INCORPORATED	Iss dt. 1/1/99	Rev dt. 1/1/99		Pg.
QUALITY OPERATING PROCEDURE	Sign	Rev # 0		20/23
REQUIREMENT: QUALITY SYSTEM MANUAL SEC: 4.16				
SUBJECT: HANDLING AND RETENTION OF QUALITY RECORDS				
QOP 002 DOCUMENT REQ'D: All records of the Quality System				

the respective subcontractor shall be part of his/her PO Folder. Only product related Purchase Orders are regulated. Others are optional. All the retained copies of the product related Purchase Orders shall show evidence of the Quality Engineer's release approval.

3.05   **The Quality Engineer shall:**

a) Maintain and retain **Form A-001** – Purchase Order Review Sheet – attached to the PO in the respective customer's Inspection Product Folder. Whenever a revised one is issued due to PO amendments, it shall be kept together with the same PO package;

b) Maintain and retain **Forms A-018 and A-019** – Inspection Report and Continuation Sheet – in the respective customer's Inspection Product Folder;

c) Maintain and retain **Form A-033** – Operator Product Verification Record – in the respective customer's Inspection Product Folder;

d) Maintain and retain **Form A-013** – Acceptance Material Ticket – on the verified and accepted product until that product is released for additional processing and/or shipment. Then, it shall be destroyed;

e) Maintain and retain **Form A-014** – Hold Ticket – on the product until the administrative resolution of internal or external issues has been satisfactorily completed. Then, it shall be destroyed;

f) Maintain and retain **Form A-006** – Non-conforming Material Report (NCMR) – in its own File Folder. Attach to it all other documents that relate to the same nonconformance. Keep the forms in sequential order according to serial number designation. (They may be stacked in binders to keep bundles in order.);

g) Maintain and retain **Form A-007** – Non-conforming Material Report Register – in front of the latest group of NCMRs in order to keep track of running serial numbers. Serial numbers once issued shall not be repeated. Serial numbers shall be sequentially arranged in every folder (binder);

h) Maintain and retain **Form A-012** – Rejected Material Ticket – on all the nonconforming product(s) (lot by lot) pending final resolution. When a nonconforming product has been dispositioned and the disposition has been satisfactorily carried out, the accepted product shall have a new inspection status indicated and documented. Then, the "Rejected Material Ticket" shall be destroyed. There are times when this ticket shall be left on a product during shipment in order to identify its status to the customer for his determination of further disposition (waiver cases);

MINFOR INCORPORATED	Iss dt. 1/1/99	Rev dt. 1/1/99	Pg.
QUALITY OPERATING PROCEDURE	Sign	Rev # 0	21/23
REQUIREMENT: QUALITY SYSTEM MANUAL SEC: 4.16			
SUBJECT: HANDLING AND RETENTION OF QUALITY RECORDS			
QOP 002 |DOCUMENT REQ'D: All records of the Quality System			

i) Maintain and retain **Form A-024** – Waiver Request attached to identified NCMR in its File Folder (binder) after customer has responded to request. When required, record the presence of this form on the shipper and attach a copy to the shipping documents when the product is released for shipment to the customer;

j) Maintain and retain **Form A-023** – Rework/Repair Record – with the identified NCMR in its File Folder (binder) after the rework/repair activity has been satisfactorily verified and accepted;

k) Maintain and retain **Form A-002** – Maintenance Record – on the individual processing equipment. When a form is completely filled in, a new one shall be issued and the old one shall be kept under its own File Folder (binder) as long as required. Then, destroyed;

l) Maintain and retain one copy of **Form A-015** Record of Received Materials – in the respective customer's Inspection Product Folder. Send the original to the Contract Administrative Secretary and send the second copy to Engineering;

m) Maintain and retain the "Quality" copy of **Form A-017** – Purchase Order (internal) – in the respective Supplier Product Folder after the Quality Engineer has approved and released the original back to the Purchasing Agent;

n) Maintain and retain **Form A-020** – Calibration and Status Record – in its own File Folder (binder);

o) Maintain and retain **Form A-025** – Shipping Log – in its own File Folder (binder). The application of this form is optional;

p) Maintain and retain **Form A-026** – Supplier Survey Questionnaire – in the identified Supplier Product Folder after the Quality Engineer has reviewed and approved it;

q) Maintain and retain **Form A-027** – Material Identification Tag – on a received or processed product, stored or in-process, which must be identified in order to facilitate further processing. Once accomplished, the tag shall be destroyed;

r) Maintain and retain a copy of **Form A-029** – Certificate of Conformance – in the respective customer's Inspection Product Folder. (Attach it to the other inspection documents relating to the same product.); Send original with shpm't.

s) Maintain and retain **Form A-030** – Receiving Log – in its own File Folder (binder). Application of this form is optional;

t) Maintain and retain **Form A-032** – Inspection Stamp Control – in its own File Folder (binder);

u) Maintain and retain **Form A-035** – Supplier Survey Form (short form) – in the Supplier Product Folder after the Quality Engineer has approved it;

MINFOR INCORPORATED	Iss dt. 1/1/99	Rev dt. 1/1/99	Pg.
QUALITY OPERATING PROCEDURE	Sign	Rev # 0	22/23
REQUIREMENT: QUALITY SYSTEM MANUAL SEC: 4.16			
SUBJECT: HANDLING AND RETENTION OF QUALITY RECORDS			
QOP 002 |DOCUMENT REQ'D: All records of the Quality System			

v) Maintain and retain **Form A-036** – Source Inspected Product Approval – in the respective (subcontractor) Supplier Product Folder after the received product has been identified and verified to the said form's requirements;

w) Maintain and retain **Form A-037** – Supplier Corrective Action Request – attached to the applicable NCMR. Place a copy of this form in the respective (subcontractor) Supplier Product Folder, as appropriate. (Customer SCARs should be filed in the Customer Complaint Folder);

x) Maintain and retain **Form A-038** – Customer Complaint and Evaluation Report under the identified Customer Complaint File after all the questions and problems documented on this form have been satisfactorily resolved and the customer notified. (Approval signature on this form means satisfactory resolution.);

y) Maintain and retain **Form A-008** – Job Traveler – after final inspection and product release in the respective customer's Inspection Product Folder.

z) Maintain and retain the blanks of **Form A-003** – Quality Operating Procedure Blank – with all the other original forms in the Master Folder. Issue copies of this form to all departments as required. (This form has a multipurpose application. First, it is the master sheet on which the process instructions are written in the Quality Operating Procedures. Second, it should be used by Quality and other departments to accomplish documented internal correspondence. When executed, a copy of it should be filed according to the subject matter it covered.)

3.06   File Management

Process owners identified under the above paragraphs shall be responsible for maintaining the assigned file folders according to the following arrangement:

a) identify the files according to the subject matter and as noted above under each form's application;

b) arrange the documents in order of application. Group the documents in sequential order as they apply to a particular product, process, or assignment;

c) provide easy access to the documents and allow authorized personnel to review them;

d) separate the obsolete documents from the active documents within the same folder. When the folders become unmanageable, remove the obsolete documents that are not needed for background information and place them in storage. Store the documents in containers (boxes) and identify them in sequential order so that their contents can be traced back to from where they have been removed;

e) provide safe location and protection from damage and unauthorized handling.

MINFOR INCORPORATED	Iss dt. 1/1/99	Rev dt. 1/1/99	Pg.
QUALITY OPERATING PROCEDURE	Sign	Rev # 0	23/23
REQUIREMENT: QUALITY SYSTEM MANUAL SEC: 4.16			
SUBJECT: HANDLING AND RETENTION OF QUALITY RECORDS			
QOP 002	DOCUMENT REQ'D: All records of the Quality System		

3.07   **Responsibility**

Responsibility to enforce the retention, distribution, control, and maintenance of all the identified records shall be assigned to the Management Representative.

OVERVIEW TO QOP 003
(CONTROL OF PURCHASES)

How often we find that the Purchasing Department is out of the quality loop because of its independent stature. Then the purchased item comes in and is rejected. The inspection report indicates that the product was made by somebody who had absolutely no knowledge as to what kind of product quality the customer required. Now, you have to rework it at your expense, for you are out of time to meet the scheduled delivery requirement. Within a process-approach quality management system, the Purchasing Department becomes part of the team. After all we work for the same company and the same customer. Don't we? So let's bring this sacred cow into the fold in order that we all know our customers' requirements for product quality. QOP 003 is the procedure that requires the Purchasing Agent and the Subcontractor to understand the product quality requirements expected to be realized in purchased products. This procedure is packed with preventive quality requirements in order that we don't buy junk that would end up causing problems and wipe out any profit we calculated to make on the project.

MINFOR INCORPORATED	Iss dt. 1/1/99	Rev dt. 1/1/99	Pg.
QUALITY OPERATING PROCEDURE	Sign	Rev # 0	1/9
REQUIREMENT: QUALITY SYSTEM MANUAL SEC: 4.6			
SUBJECT: CONTROL OF PURCHASES (Internal)			
QOP 003 DOCUMENT REQ'D: QOP 003, Form A-017			

QUALITY OPERATING PROCEDURE 003

CONTROL OF PURCHASES
(internal)

CONTENTS   Page

Form A-003 Rev 0 1999

MINFOR INCORPORATED	Iss dt. 1/1/99	Rev dt. 1/1/99		Pg.
QUALITY OPERATING PROCEDURE	Sign	Rev #	0	2/9
REQUIREMENT: QUALITY SYSTEM MANUAL SEC: 4.6				
SUBJECT: CONTROL OF PURCHASES (internal)				
QOP 003 DOCUMENT REQ'D: QOP 003, Form A-017				

REVISION HISTORY

Rev Date	Rev No	Description	Approval

MINFOR INCORPORATED	Iss dt. 1/1/99	Rev dt. 1/1/99	Pg.
QUALITY OPERATING PROCEDURE	Sign	Rev # 0	3/9
REQUIREMENT: QUALITY SYSTEM MANUAL SEC: 4.6			
SUBJECT: APPROVAL AND ISSUE OF PURCHASE ORDERS			
QOP 003 DOCUMENT REQ'D: QOP 003, Form A-017			

SECTION ONE

1.0   PURPOSE

Define the quality requirements for reviewing and approving the issuance of product related internal Purchase Orders. (Job quoting may be a preliminary requirement.)

2.0   APPLICATION

The provisions herein apply to the issuance of internal Purchase Orders for product related materials, processes (special) and other subcontracted services, which products will be delivered in compliance with the quality requirements of customers' purchase orders.

3.0   PROCEDURE

3.01   Issuance of Internal Purchase Orders

The Purchasing Agent shall:

1. Review the customer's Purchase Order requirements regarding subcontractor activities. Review Form A-001 (Purchase Order Review Sheet) issued by the Quality Engineer to regulate product quality requirements for the contracted product. (These documents are on file in the Contracts Department under the customer's name.) Review the Master Production Scheduling (MPS). Plan your procurement activities in line within the provisions of these documents; Now:
2. Obtain form A-017 (Purchase Order, Internal, pg 472) and allocate control number (if don't have one);
3. Fill in all the information in the preprinted sections (the addresses, Phone/Fax numbers, order and due dates, ship via, terms, etc.);
4. Define line item description and other contract provisions, as required. Ensure accuracy and completeness of all the technical information;
5. Review the recorded information for clarity. Sign and date Purchase Order;
6. Forward PO to the Quality Engineer.

Note 1. Include "Right of Entry" provision, as required. **(AS9000, par. 4.6.2 "Note")**

Note 2. Maintain document flow-down to ensure that subcontractors control those key characteristics (special processes) that are not verifiable in the received products, as required.**(AS9000, par. 4.6.5)**

Note 3. Delegated responsibility regarding product verification and listing shall be defined in the provisions of the Purchase Order. **(AS9000, par. 4.6.4.4)**

MINFOR INCORPORATED	Iss dt. 1/1/99	Rev dt. 1/1/99		Pg.
QUALITY OPERATING PROCEDURE	Sign	Rev # 0		4/9
REQUIREMENT: QUALITY SYSTEM MANUAL SEC: 4.6				
SUBJECT: APPROVAL AND ISSUE OF PURCHASE ORDERS				
QOP 003	DOCUMENT REQ'D: QOP 003, Form A-017			

3.02 Review and Approval of Purchase Orders by the Quality Engineer

The Quality Engineer shall:

1. Verify accuracy, clarity, and completeness of the technical information for each line item in the PO issued by the Purchasing Agent. Verify the latest revision levels of the noted documents. If incorrect, return the PO to the Purchasing Agent for correction;
2. When satisfied with the content information, record in the PO the Purchase Quality Provisions for each line item, including Source Inspection, as required. Sign and date the PO;
3. Retain the "Quality" copy of the PO. Return the rest to the Purchasing Agent;
4. File the retained copy in the Supplier Product Folder.

Note 4. The above given steps uniformly apply to any PO amendments as well.

Note 5. Special Processes must comply with the customer imposed PO requirements. **(AS9000, par. 4.6.2d)**

3.03 Releasing Purchase Orders

The Purchasing Agent shall:

1. Send the "Vendor" copy of the PO to the subcontractor. Ensure that the specified documents listed in the PO have been enclosed;
2. Distribute the other copies of the PO, as required (accounting, receiving, etc.);
3. Maintain his/her own PO filing system to ensure document control on subcontractors;
4. Interact with Production Planing, Engineering, and Quality, as required.

Note 6. The above given steps uniformly apply to PO amendments as well.

3.04 Maintenance of Supplier Product Folders

The Quality Engineer shall:

1. Set up the suppliers' product folders on those subcontractors with whom the Company has established a contractual relationship;
2. Keep the suppliers' product folders current;
3. Include in the suppliers' product folders the following records, as required:

   ✿ Self-survey, reviewed and approved (Forms A-026 or A-035)
   ✿ Copy of purchase orders and amendments (Form A-017)
   ✿ The subcontractor sent inspection reports
   ✿ NCMR related documents (copy of Form A-006)
   ✿ Corrective action report responses (Form A-037 (SCAR))
   ✿ Product and quality related correspondence
   ✿ Source Inspection Reports (Form A-036 (SAR))

MINFOR INCORPORATED	Iss dt. 1/1/99	Rev dt. 1/1/99	Pg.
QUALITY OPERATING PROCEDURE	Sign	Rev # 0	5/9
REQUIREMENT: QUALITY SYSTEM MANUAL SEC: 4.6			
SUBJECT: APPROVAL AND ISSUE OF PURCHASE ORDERS			
QOP 003 |DOCUMENT REQ'D: QOP 003, Form A-017			

Note 7. AS9000, paragraph 4.6.1 "Note," provision is a standard requirement.

4.0   Control of Purchase Order Amendments

4.01   Issuance of Amendments

Note 8. Preliminary assessment of terms and conditions should be discussed with sub-contractors prior to releasing PO amendments. Job quoting may be a requirement.

The Purchasing Agent shall:

1. Obtain current PO package from subcontractor's file;
2. Review and determine in the PO which line item has been impacted by the required amendment;
3. Revise the text in the PO in line with the amendment requirements. Ensure correctness of the technical data. Review the revised PO for accuracy, clarity, and completeness. Sign and date the revised PO;
4. Send the revised PO with all the related technical documents attached to the Quality Engineer.

4.02   Review and Approval of the PO Amendments by the Quality Engineer

The Quality Engineer shall:

1. Review and approve the PO amendments in the same manner as done with the originally issued purchase orders stated above in paragraph 3.02;
2. Maintain the Suppliers' Product Folders regarding PO amendments in the same manner as stated above in paragraph 3.04.

4.03   Releasing the Amended Purchase Orders

The Purchasing Agent shall:

1. Release the amended purchase orders by following the release steps of the standard purchase orders as defined under paragraph 3.03 above;
2. Follow up implementation schedule and related communication with the subcontractors;
3. Maintain subcontractors' file regarding amendments, as required;
4. Interact with Production Planning, Engineering, and Quality, as required.

MINFOR INCORPORATED	Iss dt. 1/1/99	Rev dt. 1/1/99		Pg.
QUALITY OPERATING PROCEDURE	Sign	Rev #	0	6/9
REQUIREMENT: QUALITY SYSTEM MANUAL SEC: 4.6.2				
SUBJECT: EVALUATION OF SUBCONTRACTORS				
QOP 003 DOCUMENT REQ'D: QOP 003, Form A-026, and A-035 (optional)				

SECTION TWO

1.0   PURPOSE

Define the quality requirements regarding evaluation of subcontractors (Suppliers).

2.0   APPLICATION

The results of evaluation shall apply to the selection and approval of subcontractors. The evaluation process shall be based on subcontractors' capability to deliver acceptable products, and on evidence of documented procedures of quality requirements. **(AS9000, par. 4.6.2 "NOTE")**

3.0   PROCEDURE

3.01   Selection of Suppliers (Subcontractors)

The Purchasing Agent shall:
1. Select the suitable subcontractors.
   a) on established subcontractors, base suitability on the already existing quality records from the Supplier Product Folders located in Quality. Observe previously demonstrated capability, product quality and delivery history and act accordingly;
   b) on new subcontractors, base the initial suitability by reviewing their available background profiled on the Internet, or any records available from the local Better Business Bureau, or recommendations from customers and business associates.
2. Request Quality Engineering to determine subcontractors' capability by survey, as required;
3. Ensure that procurement activities are in line with production scheduling requirements.
Note 1.   In urgent situations, the Purchasing Agent should interact with Quality to evaluate and approve subcontractors at the same time internal Purchase Orders are reviewed for approval. "Blind" approval should not be accepted.

3.02   Evaluation of Suppliers (Subcontractors)

The Quality Engineer shall:
1. Review the chosen subcontractors selected by the Purchasing Agent regarding evaluation and approval;
2. Determine whether the selected subcontractor(s) are established or new suppliers;

MINFOR INCORPORATED	Iss dt. 1/1/99	Rev dt. 1/1/99		Pg.
QUALITY OPERATING PROCEDURE	Sign	Rev #	0	7/9
REQUIREMENT: QUALITY SYSTEM MANUAL SEC: 4.6.2				
SUBJECT: EVALUATION OF SUBCONTRACTORS				
QOP 003 DOCUMENT REQ'D: QOP 003, Form A-026, and A-035 (optional), A-037				

3. Prepare for the evaluation and approval process and carry out the requirements accordingly:

    a) do on-site survey
    b) or mail-in survey

3.02.1 On-site Survey

The Quality Engineer shall:
1. Contact subcontractor's Quality and arrange the date for the on-site survey;
2. Review the project specifics that would be required for capability determination in doing the survey at the subcontractor;
3. Prepare the necessary documents to take in order to facilitate the survey (on Form A-026 or A-035, whichever is more suitable). Go to the subcontractor as scheduled;
4. Conduct the quality survey by following the questions listed on the selected survey form. Check off each item according to finding. Take notes as required;
5. Review the survey results with the subcontractor's Quality Representative. Point out areas of non-conformance, if any, and inform the subcontractor that approval is dependent on corrective action. Or inform the subcontractor that there have been no violations found. Prepare to conclude the survey;
6. Sign and date the survey form and give a copy to the subcontractor's representative;
7. Back home, submit the completed survey form to supervision for final approval (auditing) and issue trip report on survey results to all concerned (Purchasing);
8. Make up the Supplier Product Folder as required and file in it the relevant survey related documents, including the trip report.

Note 2. Issuance and approval of purchase order(s) to subcontractors with outstanding corrective action requirements should not be done until the identified noncompliance has been corrected and implemented. Issue Form A-037 (Supplier Corrective Action Request, pg 496) in order to verify compliance.

3.02.2 Mail-in Survey

The Quality Engineer shall:
1. Send (mail or fax) Survey Request (Form A-026) to the selected subcontractor. State in the cover letter the reason and contingency for the survey requirement;
2. Review the returned Survey Request. Sign and date it;
3. Forward the signed Survey Request to supervision for approval review;
4. Notify all concerned (Purchasing) on approval or disapproval (on Form A-003);

MINFOR INCORPORATED	Iss dt. 1/1/99	Rev dt. 1/1/99	Pg.
QUALITY OPERATING PROCEDURE	Sign	Rev # 0	8/9
REQUIREMENT: QUALITY SYSTEM MANUAL SEC: 4.6.2			
SUBJECT: EVALUATION OF SUBCONTRACTORS			
QOP 003 |DOCUMENT REQ'D: QOP 003, Form A-026, and A-035 (optional), A-037			

5. File the approved Survey Request in the Supplier Product Folder.

Note 3. Issuance and approval of purchase order(s) to subcontractors with outstanding corrective action requirements should not be done until the identified noncompliance has been corrected and implemented. Issue Form A-037 (Supplier Corrective Action Request) in order to verify compliance.

3.03 Subcontractor Rating

General

Subcontractors are expected to comply with the provisions of Purchase Orders issued to them. These Purchase Orders contain stipulated provisions relating to general and specific product related requirements. The Quality Provisions are just one component of the total requirements. In a rating process, the violations of the imposed PO provisions should be compared against a predetermined criteria (penalty points) made up according to the business interest of the Company. The predetermined criteria should be the standard rule against which subcontractors are rated and after "so many" violations disqualified. (In small businesses, subcontractors are relatively few and the rating system is based more on arbitrary decision than on an established rating system. Accordingly, the arbitrary decision making is heavily based on two factors: delivery and product quality capability. In tune with this approach, the following process instructions should be subject to individual consideration.)

Implementation

The Quality Engineer shall:

1. Compile a list of all the active subcontractors;

2. Design a form (uncontrolled) with only two entries:

(Uncontrolled Form)	**VENDOR RATING**	
Vendor's Name		Violations
1. The Hatchet Company		1 2 3 4 5 6 7 8 9 10

3. Circle the number of violations progressively each time the same vendor had been issued a SCAR (Supplier Corrective Action Request);

4. Interact with management in order to determine at which point to "pull the plug" on a subcontractor;

5. Document the result of management's decision on Form A-003;

MINFOR INCORPORATED	Iss dt. 1/1/99	Rev dt. 1/1/99		Pg.
QUALITY OPERATING PROCEDURE	Sign	Rev #	0	9/9
REQUIREMENT: QUALITY SYSTEM MANUAL SEC: 4.6.2				
SUBJECT: EVALUATION OF SUBCONTRACTORS				
QOP 003 DOCUMENT REQ'D: QOP 003, Form A-026, and A-035 (optional), A-037				

6. Notify the subcontractor of the decision;

7. File all the related papers in the Supplier Product Folder.

Note 4. Consideration should be given prior to disqualifying subcontractors to the availability of alternate suppliers.

OVERVIEW TO QOP 004
(CONTROL OF CUSTOMER SUPPLIED PRODUCT)

Customer supplied products could be just about anything that an organization may need in supporting the contracted product's realization processes. The presumption that whatever the customer may provide in supporting product realization is already a qualified product, because the customer checked it before releasing it, is as far from the truth as catching a 10 pound bass on dry land. Of course, not every customer falls under this scenario. The customer's previously demonstrated capability should be the yardstick by which we should take any chances in accepting the supplied product(s). Without controlling customer-supplied products, we are inviting the same type of problems as accepting unverified products from subcontractors. Just when we are going through the first production piece verification, we discover the problems. This is another area where any profit we wanted to make on the job had evaporated. Sometimes I wonder where customer satisfaction begins, now that I am buying product support from the customer. Anyway, QOP 004 is giving you the tools in order to control customer supplied products, similarly as you would buy products from anybody else. It's a fair deal.

MINFOR INCORPORATED	Iss dt. 1/1/99	Rev dt. 1/1/99		Pg.
QUALITY OPERATING PROCEDURE	Sign	Rev #	0	1/5
REQUIREMENT: QUALITY SYSTEM MANUAL SEC: 4.7				
SUBJECT: CONTROL OF CUSTOMER SUPPLIED PRODUCT				
QOP 004|DOCUMENT REQ'D: QOP 004, Forms A-006, A-012, A-015, A-027				

QUALITY OPERATING PROCEDURE 004

CONTROL OF CUSTOMER SUPPLIED PRODUCT

CONTENTS   **Page**

Form A-003 Rev 0 1999

MINFOR INCORPORATED	Iss dt. 1/1/99	Rev dt. 1/1/99		Pg.
QUALITY OPERATING PROCEDURE	Sign	Rev #	0	2/5
REQUIREMENT: QUALITY SYSTEM MANUAL SEC: 4.7				
SUBJECT: CONTROL OF CUSTOMER SUPPLIED PRODUCT				
QOP 004	DOCUMENT REQ'D: QOP 004, Forms, A-006, A-012, A-015, A-027			

REVISION HISTORY

Rev Date	Rev No	Description	Approval

MINFOR INCORPORATED	Iss dt. 1/1/99	Rev dt. 1/1/99		Pg.
QUALITY OPERATING PROCEDURE	Sign	Rev #	0	3/5
REQUIREMENT: QUALITY SYSTEM MANUAL SEC: 4.7				
SUBJECT: CONTROL OF CUSTOMER SUPPLIED PRODUCT				
QOP 004	DOCUMENT REQ'D: QOP 004, Forms A-006, A-012, A-015, A-027			

1.0   PURPOSE

To control customer supplied products in accordance with the quality provisions of the Purchase Order and this procedure.

2.0   APPLICATION

Customer supplied product determines the processing schedule to furnish goods within delivery requirements of the contract. As such, the quality of the supplied product bears a substantial impact to comply with that requirement. The application of this procedure shall ensure the expedient identification, verification, documentation and release of the customer supplied product.

3.0   PROCEDURE

3.01   Receiving

The Receiver shall:
1. Receive the delivered product. Sign delivery papers as required;
2. Verify that the delivered product is accurately reflected in the customer's Packing Slip;
3. Record on Form A-015 (Record of Received Materials, pg 470) the product related information as requested in the form. Ensure accuracy and completeness. Sign and date it. Put this form on your desk for now;
4. Fill out Form A-027 (several may be required, pg 486), indicating the received product's identification according to the form's description requirement. Secure Form A-027 onto the product, according to the quantities, lot/batch, S/N, so that the shipment can be identified and located as one controlled entity;
5. Place the entire shipment in a safe area in order to prevent unauthorized removal;
6. Take now Form A-015 and all the documents sent by the customer and turn them over to the Quality Department.

3.02   The Receiving Inspector shall:
1. Accept the documents from the Receiver and schedule the identification, verification and release requirements of the received product by reviewing Form A-015, the customer's PO, and Form A-001(both the PO and Form A-001 are located in the Inspection Product Folder);
2. Determine the extent of the verification, identification, and release requirements:

   a) base the dimensional verification and documentation requirement on the customer's previously demonstrated product quality history. Accordingly, waive or impose dimensional verification and documentation;

MINFOR INCORPORATED	Iss dt. 1/1/99	Rev dt. 1/1/99	Pg.
QUALITY OPERATING PROCEDURE	Sign	Rev # 0	4/5
REQUIREMENT: QUALITY SYSTEM MANUAL SEC: 4.7			
SUBJECT: CONTROL OF CUSTOMER SUPPLIED PRODUCT			
QOP 004 DOCUMENT REQ'D: QOP 004, Forms A-006, A-012, A-015, A-027			

    b) base any product serialization and lot/batch control on two factors: **customer's requirement, internal requirement;**

    c) on customer supplied raw materials, the provisions of paragraph 4.10.2.4 (pg 283) of the QSM **(also SAE AS 9000 par. 4.10.2.4)** shall not be binding unless specifically stated in the customer's PO.

3. Verify the product as identified by the Receiver on Form A-027. Perform dimensional inspection and document the measurement results onto Form A-018, as required.

4. Mark and/or serialize the product (the part(s), lot(s)/batch(s) based on 2b above (also refer to QOP 005, paragraph 3.01, step 2, pg 354). Record product marking and/or serialization onto Form A-015 and on A-001, as required;

    a) In case of missing, damaged, or otherwise rejected product, follow the work instructions under paragraph 3.03 below;

5. Accept the received product. Stamp, or sign and date Forms A-015 and A-027 (Material Identification Tag). Release the product by notifying the Receiver;

6. Make two copies of Form A-015. File one copy together with other related documents (such as certs.) in the Inspection Product Folder;

7. Send the original of Form A-015 and the Packing Slip to Contracts. Send the second copy of Form A-015 to Engineering. Form A-027 stays on the released product.

3.03   Product Rejection

The Receiving Inspector shall:

1. Fill out Form A-012 (Rejected Material Ticket, pg 467) and attach it to the product. Fill out Form A-006 (Non-conforming Material Report) and keep all documents attached to this form, pending MRB review and disposition by customer. Keep the product segregated as practical.

2. Provide information copy of Form A-006 (NCMR) to Contract Administration;

3. Send report to customer's Quality (a copy of Form A-006) to show the rejected status of the supplied product **(SAE AS9000, par. 4.13.2.4);**

4. Record the customer's disposition onto Form A-006 (NCMR) according to the agreement reached between Contracts and the Customer;

5. Carry out the customer's disposition according to the instructions provided. Refer to QOP 009 in cases where rework/repair (pg 412), or other actions in handling nonconforming products are explained;

6. Close out Form A-006 (NCMR) as per instructions in QOP 009, Section Two, paragraph 3.01.7 (pg 416). Maintain file control over the related documents, both in the Inspection Product Folder and the NCMR Folder, as required.

MINFOR INCORPORATED	Iss dt. 1/1/99	Rev dt. 1/1/99	Pg.
QUALITY OPERATING PROCEDURE	Sign	Rev # 0	5 /5
REQUIREMENT: QUALITY SYSTEM MANUAL SEC: 4.7			
SUBJECT: CONTROL OF CUSTOMER SUPPLIED PRODUCT			
QOP 004	DOCUMENT REQ'D: QOP 004, Forms A-006, A-012, A-015, A-027		

3.04   **Miscellaneous Products Supplied by the Customer**

General

Customer supplied products may involve other than raw materials or semi-finished products. The customer's Purchase Order and/or amendments thereof should provide the necessary information regarding furnished products. Quality shall control and maintain documentation according to the type of product(s) supplied by the customer, which may involve:

a) production tooling;
b) fixtures;
c) inspection equipment;
d) software;
e) semi-finished product;
f) dropped-off product from the customer's other subcontractors;
g) other things (like shipping containers)

3.04.1   **The Receiver shall:**

Follow the provisions of paragraph 3.01 above, on page 348, from steps 1 through 6 in receiving miscellaneous products, as applicable.

3.04.2   **The Receiving Inspector shall:**

1. Follow the provisions of paragraphs 3.02 and 3.03 above, on pages 348 and 349, to the extent applicable, in verifying, documenting, and releasing the miscellaneous products;
2. Maintain file control over the related documents, both in the Inspection Product Folder and NCMR Folder, as required.

3.05   **Interaction by Contract Administration**

The Administrative Secretary shall:

1. Interact with Quality, Engineering, and the Customer regarding the resolution of problems;
2. Maintain problem-related documents (correspondence) in the customer's PO folder.

OVERVIEW TO QOP 005
(PRODUCT IDENTIFICATION AND TRACEABILITY)

Product identification is much like your birth certificate. It tells you something but not everything about yourself. Product-related documents do the same thing. This is the way we identify some very basic earmarks about the product we are putting through the various processes. The part, the serial, the lot or batch numbers, all serve to identify the product. Sometime, we don't mark the product, so we identify it on the document accompanying it. Then, we attach tags or labels to the product. We also identify the product on the Job Traveler and on the Operation Sheet. The customer does it on the Purchase Order and on the product specification. Now, we can not only match the product to its documents, but we can also trace it on the production floor, in the stockroom or even in the field. This is what product identification is all about. QOP 005 tells you how to control the process of product identification and traceability.

MINFOR INCORPORATED	Iss dt. 1/1/99	Rev dt. 9/2/99	Pg.
QUALITY OPERATING PROCEDURE	Sign	Rev # 1	1/5
REQUIREMENT: QUALITY SYSTEM MANUAL SEC: 4.8			
SUBJECT: PRODUCT IDENTIFICATION AND TRACEABILITY			
QOP 005	DOCUMENT REQ'D: Forms A-001, A-008		

QUALITY OPERATING PROCEDURE 005

PRODUCT IDENTIFICATION AND TRACEABILITY

CONTENTS   **Page**

Product Identification and Traceability

Form A-003 Rev 0 1999

MINFOR INCORPORATED	Iss dt. 1/1/99	Rev dt. 9/2/99		Pg.
QUALITY OPERATING PROCEDURE	Sign	Rev #	1	2/5
REQUIREMENT: QUALITY SYSTEM MANUAL SEC: 4.8				
SUBJECT: PRODUCT IDENTIFICATION AND TRACEABILITY				
QOP 005 DOCUMENT REQ'D: Forms A-001, A-008				

REVISION HISTORY

Rev Date	Rev No	Description	Approval
9/2/99	1	Completely rewritten to define part, batch, and lot identification marking controls	AG

MINFOR INCORPORATED	Iss dt. 1/1/99	Rev dt. 9/2//99		Pg.
QUALITY OPERATING PROCEDURE	Sign	Rev #	1	3/4
REQUIREMENT: QUALITY SYSTEM MANUAL SEC: 4.8				
SUBJECT: PRODUCT IDENTIFICATION AND TRACEABILITY				
QOP 005│DOCUMENT REQ'D: Forms A-001, A-008				

1.0   PURPOSE

To maintain control over the process of product identification and traceability through the requirements of documented procedures.

2.0   APPLICATION

This procedure shall apply to product identification and traceability for the issuance of documents and the enforcement of marking.

3.0   PROCEDURE

Product identification and traceability shall be enforced as stated in the customer's Purchase Order for two reasons:

1) internal document control for product traceability
2) implementation of customer flow-down requirements

3.01   Internal Control

General

1. The primary controlling method for document and product identification shall be the customer's Purchase Order imposed product part number and its revision level. Additional controls, such as individual serial numbering, batch or lot control, shall be imposed both in the related documents and on the product itself, as required.
2. Parts serialization, batch or lot control, when imposed, shall begin from the receipt of product(s), or as otherwise determined by the processing requirement. This shall form the basis also for identification later when splitting quantities for multiple processing is required.
3. Follow-on marking and documentation when, through processing, the original marking has been removed, shall be identically re-marked throughout processing.
4. Process owners accountability for product identification and record keeping for the parts they have done shall be enforced and tracked by supervision.
5. Identification and documentation of the products and/or services, ordered on internally issued Purchase Orders, shall conform to the customer's Purchase Order requirements.
6. Departments issuing product-related documents shall enforce the identification and documentation requirements according to the customer's Purchase Order.
7. Quality shall flow down the customer's specific product marking and traceability requirements in the Purchase Order Review Sheet, Form A-001.

MINFOR INCORPORATED	Iss dt. 1/1/99	Rev dt. 9/2/99		Pg.
QUALITY OPERATING PROCEDURE	Sign	Rev #	1	4/5
REQUIREMENT: QUALITY SYSTEM MANUAL SEC: 4.8				
SUBJECT: PRODUCT IDENTIFICATION AND TRACEABILITY				
QOP 005 DOCUMENT REQ'D: Forms A-001, A-008				

3.01.1 Issuance of Product Related Documents

Contracts, Engineering, Purchasing, Planning, and Manufacturing shall:

1. Record on the applicable documents prior to issuance, the customer's name, the product's part number and revision level, and the serial, batch or lot identifying numbers. Record other information in addition, as required;
2. Update all the documents affecting the product when customers issue Purchase Order Amendments as defined in QOP 001, Section Four (pg 307).

3.01.2 Quality Provisions for Ancillary Requirements

The Quality Engineer shall:

1. Ensure that the product-related inspection documents have been identified with the customer's name, the product's part number and revision level, and the serial, batch or lot identifying numbers, as applicable, including rejected products;
2. Ensure that the NCMR (Non-conforming Material Report) serial number has been recorded on every product-related document, identifying the individual piece, batch or lot that has been rejected;
3. Ensure that no customer related product has been staged, stored, or in process within the confines of controlled areas without identifying document.
4. Ensure that materials, purchased to satisfy a customer based purchase order requirement, have the test pieces identified and documented, and the materials have been preserved, protected, and stored in order that traceability to the certified identity of the product can be adequately verified. Furthermore, ensure that analysis to validate certifications have been planned, controlled and documented, as required. **(SAE AS9000, par. 4.10.2.4)**

3.02 Customer Flow-down Requirement

The Quality Engineer shall:

1. Define product traceability and marking requirements in Form A-001 as required by the customer's PO and by internal controls;
2. Enforce flow-down description of product traceability and marking in the Job Traveler and Operation Sheets, as applicable.

The Process Engineer shall:

1. Define product traceability and marking requirements in the Job Traveler and Operation Sheets as required by the customer's PO and Form A-001;
2. Issue detailed work instruction and sketch as to show where product marking should

MINFOR INCORPORATED	Iss dt. 1/1/99	Rev dt. 9/2/99	Pg.
QUALITY OPERATING PROCEDURE	Sign	Rev # 1	5/5
REQUIREMENT: QUALITY SYSTEM MANUAL SEC: 4.8			
SUBJECT: PRODUCT IDENTIFICATION AND TRACEABILITY			
QOP 005	DOCUMENT REQ'D: Forms A-001, A-008		

be located and what methods, tools and materials be applied to accomplish it;

3. Incorporate any PO Amendments impacting product marking into the Job Traveler and Operation Sheets, as required. Maintain document change control regarding PO Amendments as defined in QOP 001, Section Four, par 3.11, step 2, (pg 309).

OVERVIEW TO QOP 006
(INSPECTION AND TEST CONTROL)

This QOP is the longest of the quality operating procedures and it is by design that way. Instead of having a dozen or more procedures written separately on the different inspection assignments needed to support product quality enforcement, I have combined them all under one control. During my forty years in the field of Quality, I found that the various inspection functions were documented in such a fragmented, confusing manner that it took longer to find the procedure covering a specific type of inspection than to do the work itself. This was one reason why I combined them all under one directory. The second reason was to cut down on the volume of documents burdening the quality system, and the third was to maintain linkage to one QOP instead of a dozen. Now, anything you want to find regarding inspections will take you a second to find under its own index.

This consolidated QOP is the first most important controlling document of the quality management system, for it deals with verification, acceptance, documentation and release of products to the customer. It has seven separate sections, each dedicated to control one or more inspection function, and it covers the entire processing cycle of a contracted product, including customer and subcontractor source inspection. Please look up the index, next page, to discover how extensively we cover the field of inspections.

MINFOR INCORPORATED	Iss dt. 1/1/99	Rev dt. 9/3/99	Pg.
QUALITY OPERATING PROCEDURE	Sign	Rev # 1	1/32

REQUIREMENT: QUALITY SYSTEM MANUAL SEC: 4.10

SUBJECT: INSPECTION AND TEST CONTROL

QOP 006|DOCUMENT REQ'D: This procedure

QUALITY OPERATING PROCEDURE 006

INSPECTION AND TEST CONTROL

CONTENTS   Page

Form A-003 Rev 0 1999

MINFOR INCORPORATED	Iss dt. 1/1/99	Rev dt. 9/3/99	Pg.
QUALITY OPERATING PROCEDURE	Sign	Rev # 1	2/32
REQUIREMENT: QUALITY SYSTEM MANUAL SEC: 4.10			
SUBJECT: INSPECTION AND TEST CONTROL			
QOP 006|DOCUMENT REQ'D: This procedure			

QUALITY OPERATING PROCEDURE 006

INSPECTION AND TEST CONTROL

Form A-003 Rev 0 1999

MINFOR INCORPORATED	Iss dt. 1/1/99	Rev dt. 9/3/99	Pg.
QUALITY OPERATING PROCEDURE	Sign	Rev # 1	3/32

REQUIREMENT: QUALITY SYSTEM MANUAL SEC: 4.10
SUBJECT: INSPECTION AND TEST CONTROL
QOP 006|DOCUMENT REQ'D: This procedure

QUALITY OPERATING PROCEDURE 006

INSPECTION AND TEST CONTROL

CONTENTS   Page

Form A-003 Rev 0 1999

MINFOR INCORPORATED		Iss dt. 1/1/99	Rev dt. 9/3/99		Pg.
QUALITY OPERATING PROCEDURE		Sign	Rev # 1		4/32
REQUIREMENT: QUALITY SYSTEM MANUAL SEC: 4.10					
SUBJECT: INSPECTION AND TEST CONTROL					
QOP 006	DOCUMENT REQ'D: This procedure				

REVISION HISTORY

Rev Date	Rev No	Description	Approval
9/3/99	1	In Section Five, under paragraph 3.02.2, step 4 had been added.	AG

MINFOR INCORPORATED	Iss dt. 1/1/99	Rev dt. 9/3/99	Pg.
QUALITY OPERATING PROCEDURE	Sign	Rev # 1	5/32
REQUIREMENT: QUALITY SYSTEM MANUAL SEC: 4.10			
SUBJECT: INSPECTION AND TEST CONTROL			
QOP 006|DOCUMENT REQ'D: This procedure			

GENERAL REQUIREMENTS

1.0   APPLICATION

This QOP 006 shall be applicable to enforce inspection and test requirements in the following areas:

1. First Piece Inspection
2. First Article Control
3. Process Control
4. Final Inspection
5. Receiving Inspection
6. Customer Source Inspection
7. Source Inspection at Subcontractors

2.0   DEFINITIONS   (used in controlling manufacturing processes)

1. **Controlled Process** – a defined activity undertaken to meet the requirement of a specification or work instruction after the approval of initial setup.
2. **Deviation** – a specific customer authorization issued prior to the processing of a product to allow departure from a defined design requirement for a specific number of units for a specific duration.
3. **First Article Inspection** – the complete inspection and test of a processed product made under process control wherein all measurements and test results, including special processes and workmanship, have been fully documented as required by a customer's purchase order.
4. **First Piece Inspection** -- the complete inspection and test of a product's phase operation, which may be a single or multiple process, combined and carried out under one setup.
5. **Process Owner** – any employee trained to carry out a defined process in order to meet a specified requirement.
6. **Product non-conformance** – is any condition that violates the requirement of a specification, process, or procedure.
7. **Repair** – relates to a non-conforming product, which cannot be further processed to meet a specified requirement without written approval from the customer.
8. **Rework** – relates to a non-conforming product that can be reprocessed under defined conditions to meet a specified drawing requirement.
9. **Waiver** – a written request sent to a customer to dispose of a non-conforming product.

MINFOR INCORPORATED	Iss dt. 1/1/99	Rev dt. 9/3/99	Pg.
QUALITY OPERATING PROCEDURE	Sign	Rev # 1	6/32
REQUIREMENT: QUALITY SYSTEM MANUAL SEC: 4.10			
SUBJECT: INSPECTION AND TESTING CONTROL			
QOP 006|DOCUMENT REQ'D: This procedure			

3.0   PROCESS OWNER REQUIREMENTS

Process owners (and inspectors) performing production assignments shall observe the following:

1. that the measuring equipment being used is calibrated and periodically re-verified;
2. that the customer's product drawing number and revision level is verified against the Job Traveler and the program software. That notification to supervision is given when inconsistency is discovered;
3. that the first production piece is submitted for inspection verification and approval on all new setups identified in the Job Traveler;
4. that the production processes on customers' products are not carried out without work results verification and documentation.
5. that each completed operation as indicated in the Job Traveler is stamped, or signed and dated, and that all quantities have been accounted for.

MINFOR INCORPORATED	Iss dt. 1/1/99	Rev dt. 9/3/99	Pg.
QUALITY OPERATING PROCEDURE	Sign	Rev # 1	7/32
REQUIREMENT: QUALITY SYSTEM MANUAL SEC: 4.10.3			
SUBJECT: FIRST PIECE INSPECTION			
QOP 006|DOCUMENT REQ'D: Forms A-006, A-008, A-012, A-013, A-017, A-018, A-023, A-033, A-037			

SECTION ONE

1.0   PURPOSE

Enforce quality provisions in controlling the verification, documentation, and approval of the first production unit. **(see SAE AS9000, paragraph 4.10.5.1)**

2.0   APPLICATION

The provisions of this procedure shall apply to:

1. First production piece verification and approval;
2. First piece rework/repair verification and approval;
3. First piece verification and approval of subcontracted work, excluding special process.

3.0   PROCEDURE

3.01   First Production Piece Verification and Approval

The Process Owners shall:
1. Ensure that the measuring equipment is suitable for the purpose and is calibrated;
2. Verify revision level of processing related documents to Job Traveler requirement;
3. Measure the work-piece and document the work results on Form A-033, as required;
4. Review the requirements to ensure that what is specified has been measured and the work results accurately recorded. Keep product and its document(s) together;
5. Sign the Job Traveler. Then, submit the product with its documents to inspection for first piece verification and approval or
6. Reject the product if a specified requirement is out of tolerance and it cannot be readjusted to meet the requirement. Notify your supervisor.

The Supervisor shall:
1. Monitor processing to ensure first production piece approval. Interact with process owner when the first production piece had been rejected;
2. Process NCMR through Quality for the rejected product, as required;
3. Implement corrective action, as required;
4. Determine production continuation (while first piece inspection approval is pending). Base this decision on risk involved, the complexity of the product and the skill level of the operator. Review job completion, according to the Job Traveler, then sign Form

MINFOR INCORPORATED	Iss dt. 1/1/99	Rev dt. 9/3/99	Pg.
QUALITY OPERATING PROCEDURE	Sign	Rev # 1	8/32
REQUIREMENT: QUALITY SYSTEM MANUAL SEC: 4.10.3			
SUBJECT: FIRST PIECE INSPECTION			
QOP 006 DOCUMENT REQ'D: Forms A-006, A-008, A-012, A-013, A-017, A-018, A-023, A-033, A-037			

A-033 (pg 492) under the "Audit" line;

5. Interact with all departments in matters involving processing problems affecting product quality, tooling, equipment, maintenance, scheduling, and performance requirements.

The Quality Inspector shall:

1. Identify the submitted product to the Job Traveler. Then obtain own drawing and related documents (amendment) from the Inspection Product Folder, as required;
2. Verify part number and revision level of your documents against process owner's documents. They must match. Work out differences, as required;
3. Ensure that your measuring equipment is suitable and is calibrated;
4. Measure the work-piece either in restrained or free state, whichever is required;
5. Compare the measurement results to the operator's recorded results on Form A-033; (Separate recording onto Form A-018, pg 473, may not be required)
6. Tag the first production piece with form A-013 (Acceptance Material Ticket) when all the requirements have been met. Stamp, or sign and date the Job Traveler. Do the same on Form A-033;
7. Notify the supervisor of first piece acceptance and the approval of the process. Return the accepted and tagged first piece to the process owner, as required, or
8. Advise the supervisor if the first production piece was rejected and that the process was not approved. If rework cannot be accomplished, process NCMR, as required. Resubmission for first piece approval shall be required. Resume first piece verification from step "4" above when resubmission takes place;
9. Maintain control over filing the documents in line with QOP 002, Sec Three (pg 328).

3.02   First Piece Rework/Repair Verification and Approval

Process and Quality Engineers shall observe the following:

1. When rework is required while processing is still in progress, irrespective of before or after first production unit approval, the non-conforming product may be reprocessed on the authority of the Job Traveler without rework instruction, issued on Form A-023;
2. When rework is required after processing has been completed and the setup dismantled, the non-conforming product shall be reprocessed to the requirement of work instruction, issued on Form A-023 (pg 479);
3. When more than one piece is required to be reworked/repaired, the first reworked/repaired unit shall be submitted to inspection for verification and approval;
4. With one piece rework/repair, the setup shall not be dismantled until the reworked/repaired product has been verified and accepted by authorized personnel;
5. All NCMR based rework/repair shall have release approval by final inspection;

MINFOR INCORPORATED	Iss dt. 1/1/99	Rev dt. 9/3/99	Pg.
QUALITY OPERATING PROCEDURE	Sign	Rev # 1	9/32

REQUIREMENT: QUALITY SYSTEM MANUAL SEC: 4.10.3

SUBJECT: FIRST PIECE INSPECTION

QOP 006 | DOCUMENT REQ'D: Forms A-006, A-008, A-012, A-013, A-017, A-018, A-023, A-033, A-037

The Process Owner shall:

1. Determine rework/repair requirement. Follow the instructions. Rework/ repair the product, as required.
2. Verify the reworked/repaired unit to the specified requirement and record the measurement result(s) either on the first worksheet (A-033) or onto Form A-023, whichever is required. Sign and date the document showing the work results;
3. Submit the reworked/repaired product with its documents to inspection for verification and product release approval;
4. Report processing and/or document problems to your supervisor.

The Quality Inspector shall:

1. Review the rework/repair requirements from Form A-023. Keep track of the identified NCMR. Determine whether rework/ repair quantity is one or more piece(s);
2. Do final release verification and approval when rework/repair quantity is one piece. Do first piece verification and approval when the rework/repair quantity is more than one piece. Sign and date Form A-023 after the reworked/repaired piece(s) have been verified and accepted;
3. Release the product according to the identified NCMR requirement. (The NCMR number is recorded on Form A-023). Keep the product tagged (A-012), as required;
4. Maintain document retention according to QOP 002, Section Three (pg 328);
5. When the reworked/repaired product is rejected and it cannot be corrected through additional processing, issue a new NCMR (see pg 410).

3.03   First Piece Approval of Subcontracted Work, Excluding Special Process

The Receiver shall:

1. Sign document(s) and receive the product sent by the subcontractor. (If product had been hand carried, there may be nothing to sign.);
2. Turn the product and the attached document(s) over to the Quality Department.

The Quality Inspector shall:

1. Identify the subcontractor and obtain the internally issued PO from the Supplier Product Folder;
2. Determine the Quality Provisions from the PO as additional requirements to the terms and conditions of the PO;
3. Verify the product to the PO and the specification (drawing) requirements. Record measurement results onto Form A-018. Sign and date Form A-018;
4. Notify Purchasing (or Subcontractor's Quality) on verification results:
   a) If the product had been rejected, require resubmission of another first piece;

MINFOR INCORPORATED	Iss dt. 1/1/99	Rev dt. 9/3/99	Pg.
QUALITY OPERATING PROCEDURE	Sign	Rev # 1	10/32
REQUIREMENT: QUALITY SYSTEM MANUAL SEC: 4.10.3			
SUBJECT: FIRST PIECE INSPECTION			
QOP 006 DOCUMENT REQ'D: Forms A-006, A-008, A-012, A-013, A-017, A-018, A-023, A-033, A-037			

b) If the product had been accepted, authorize production;

c) Send the product with inspection results (A-018) back to the subcontractor, as required. Tag the product before returning it, either with Form A-013 (Acceptance Material Ticket, pg 468), or with Form A-012 (Rejected Material Ticket, pg 467), whichever is applicable;

d) Issue SCAR on Form A-037 (Supplier Corrective Action Request), as required;

5. Retain Form A-018 in the Supplier Product Folder. Return the folder to its original file;

6. Repeat steps 1 through 5 again when product resubmission takes place.

MINFOR INCORPORATED	Iss dt. 1/1/99	Rev dt. 9/3/99	Pg.
QUALITY OPERATING PROCEDURE	Sign	Rev # 1	11/32
REQUIREMENT: QUALITY SYSTEM MANUAL SEC: 4.10			
SUBJECT: FIRST ARTICLE CONTROL (qualification process)			
QOP 006	DOCUMENT REQ'D: A-001, A-006, A-008, A-012, A-013, A-017, A-018, A-023, A-033, A-037		

SECTION TWO

1.0 PURPOSE

Enforce the quality provisions in controlling First Article related assignments for internal and external customer requirements. (Internal customer requirements are marked with a star. External customers include both, as required.) This is a first article qualification process control for a single unit. For multiunit, becoming an assembly, the same provisions apply, except in the plural.

2.0 APPLICATION

Provisions of this procedure shall apply to controlling:
☆ 1. phases of the process assignments
☆ 2. inspection and documentation
☆ 3. customer source inspection
☆ 4. packaging and delivery
☆ 5. document retention

3.0 PROCEDURE

General Requirements

☆ 1. First Article quality requirements shall be detailed in the Purchase Order Review Sheet, Form A-001 (pg 455);
☆ 2. The Job Traveler (A-008, pg 463) shall outline the processing steps in sequential order, including subcontracting, in order to control an orderly process realization;
☆ 3. The Quality Engineer shall enforce process controls in line with the requirements stated in the Job Traveler and the Purchase Order Review Sheet;
☆ 4. All the provisions of the quality system procedures shall be binding on all departments carrying out the First Article objectives, as required.

4.0 Issuance of Internal Product Quality Requirements

The Quality Engineer shall:
☆ 1. Review the customer's PO and product specifications in order to evaluate, plan, and formulate the internal quality requirements of the First Article unit to control it during its entire processing cycle;
☆ 2. Determine project specific quality requirements for materials, in-house production

MINFOR INCORPORATED	Iss dt. 1/1/99	Rev dt. 9/3/99	Pg.
QUALITY OPERATING PROCEDURE	Sign	Rev # 1	12/32
REQUIREMENT: QUALITY SYSTEM MANUAL SEC: 4.10.1			
SUBJECT: FIRST ARTICLE INSPECTION (qualification process)			
QOP 006 DOCUMENT REQ'D: A-001, A-006, A-008, A-012, A-013, A-017, A-018, A-023, A-033, A-037			

processes and subcontracted operations in order to summarize the detail flow-down requirements. (Refer to the customer's PO Quality Provisions);

☆ **3.** Write down these requirements onto Form A-001 (Purchase Order Review Sheet). If required, do the following:

   a) put together a supplemental Quality Plan, using the blank form of A-003 pg 457. Do the paragraph numbering and sequencing of events as you see it done, for instance, in this QOP;

   b) write in the body of the "Procedure," paragraph by paragraph, the various steps needed to impose the specific requirements. Don't write the quality operating procedures over. Just cite "as stated in QOP —, Section —, paragraph —";

☆ **4.** Issue the completed Form A-001 to Engineering and Contracts. Attach the Quality Plan, as required;

☆ **5.** Keep all the Fist Article documents (A-001, PO, dwg., etc.) in one folder and locate it centrally as you and others will frequently need to refer to it later. (This folder will be the First Article Inspection Product Folder.) Mark the folder "DEVELOPMENT."

5.0 Issuance of the First Article Production Folder

The Project Engineer shall:

☆ **1.** Review the customer's PO and product specifications in order to evaluate, plan, and formulate the processing related requirements of the First Article unit to control it during its entire production cycle. Review Form A-001 sent to you by the Quality Engineer and incorporate the quality requirements into the processing related documents (A-008), as required;

☆ **2.** Determine the project related requirements for materials, in-house production processes and subcontracted operations in order to incorporate the detail flow-down requirements of the First Article into the processing layout plans. Also determine the following:

   a) phase operational requirements for equipment, tools and fixtures, special gages, programming, and qualified personnel;

   b) processing capability for each phase operation. If required, invoke the "developmental" process controls stated under QOP 012 , Sec Three (pg 449), for proving out the process techniques before issuing the final processing related procedures;

☆ **3.** Interact with the Quality Engineer in order to ensure that the processing related documents (A-008) and the product's inspection requirements (A-001) are properly linked to control the First Article throughout its processing cycle;

☆ **4.** Interact with the department heads to ensure adequate planning and coordination regarding purchasing, scheduling, production, maintenance, training and other related matters;

MINFOR INCORPORATED	Iss dt. 1/1/99	Rev dt. 9/3/99	Pg.
QUALITY OPERATING PROCEDURE	Sign	Rev # 1	13/32
REQUIREMENT: QUALITY SYSTEM MANUAL SEC: 4.10.1			
SUBJECT: FIRST ARTICLE INSPECTION (qualification process)			
QOP 006|DOCUMENT REQ'D: A-001, A-006, A-008, A-012, A-013, A-017, A-018, A-023, A-033, A-037			

☆ **5.** Bring together and formalize the processing related procedures, as required – (setup sheet, tool and gage sheets, cut sheet, programming, sketches, operation sheets and the job traveler);

☆ **6.** Review and approve these controlling procedures. Issue the "Developmental" Production Folder, as required. Mark the folder as such.

☆ **7.** After successful process qualification and final inspection approval, formalize the developmental procedures and release them in the applicable Production Folders as standard operating work instruction, as required.

6.0   First Article Process Control

The Quality Engineer shall:

☆ **1.** Identify the process control specifics from the already issued Job Traveler and Purchase Order Review Sheet and write them down on a blank form of A-003 and title it "In-process Surveillance and Follow-up." Then, determine the number and title of the various forms required to record the work results in the different areas of assignments. Create a matrix right on form A-003 and use it as your checklist as you do the process review;

☆ **2.** Now, go to Section Three of this QOP and determine the process control requirement for each form – what it stands for, where it is used, and who is the process owner;

☆ **3.** Carry out the process review in each department, as required. Review and verify whether each process owner has carried out the work result documentation (recording) in the assigned form(s), as required. Note the reviewed results on form A-003;

☆ **4.** Carry out on spot corrective action where infringements are minor. Note this also on form A-003. For major product quality violations, go to paragraph 7.0 below;

☆ **5.** Sign and date each form under the "Audit" line. Sign and date form A-003 also;

☆ **6.** Enter all the corrective action related instructions you did onto Form A-016 (Training Log, pg 471);

☆ **7.** Retain the training log sheets in the Training Log Folder. Retain Form A-003 in the first article developmental folder.

7.0   Handling First Article Non-conforming Products

The Quality Engineer shall:

☆ **1.** Review the applicable Inspection Report (A-018) indicating the recorded evidence of non-compliance (rejection data);

☆ **2.** Process Form A-006 (NCMR) in line with QOP 009, Sec Two (pg 410), as required;

☆ **3.** Submit the non-conforming product with all its document attachment to MRB for review and disposition in accordance with QOP 009, Sec Two (pg 411);

MINFOR INCORPORATED	Iss dt. 1/1/99	Rev dt. 9/3/99		Pg.
QUALITY OPERATING PROCEDURE	Sign	Rev #	1	14/32
REQUIREMENT: QUALITY SYSTEM MANUAL SEC: 4.10.1				
SUBJECT: FIRST ARTICLE INSPECTION (qualification process)				
QOP 006|DOCUMENT REQ'D: A-001, A-006, A-008, A-012, A-013, A-017, A-018, A-023, A-033, A-037				

☆ **4.** Interact with the department heads and others in carrying out the MRB disposition. (The MRB disposition may involve the customer through the waiver request on Form A-024, pg 480 and rework/repair on Form A-023, pg 479);

☆ **5.** Interact with those departments required to implement corrective action, as required;

☆ **6.** Follow up the corrective action in order to determine effective implementation;

☆ **7.** Close out the NCMR (A-006) in line with QOP 009, Sec Two, par 3.01.7 (pg 416);

☆ **8.** Retain a copy of the First Article related MRB documents in the First Article Developmental Product Folder (customer or in-house, whichever the case).

8.0   Compiling the First Article Inspection Report

Note 1.   The First Article Inspection Report shouldn't be put together until all the required processing steps have been completed and the product verified, approved, and released from final inspection.

Note 2.   The First Article Inspection Report consists of all those documents that were issued to control the various processing operations including subcontracting, special processes and any MRB activity, proving work results documentation in compliance with requirements.

The Quality Engineer shall:

1. Bring together the Production Folder, First Article Inspection Product Folder, and the Supplier Product Folder. (These three folders should contain all the First Article records substantiating the processing and product quality control records);

2. Find the Job Traveler and Form A-001. Review both documents and ensure that all required operations had been signed off. If omissions have been found, determine the impact and process an NCMR, as required, and execute it in accordance with QOP 009, Section One and Two. (You may omit this step if forms A-008 and A-001 were already reviewed under paragraph 6.0 above.) Begin gathering the documents needed to make up the First Article Inspection Report (FAIR);

3. Start with the Job Traveler. The first operation should be "Material Release" (see parts list, as required). Find the material's certification. This document should be your fourth record in the "FAIR" (First Article Inspection Report) stack. (The first document is the Customer's PO. The second one is the Job Traveler. The third one is Form A-001);

Note 3.   When the customer supplies the material, it usually doesn't provide the certification with it. The customer's PO should specify what has been supplied.

4. Go step by step following the Job Traveler, pull together from the folders all those records substantiating work results documentation, including subcontracting and special processes. Place the documents in sequential order. Place any NCMR related

MINFOR INCORPORATED	Iss dt. 1/1/99	Rev dt. 9/3/99	Pg.
QUALITY OPERATING PROCEDURE	Sign	Rev # 1	15/32
REQUIREMENT: QUALITY SYSTEM MANUAL SEC: 4.10.1			
SUBJECT: FIRST ARTICLE INSPECTION (qualification process)			
QOP 006	DOCUMENT REQ'D: A-001, A-006, A-008, A-012, A-013, A-017, A-018, A-023, A-033, A-037		

documents where the non-conformance occurred. The last document should be the Final Inspection Report (A-018). (The Certificate of Conformance should be issued as part of the shipping documents, unless otherwise arranged);

5. Remove the staples from the documents, as required. Make as many copies of the collection as needed. Then, sort the copies in sequential order to make up the individual stacks;
6. Take the originals and staple them back together as they were before and put them back into the folders from where you removed them;
7. Make up the index page indicating the sequential order of the documents – title and subject. Numbering the pages is optional. Title the index page "FAIR" and show the customer's name, PO and part number on it.
8. Put one index page on top of each stack and bind each stack in a folder, suitable for the occasion.
9. Locate the First Article unit. (Ensure that Form A-013, Acceptance Material Ticket, is on it. If not, check completion of Final Inspection. Also, if there were customer authorized rework/repair done, Form A-012 should also be on it as well);
10. Put a copy of the "FAIR" onto the First Article unit and move it to a clean room. Hold the original folders as well as the additional "FAIR" copies close by the product until the customer's source inspection is over. (The Source Representative will need the product specification documents (drawings) and other related documents from the Production or the Inspection Folders);

9.0   Preparation for Customer Source Inspection

The Quality Engineer shall:
1. Contact the customer's quality department and request First Article Source Inspection service. Agree on time and date;
2. Make arrangement for the following while waiting for source inspection:
    a) have the first article unit cleaned;
    b) have the measuring equipment set up to facilitate source inspection. Check the calibration stickers for validity (due date);
    c) have the place cleaned and organized;
    d) provide assistance to the customer's representative as needed.

10.0   Preparation for First Article Delivery

The Quality Engineer shall:
1. Review and approve (as required) all shipping related documents issued by authorized

MINFOR INCORPORATED	Iss dt. 1/1/99	Rev dt. 9/3/99		Pg.
QUALITY OPERATING PROCEDURE	Sign	Rev #	1	16/32
REQUIREMENT: QUALITY SYSTEM MANUAL SEC: 4.10.1				
SUBJECT: FIRST ARTICLE INSPECTION (qualification process)				
QOP 006 DOCUMENT REQ'D: A-001, A-006, A-008, A-012, A-013, A-017, A-018, A-023, A-033, A-037				

    personnel. Ensure listing on the Packing Slip of all documents that are required to be sent to the customer with the shipment;

2. Issue Certificate of Conformance (Form A-029, pg 488);
3. Verify packaging to the customer's PO and A-001 requirement;
4. Place the First Article Inspection Report, the "C of C," Form A-038 (pg 497) and the customer's Source Acceptance Report (SAR) in the "Documents Enclosed" envelope and secure it to the inside or outside of the product container (box, skid, etc.) as required. Review "Bill of Lading" as required;
5. Release the product for shipment as directed by the customer.

11.0   Document Retention

☆ **General**

The First Article documents are not standard production documents due to the nature of project specific handling and controls imposed externally by the customer's Purchase Order, or internally for processing qualification requirements. Although the controlling documents had been derived from the established quality system's procedures, the purpose for which they were applied made them isolated under First Article sectioning. Therefore, retention of these documents should not be intermixed with the standard production and quality records. For internal document retention, file the records under "Developmental Documents." For external requirement, file the records as agreed with Engineering.

The Quality Engineer shall:
1. Interact with Engineering and determine the storage location of the First Article documents;
2. Place the Production Folder, the First Article Inspection Product Folder, the Supplier Product Folder and the extra copies of the FAIR in a container suitable for storage;
3. Place a label on the container, indicating "FAIR," the customer's name and product part number, and the container's contents. Place the container in storage, as required.

Note 4. Should the First Article unit be released by the customer for production, the stored documents would facilitate prompt reestablishment of all the past performance and product quality requirements (for a fixed process control – ESA).

MINFOR INCORPORATED	Iss dt. 1/1/99	Rev dt. 9/3/99	Pg.
QUALITY OPERATING PROCEDURE	Sign	Rev # 1	17/32
REQUIREMENT: QUALITY SYSTEM MANUAL SEC: 4.9			
SUBJECT: PROCESS CONTROL (Performance Surveillance)			
QOP 006│DOCUMENT REQ'D: All the forms listed and controlled in the Quality System			

SECTION THREE

1.0   PURPOSE

Maintain performance surveillance over those activities affecting the quality system's objectives and requirements.

2.0   APPLICATION

Performance surveillance shall apply to the following document handling, recording, and processing related activities: (Mistake-Proofing Control)
1. material control
2. tools, fixtures, and processing equipment
3. processing related documents
4. dimensional verification
5. special processes
6. final inspection
7. workmanship and other processing related requirements
8. non-conforming materials
9. product release
10. customer returns
11. corrective action and follow-up
12. contracts, engineering, and purchasing

3.0   PROCEDURE

General Requirements
1. Supervision shall be responsible for processing review performance results of the activities listed under paragraph 2.0 above in order to control those activities to ensure compliance with the requirements of documented procedures. The applicable forms shall be acknowledged by the reviewer by his/her signature and date at the time the performance review was accomplished.
2. The supervisor shall have the authority to implement on-spot corrective action with the process owner(s) who are assigned under his/her responsibility. For minor violations where only instructional methods have been used to correct documentation, the supervisor shall apply Form A-016 (Training Log Sheet) to record the activity. For major violations impacting product quality, the supervisor shall apply the NCMR process (on Form A-006) in line with QOP 009, Sec One (pg 408) and Two (pg 410), as required.

MINFOR INCORPORATED	Iss dt. 1/1/99	Rev dt. 9/3/99	Pg.
QUALITY OPERATING PROCEDURE	Sign	Rev # 1	18/32
REQUIREMENT: QUALITY SYSTEM MANUAL SEC: 4.9			
SUBJECT: PROCESS CONTROL (Performance Surveillance)			
QOP 006 DOCUMENT REQ'D: All the forms listed and controlled in the Quality System			

3.01 Material Control

The Inspection Supervisor, or the Quality Engineer shall:

1. review the process owner's performance to ensure that the product's identity (by part number and description) and the related identifiers (customer, S/N, lot/batch number) have been accurately recorded on the applicable forms in the controlled areas, as required:

   a) at Receiving (on Forms **A-015 and A-027**) as stated in QOP 006, Section Five, paragraph 3.02.1 (pg 383);

   b) during processing (on Forms **A-008 and A-033**), as required;

   c) when product is non-conforming (on Form **A-006**) as stated in QOP 009, Sec Two.

3.02 Tools, Fixtures, and Equipment

The Production Supervisor shall:

Review that tools, fixtures, and processing equipment are adequately maintained as indicated on Form **A-002** and available upon the release of the Production Folders to manufacturing. See QOP 012, Sec One, par 3.03.2 (pg 440). **(SAE AS9000, par 4.9.2)**

3.03 Processing Related Documents

The Production Supervisor shall:

1. review before the start of production that processing related documents have been authorized, adequately completed, incl. workmanship and safety notes. Verify their presence in Production Folder(s) at processing. Review the validity of revisions in the applicable documents as stated in QOP 001, Sec Two, par 3.03, step 5, (pg 302):
   a) in the job traveler (**A-008**) and operation sheet, as required;
   b) in the setup sheet, tool sheet, gage sheet, program print-out, cut sheet, drawings and other related specifications, as required;

3.04 Dimensional Verification

The Production Supervisor shall:

Review that process owners had properly recorded onto Forms **A-033** the work results after product verification of the first piece production unit as stated in QOP 006, Section One, par 3.01, step 3, (pg 364), and thereafter for the production pieces as stated in the QSM, par 4.9c (pg 281), as required. Review that process owners have signed the Job Traveler (**A-008**) under the designated operation(s) as required by the same quality provisions. Take C/A as needed.

MINFOR INCORPORATED	Iss dt. 1/1/99	Rev dt. 9/3/99	Pg.
QUALITY OPERATING PROCEDURE	Sign	Rev # 1	19/32
REQUIREMENT: QUALITY SYSTEM MANUAL SEC: 4.9			
SUBJECT: PROCESS CONTROL (Performance Surveillance)			
QOP 006 DOCUMENT REQ'D: All the forms listed and controlled in the Quality System			

3.05   Special Process

The Inspection Supervisor shall:

1. review that special process requirements have been transcribed, as required, from the customer's PO onto Form **A-001** as stated in QOP 001, Sec Two, par 3.02 (pg 302) and from there onto the Job Traveler, Form **A-008**, as stated in QOP 001, Sec Two, par 3.03 in order to implement the specification and process flow-down requirements;
2. review that the internally issued Purchase Orders (**A-017**) have also imposed the same flow-down requirements on the subcontractor(s) for the identified product(s) as stated in QOP 003, Sec One, par 3.02, Note 5, (pg 339). **(see SAE AS9000, par 4.9.1)**

3.06   Final Inspection

The Inspection Supervisor shall;

1. review that the first piece product verification and approval activities have been carried out, recorded, and signed for by inspection personnel on Forms **A-018** and **A-008** as stated in QOP 006, Section One (pg 364);
2. review final product verification and acceptance on Forms **A-018, A-013, and A-008** as stated in QOP 006, Section Four (pg 380). Review the inspectors' job performance as stated in QOP 006, Section Four, all paragraphs apply;
3. review the tracking and recording of the accepted and rejected quantities as described in the same quality provisions. **(see SAE AS9000, par 4.9h and 4.9i)**

3.07   Workmanship and Other Processing Related Requirements

The Production Supervisor shall:

1. review that workmanship and other processing related requirements have been completed, verified and documented as per work instructions in the Job Traveler and Operation Sheet. Refer to QOP 001, Section Two, par 3.03, step 2 (pg 302);
2. review that the completed work pieces have been cleaned and foreign objects (chips, dirt) removed and preservation is maintained to the same provisions of QOP 001. **(see SAE AS9000, par 4.9j)**

3.08   Non-conforming Materials

The Inspection Supervisor shall:

1. review that non-conforming products have been identified and documented throughout

MINFOR INCORPORATED	Iss dt. 1/1/99	Rev dt. 9/3/99	Pg.
QUALITY OPERATING PROCEDURE	Sign	Rev # 1	20/32
REQUIREMENT: QUALITY SYSTEM MANUAL SEC: 4.9			
SUBJECT: PROCESS CONTROL (Performance Surveillance)			
QOP 006 DOCUMENT REQ'D: All the forms listed and controlled in the Quality System			

the entire processing areas as stated in QOP 009, Section One and Two. Identification requires the application of Form **A-012** and documentation requires the application of Form **A-006**;

2. review that the NCMR number has been transcribed onto the other processing related documents such as Forms **A-008, A-009, A-018, A-012, A-024, A-023 and A-037** to ensure the recognition that the product had NCMR history;

3. review non-conforming products have not been intermixed with conforming products.

3.09   Product Release

The Inspection Supervisor shall:

1. review that product releases have complied with documented procedures:
    a) at Receiving **(A-015)** as stated in QOP 006, Section Five, paragraph 3.01 and 3.02 (pg 383);
    b) during in-process **(A-008)** as stated in QOP 006, Sec One, par 3.01 (pg 364);
    c) at MRB **(A-006)**, as stated in QOP 009, Sec Two, par 3.01.7 (pg 416);
    d) at shipping **(A-018, A-013, A-028, and A-029)** as stated in QOP 006, Section Four, paragraph 3.02 (pg 381) and 3.03 (pg 382).

3.10   Customer Returns

The Inspection Supervisor shall:

Review that customer returned materials have been handled in accordance with documented procedures as defined in QOP 009, Section Three (pg 417). All the provisions apply.

3.11   Corrective Action and Follow up

The Inspection Supervisor shall:

1. review to verify that the Management Representative has accepted the effective implementation of corrective action as stated in QOP 009, Section Two, paragraph 3.01.6 (pg 415);

2. review closure of the applicable NCMR **(A-006)** as stated in QOP 009, Section Two, paragraph 3.01.7 (pg 416) in order to maintain continuous improvement built into the quality system's procedures.

3.12   Contracts, Engineering, and Purchasing Activities

3.12.1 The Contracts Administrator shall:

MINFOR INCORPORATED	Iss dt. 1/1/99	Rev dt. 9/3/99	Pg.
QUALITY OPERATING PROCEDURE	Sign	Rev # 1	21/32
REQUIREMENT: QUALITY SYSTEM MANUAL SEC: 4.9			
SUBJECT: PROCESS CONTROL (Performance Surveillance)			
QOP 006 DOCUMENT REQ'D: All the forms listed and controlled in the Quality System			

1. review to verify that the Administrative Secretary properly managed the handling, recording and filing of the customers' Purchase Orders and Amendments as stated in QOP 001 Section One, Two, Three, and Four (pg 299,301,304, 307);
2. review that work results recorded on Forms **A-004, A-005, A-011, and A-031** adequately demonstrate compliance to the requirements of the same quality operating procedures.

3.12.2 The Engineering Supervisor shall:

1. review that the Process Engineer has adequately carried out: the product related planning; the determination of the product's processing layout; the formulation, review, approval, and issuance of the product's processing related procedures; and the follow-up activities as stated in QOP 001, Section One, Two, Three, and Four;
2. review that the Process Engineer has effectively carried out and complied with the internal document change provisions of QOP 001, Section Four (pg 307), in the application and execution of Form **A-009**;
3. review that the Process Engineer has adequately carried out and complied with the provisions of QOP 009, Section One, Two, and Three regarding the review, disposition, and determination of corrective action on non-conforming products;
4. review that the Process Engineer had adequately complied with the provisions of QOP 002, Sec Three, par 3.03 (pg 330) in maintaining document retention;
5. review that the Process Engineer has carried out the engineering related performance requirements stated in QOP 012, Section One, Two, and Three (pg 438,442,449).

3.12.3 The Purchasing Supervisor shall:

Review that the Purchasing Agent has adequately complied with the provisions of QOP 003, Sec One and Two, in carrying out the preliminary investigation in the selection of subcontractors and the subsequent review, approval, and issuance of internal Purchase Orders to the subcontractors in collaboration with the cognizant Quality Eng (pg 338).

3.13   Audit Compliance Verification and Documentation

3.13.1 The Management Representative shall:

Audit, the supervisory review requirements imposed in this Process Control procedure and those that are imposed in QOP 010 (pg 421) according to the Audit Schedule, Form **A-021** (pg 477). Record the audit results onto Form **A-022** (pg 478).

Note 5: The Management Representative shall not audit those areas for which he/she has direct work assignment responsibility as in paragraph 3.11 above.

MINFOR INCORPORATED	Iss dt. 1/1/99	Rev dt. 9/3/99		Pg.
QUALITY OPERATING PROCEDURE	Sign	Rev #	1	22/32
REQUIREMENT: QUALITY SYSTEM MANUAL SEC: 4.9				
SUBJECT: PROCESS CONTROL (Performance Surveillance)				
QOP 006 DOCUMENT REQ'D: All the forms listed and controlled in the Quality System				

3.13.2 Any Manager Except The Management Representative shall:

Audit, independently, that the Management Representative has carried out and complied with the requirements of QOP 010 (pg 421). Record the independent audit results on Form **A-022.** Title it: "INDEPENDENT AUDIT VERIFICATION."

MINFOR INCORPORATED	Iss dt. 1/1/99	Rev dt. 9/3/99	Pg.
QUALITY OPERATING PROCEDURE	Sign	Rev # 1	23/32
REQUIREMENT: QUALITY SYSTEM MANUAL SEC: 4.10.4			
SUBJECT: FINAL INSPECTION AND TESTING			
QOP 006|DOCUMENT REQ'D: A-001, A-006, A-008, A-012, A-013, A-014, A-018, A-028, A-029, A-033			

SECTION FOUR

1.0   PURPOSE

Maintain control over final inspection and testing.

2.0   APPLICATION

This procedure shall apply to the verification, documentation, acceptance, and release of a completed product in compliance with the customer's Purchase Order requirements.

3.0   PROCEDURE

General Requirement

1. Final inspection and testing entail the review of past performance regarding the product's processing, verification, acceptance, and documentation. Additional verification and documentation shall be done, as necessary, in order to demonstrate complete compliance to the customer's Purchase Order requirements, prior to delivery (or storage);
2. In the absence of PO quality provisions, the internally imposed and implemented quality requirements shall be reviewed and approved prior to releasing a completed product for delivery to the customer (or to storage);
3. The above provisions equally apply to customer returned products.

3.01   Document Review

The Quality Inspector shall:

1. collect the product's processing and inspection folders, as applicable (Inspection Product Folder, Supplier Product Folder, Production Folder, and the NCMR Folder);
2. take the Job Traveler (A-008) from the Production Folder and Form A-001 (Purchase Order Review Sheet) from the Inspection Product Folder;
3. review the Job Traveler against the inspection requirements imposed in Form A-001 and establish compliance to the imposed requirements:
   a) verify the process owners' signature under each specified operation;
   b) verify the Quality Inspectors' signature where first piece inspection was required;
   c) verify any subcontracted work acceptance either on Form A-018 or on the subcontractor's form;
   d) verify the acceptance of rework/repair, documented on Form A-023, resulting from MRB and the customer's disposition (waiver A-024);
   e) verify the acceptance of customer returned product(s) as per "d" above.

MINFOR INCORPORATED	Iss dt. 1/1/99	Rev dt. 9/3/99	Pg.
QUALITY OPERATING PROCEDURE	Sign	Rev # 1	24/32
REQUIREMENT: QUALITY SYSTEM MANUAL SEC: 4.10.4			
SUBJECT: FINAL INSPECTION AND TESTING			
QOP 006 DOCUMENT REQ'D: A-001, A-006, A-008, A-012, A-013, A-014, A-018, A-028, A-029, A-033			

4. verify further that the documented records fully cover verification and acceptance of all the quantities required by the Job Traveler, (see SAE AS9000, par 4.9h and 4.9i);
5. contact the responsible supervisor if the documentation is incomplete or the records are missing in order to implement on spot corrective action as stated under paragraph 3.04 (pg 375) of Section Three of this QOP 006. (Last sentence)

3.02 Final Product Verification

The Quality Inspector shall:
1. ensure that the selected measuring equipment is calibrated;
2. verify that the part description and revision level in the controlling documents (drawing, job traveler and operation sheet, software program) accurately reflect the customer's Purchase Order part description and revision level requirement;
3. determine whether 100% or sample inspection is required. Comply with Mil-Std-105, or ANSI Z1.4, or the customer-imposed requirement, whichever is applicable, in selecting samples. Select the required samples. Then:
4. obtain a blank form of A-018 and fill in the product related information from the Job Traveler in the upper section of the form. Fill in under "Type of Inspection": Final Inspection and Testing. Now:
5. inspect the product as required and record the verification results on Form A-018. Then:
   a) verify and record the results of any special process work done, as applicable;
   b) verify and record the results of any rework/repair work done, as applicable;
   c) verify and record the results of any workmanship done (refer to drawing notes);
   d) do visual (optical) inspection and record the results, as required.
6. handle the product which may require MRB disposition in accordance with QOP 009, Section One, Two, and Three. Otherwise, interact with departmental supervision;
   Note 6: any product with customer authorized rework/repair (on waiver Form A-024) requires tagging with Form A-012 prior to shipment in addition to tag A-013.
7. stamp, or sign and date all the required documents after product acceptance;
8. follow up and ensure NCMR closure (pg 416), as required;
9. release the product on Form A-013 (Acceptance Material Ticket) for delivery to the customer (or to storage). Send a copy of Form A-018 to the Contracts Administrative Secretary to issue the Packing Slip, Form A-028 (pg 487). Advise him/her that the following documents (give the description from A-001) will have to be listed in the Packing Slip as enclosures with shipment;
10. now make the copies of those documents required to be enclosed with product shipment and put them aside 'till you receive the Packing Slip. Now:
11. put all the original documents back to their respective folders (Inspection Product Folder, Supplier Product Folder, Production Folder, NCMR Folder), except:

MINFOR INCORPORATED	Iss dt. 1/1/99	Rev dt. 9/3/99		Pg.
QUALITY OPERATING PROCEDURE	Sign	Rev #	1	25/32
REQUIREMENT: QUALITY SYSTEM MANUAL SEC: 4.10.4				
SUBJECT: FINAL INSPECTION AND TESTING				
QOP 006 DOCUMENT REQ'D: A-001, A-006, A-008, A-012, A-013, A-014, A-018, A-028, A-029, A-033				

12. retain the Job Traveler, the Operation Sheet, and Form A-033 in the Inspection Product Folder, as required. Return the Production Folder to Engineering and put the rest of the folders back to their original locations.

3.03   Product Delivery

The Quality Inspector shall:

1. determine from Form A-001 whether Customer Source Inspection is a requirement. If yes, follow Section Six of this QOP 006 (pg 386), otherwise:
2. review the Packing Slip. Ensure that the documents that will be sent to the customer with the shipment have been accurately listed. Require correction, as necessary. Then, sign the Packing Slip and release it to the Shipping Department. In the meantime:
3. issue the "C of C," when required, on Form A-029 (pg 488). Make a copy of it!
4. place the listed documents, including the original of the "C of C" and Form A-038 (Customer Complaint and Evaluation Report, pg 497) and the customer's Source Acceptance Report (SAR), when applicable, in the "Documents Enclosed" envelope (mark the envelop as such). Take the envelope and go to the Shipping Department. Now:
5. verify any handling damage, cleaning and preservation and wrapping requirements, verify product marking, as required, prior to packaging. Verify the packaging methods of the product as stipulated by the customer's PO and/or A-001. When satisfied, take the "Customer Copy" of the Packing Slip and place it inside the envelope. Seal the envelope and secure it to the inside or outside of the product container (box, skid, etc.) as required. Review "Bill of Lading" as appropriate;
6. release the product for shipment;
7. retain the "Quality Copy" of the Packing Slip in the Inspection Product Folder. Retain a copy of the "C of C" in the Certification Folder. Forward the "Contracts Copy" of the Packing Slip to the Administrative Secretary.

MINFOR INCORPORATED	Iss dt. 1/1/99	Rev dt. 9/3/99	Pg.
QUALITY OPERATING PROCEDURE	Sign	Rev # 1	26/32
REQUIREMENT: QUALITY SYSTEM MANUAL SEC: 4.10.2			
SUBJECT: RECEIVING INSPECTION AND TESTING			
QOP 006|DOCUMENT REQ'D: Forms A-001, A-006, A-012, A-015, A-017, A-018, A-027			

SECTION FIVE

1.0   PURPOSE

Maintain control over received products.

2.0   APPLICATION

This procedure shall apply to Receiving and Receiving Inspection to control:
1. customer supplied materials, tools, gages, fixtures and other product related items;
2. internally purchased, product related materials, and subcontract services which may include special processes;
3. customer returns.

3.0   PROCEDURE

3.01   Customer Supplied Materials, Tools, Gages, Fixtures and other product related items

3.01.1 The Receiver shall:
Go to QOP 004 (pg 345) and carry out the receipt of the customer supplied product(s) as stated above under paragraph 3.01.

3.01.2 The Receiving Inspector shall:
Go to QOP 004 and carry out the receiving inspection as stated under paragraphs 3.02 (pg 348), 3.03 (pg 349), and 3.04 (pg 350).

3.02   Company Purchased, Product Related Materials and Services

3.02.1 The Receiver shall:
1. Receive the delivered product. Sign delivery papers as required;
2. Verify that the delivered product is accurately reflected in the sender's Packing Slip;
3. Record on Form A-015 (Record of Received Materials, pg 470) the product related information as requested in the form. Ensure accuracy and completeness. Sign and date it. Put this form on your desk for now;
4. Fill out Form A-027 (several may be required, pg 486), indicating the received product's identification according to the form's description requirement. Secure Form A-027 onto the product, according to the quantities, lot/batch, S/N, so that the shipment can be identified and located as one controlled entity;

MINFOR INCORPORATED	Iss dt. 1/1/99	Rev dt. 9/3/99	Pg.
QUALITY OPERATING PROCEDURE	Sign	Rev # 1	27/32
REQUIREMENT: QUALITY SYSTEM MANUAL SEC: 4.10.2			
SUBJECT: RECEIVING INSPECTION AND TESTING			
QOP 006│DOCUMENT REQ'D: Forms A-001, A-006, A-012, A-015, A-017, A-018, A-027			

5. Place the entire shipment in a safe area in order to prevent unauthorized transfer;
6. Now take Form A-015 and all the documents sent by the supplier and turn them over to the Quality Department.

3.02.2 The Receiving Inspector shall:

1. Accept the documents from the Receiver and schedule the identification, verification and release requirements of the received product by reviewing Form A-015 and the internally issued PO (A-017 located in the Supplier Product Folder);
2. Determine the extent of the verification, identification, and release requirements;

   a) base the dimensional verification and documentation requirement on the Supplier's previously demonstrated product quality history. Accordingly, waive or impose dimensional verification;

   b) base any product serialization and lot/batch control on two factors: **customer's requirement, internal requirement;**

   c) note that on purchased raw materials, the provisions of paragraph 4.10.2.4 (pg 283) of the QSM **(also SAE AS 9000 par. 4.10.2.4)** may be binding as determined by the customer's PO (refer to QOP 005, paragraph 3.01.2, step 4). Act accordingly;

   SPECIAL INSTRUCTION: At this time you may release the product for urgent production requirement without any further inspection. Note this action on Form A-015 and skip step 3. Continue with steps 4,5,6,7. When the urgently released product is found non-conforming during production, recall the product and process it through the NCMR provisions of QOP 009, Section Two (pg 410).

3. Verify the product as identified by the Receiver on Form A-027. Perform dimensional inspection and document the measurement results onto Form A-018, as required.

4. *Mark or serialize the product (the part(s), lot(s)/batch(s)) based on 2b above (also refer to QOP 005, paragraph 3.01, step 2, pg 354). Record any additional product marking and/or serialization done on Forms A-015 and on A-001, as required;*

   a) in case of missing, damaged, or otherwise rejected product, follow the provisions of QOP 009, Section Two, par 3.01.4.4 (the NCMR process on RTV, pg 413).

5. Accept the received product. Stamp, or sign and date Form A-015 and A-027 (Material Identification Tag). Release the product by notifying the Receiver;

6. Make two copies of Form A-015. File one copy together with other related documents (such as certs.) in the Supplier Product Folder;

7. Send the original of Form A-015 and the Packing Slip to Contracts. Send the second copy of Form A-015 to Engineering. Form A-027 stays on the released product.

MINFOR INCORPORATED	Iss dt. 1/1/99	Rev dt. 9/3/99	Pg.
QUALITY OPERATING PROCEDURE	Sign	Rev # 1	28/32
REQUIREMENT: QUALITY SYSTEM MANUAL SEC: 4.10.2			
SUBJECT: RECEIVING INSPECTION AND TESTING			
QOP 006 DOCUMENT REQ'D: Forms A-001, A-006, A-012, A-015, A-017, A-018, A-027			

3.03   Customer Returns

3.03.1 The Receiver shall:

Go to QOP 009, Sec Three, par 3.02.1 (pg 418) and follow the stated instruction.

3.03.2 The Quality Engineer shall:

Go to QOP 009, Sec Three, par 3.02.2 (pg 418) and follow the stated instructions from steps 1 through 7.

MINFOR INCORPORATED	Iss dt. 1/1/99	Rev dt. 9/3/99		Pg.
QUALITY OPERATING PROCEDURE	Sign	Rev #	1	29/32
REQUIREMENT: QUALITY SYSTEM MANUAL SEC: 4.10.4.1				
SUBJECT: CUSTOMER SOURCE INSPECTION				
QOP 006|DOCUMENT REQ'D: Inspection and Supplier Product Folders				

SECTION SIX

1.0   PURPOSE

Maintain control over how customer source inspection is carried out.

2.0   APPLICATION

This procedure shall apply to the presentation of a verified, documented, and accepted final product, made available for customer source inspection.

3.0   PROCEDURE

Customer source inspection may be imposed in two different ways:

1. at the supplier (the Company)
2. at the subcontractor (of the Company)

3.01   Customer Source Inspection at the Supplier

The Quality Engineer shall:
1. identify the product and find the Inspection Product Folder. Review the final inspection results documented on Form A-018. Ensure that the designated product had been fully verified, documented, and accepted in compliance with the specified requirements of the customer's PO and the Quality Plan A-001 (pg 455);
2. contact the customer's Quality Department and request source inspection services. Agree upon the time and date;
3. prepare the product and the related inspection documents (Form A-018 and drawing(s)). Set up the area (in the inspection department) where the product will be situated for customer source inspection. Have on hand the Production Folder, the Supplier and Inspection Product Folders in case the Source Representative wants to verify Process Controls.
4. ensure the following:
    a) the product and source inspection area have been cleaned;
    b) the measuring equipment has been calibrated and set up to facilitate source inspection;
    c) the documents are in order and at hand to accommodate the review process;
    d) inspection personnel are made available to assist the customer's representative;

MINFOR INCORPORATED	Iss dt. 1/1/99	Rev dt. 9/3/99		Pg.
QUALITY OPERATING PROCEDURE	Sign	Rev #	1	30/32
REQUIREMENT: QUALITY SYSTEM MANUAL SEC: 4.10.4.1				
SUBJECT: CUSTOMER SOURCE INSPECTION				
QOP 006 DOCUMENT REQ'D: Inspection and Supplier Product Folders				

5. assist the customer's representative in all areas concerned. Follow up any action item to ensure product quality compliance and customer satisfaction;

6. arrange the final preparation to complete the source accepted product for delivery in accordance with the Job Traveler and/or Operation Sheet requirements, as applicable;

7. release the accepted product for shipment in accordance with the "Product Delivery" requirements of this QOP 006, Sec Four, par 3.03, steps 2 through 7 (pg 382).

3.02   Customer Source Inspection at the Subcontractor

The Quality Engineer shall:

1. interact between the customer's and the subcontractor's Quality Departments in setting up customer source inspection. Agree upon time and date.

2. prepare the subcontractor for the customer source inspection:
    a) assure that the subcontractor has final inspected, documented the work results, and accepted the product;
    b) assure that the subcontractor has set up a clean area in which to do the source inspection, has provided the appropriate measuring equipment, has organized the product related documents, has prepared (cleaned) the product for source inspection;
    c) visit the subcontractor, as appropriate, and verify that "a" and "b" above have been in fact accomplished.

3. prepare yourself for the customer source inspection at the subcontractor:
    a) review, as appropriate, the Inspection and Supplier Product Folders, the Production Folder, and the NCMR Folder in order to determine any in-house activity impacting the acceptance of the subcontracted product being readied for source inspection;
    b) organize to take with you all those documents that will be needed to substantiate any in-house processing related effort impacting the verification and acceptance of the presented product during source inspection;
    c) interact with Contracts, Purchasing and Engineering, as appropriate, to take care of internal communication and possible product support liaison during and after customer source inspection.

4. interact between the customer source representative and the subcontractor during source inspection in order to mediate inconsistencies and follow-up activities.

5. arrange the final preparation to complete the source accepted product for delivery in accordance with the Job Traveler and/or Operation Sheet requirements, as applicable.

6. release the accepted product for shipment in accordance with the "Product Delivery" requirements of this QOP 006, Sec Four, par 3.03, steps 2 through 7 (pg 382), as applicable.

MINFOR INCORPORATED	Iss dt. 1/1/99	Rev dt. 9/3/99	Pg.
QUALITY OPERATING PROCEDURE	Sign	Rev # 1	31/32
REQUIREMENT: QUALITY SYSTEM MANUAL SEC: 4.10.4.2			
SUBJECT: SOURCE INSPECTION AT THE SUBCONTRACTOR			
QOP 006	DOCUMENT REQ'D: Inspection and Supplier Product Folders, Forms A-018, A-036		

SECTION SEVEN

1.0   PURPOSE

Maintain control over source inspection at the Company's subcontractor(s).

2.0   APPLICATION

Procedures in this section shall apply to source control activities of subcontracted product(s).

3.0   PROCEDURE

General Requirement
1. Internal Purchase Orders issued to subcontractors shall state source control requirements under the quality provisions imposed by the cognizant Quality Engineer as stipulated in QOP 003, Section One, paragraph 3.02, step 2 (pg 339).
2. Requests to schedule Source Inspection shall be directed to the Quality Department by all subcontractors. Purchasing shall be required to follow the same provision.

3.01   Preparation for Source Inspection

The Quality Engineer shall:
1. contact the subcontractor's Quality Department and inquire about product readiness:
   a) find out whether final inspection has been completed, the work results documented, and the product accepted. Interact accordingly;
   b) agree upon time and date for source inspection.
2. review the Supplier Product Folder and determine the subcontractor's past performance regarding product quality, including any SCAR activity. Take note of any outstanding issues and determine whether those issues impact the product you are about to source inspect. Interact accordingly;
3. gather the necessary documents (drawing(s), forms A-018 and A-036, etc.) you need to take with you in order to carry out the source inspection. Flag the folders from which you have removed the needed documents for they will have to be returned there;
4. interact with Purchasing about your source inspection activity. Maintain communication between the home office and the subcontractor during source inspection in order to resolve any conflict impacting product quality and expected delivery;

MINFOR INCORPORATED	Iss dt. 1/1/99	Rev dt. 9/3/99		Pg.
QUALITY OPERATING PROCEDURE	Sign	Rev #	1	32/32
REQUIREMENT: QUALITY SYSTEM MANUAL SEC: 4.10.4.2				
SUBJECT: SOURCE INSPECTION AT THE SUBCONTRACTOR				
QOP 006|DOCUMENT REQ'D: Inspection and Supplier Product Folders, Forms A-018, A-036				

3.02 Source Inspection Performance

The Quality Engineer shall:

1. verify at the subcontractor's premises that the product's controlling documents (drawing(s) and any specifications (special process)) have identical part numbers and revision levels to that which are stated in the Purchase Order and any Amendments thereof;
2. verify that the subcontractor's measuring equipment is calibrated;
3. verify that the subcontractor's final inspection report has been completed and the submitted product has been in fact accepted through work results substantiation evidenced by the final inspection report;
4. subject the submitted product to measurement verification to the extent required in order to substantiate product quality compliance to the specified requirements, including workmanship. Record the measurement results on Form A-018, as required.
5. resolve any minor violations affecting product quality and documentation with the subcontractor, or issue SCAR (A-037), as appropriate, and act accordingly. Base corrective action requirement on the subcontractor's quality system. Contact the home office for further direction, as appropriate, or:
6. accept the product. Stamp, or sign and date Form A-018. Then, complete Form A-036 (Source Acceptance Report) and stamp, or sign and date it;
7. make 2 copies of each of A-018 and A-036 and give them to the subcontractor's representative. Ensure that one copy of each accompanies the product upon delivery. The other copies stay with the subcontractor for his records. Bring the originals (of A-018 and A-036) back home with you;
8. upon arrival at the home office, return all the previously removed documents to their respective folders. File the closed out forms of A-018 and A-036 in the Supplier Product Folder. Maintain the Inspection Product Folder, as required.
9. Issue your Trip Report and notification of the source inspection results on Form A-003 (blank, pg 457) and distribute it as required (to Purchasing and Production Planning).

3.03 Product Handling After Source Inspection

The Quality Engineer shall:

1. track the source inspected product and follow up on any SCAR related activity at the subcontractor until the imposed corrective action had been satisfied, as required;
2. ensure that the Job Traveler and Form A-001 requirements are followed and accomplished in order to complete any additional assignment before product delivery in accordance with this QOP 006, Sec Four, par 3.03, steps 1 through 7 (pg 382).

OVERVIEW TO QOP 007
(CONTROL OF INSPECTION, MEASURING AND TEST EQUIPMENT)

When we talk about product verification, what we really mean is to check that product with a measuring device to determine if it meets a determined tolerance limit. How often do we do this without questioning whether that measuring device we just used is good enough to do the job? So you see, that measuring device also has a tolerance limit. And then, when we confirm it to determine its accuracy, the validating equipment also has a tolerance limit. There must be someone in every shop where measuring equipment is used, to take care of the gages, or very soon we are out of business, because we can't accurately measure what we made. All this, of course, is a reminder that we must always know, ahead of time, whether our measuring and test equipment were calibrated and that they are suitable for the type of measurement we want to make. QOP 007 is the procedure that tells us how to go about controlling measuring and test equipment. This procedure controls not only company owned equipment, but also personal and customer owned or loaned measuring and test equipment.

MINFOR INCORPORATED	Iss dt. 1/1/99	Rev dt. 1/1/99	Pg.
QUALITY OPERATING PROCEDURE	Sign AG	Rev # 0	1/6
REQUIREMENT: QUALITY SYSTEM MANUAL SEC: 4.11			
SUBJECT: CONTROL OF INSPECTION, MEASURING, AND TEST EQUIPMENT			
QOP 007|DOCUMENT REQ'D: This procedure and Form A-020			

QUALITY OPERATING PROCEDURE 007

CONTROL OF INSPECTION, MEASURING, AND TEST EQUIPMENT

Form A-003 Rev 0 1999

MINFOR INCORPORATED	Iss dt. 1/1/99	Rev dt. 1/1/99	Pg.
QUALITY OPERATING PROCEDURE	Sign AG	Rev # 0	2/6
REQUIREMENT: QUALITY SYSTEM MANUAL SEC: 4.11			
SUBJECT: CONTROL OF INSPECTION, MEASURING, AND TEST EQUIPMENT			
QOP 007 DOCUMENT REQ'D: This procedure and Form A-020			

REVISION HISTORY

Rev Date	Rev No	Description	Approval

MINFOR INCORPORATED	Iss dt. 1/1/99	Rev dt. 1/1/99	Pg.
QUALITY OPERATING PROCEDURE	Sign AG	Rev # 0	3/6
REQUIREMENT: QUALITY SYSTEM MANUAL SEC: 4.11			
SUBJECT: CONTROL OF INSPECTION, MEASURING, AND TEST EQUIPMENT			
QOP 007|DOCUMENT REQ'D: This procedure and Form A-020 (pg 476)			

1.0   PURPOSE

Maintain control over inspection, measuring, and test equipment.

2.0   APPLICATION

This procedure shall apply to all types of Company and employee owned inspection, measuring, and test equipment, as well as to those under loan agreement (verbal or documented).

3.0   PROCEDURE

General Requirement

Record keeping shall be the acceptable means to demonstrate control of inspection, measuring and test equipment, including test software, and comparative references. The control of measuring and test equipment shall comply with ISO 10012 -1 and/or ANSI/NCSL Z540-1. Computerized data collection shall be allowed to maintain the records of calibration.

The built-in variability (gage maker's tolerance) of the selected measuring and test equipment shall be known in order to be able to assign that equipment to confirm a specified requirement (dimension, weight, etc.) to the defined tolerance limits. (Match the equipment to the capability requirement.) Any software used in measuring applications shall be confirmed prior to its application.

The documented accuracy of the calibration masters shall be traceable to known standards (national or international, or even to a homemade standard in exceptional cases) in order to ensure suitability for the intended confirmation requirement. The documented accuracy and repeatability of the calibration masters and comparative references shall be certified by independent authority at predetermined intervals based on the equipment manufacturers' recommendations or as otherwise specified by internal requirement.

The calibration records shall demonstrate the frequency (cycle time) of confirmations of all the identified measuring and test equipment, including process owners' equipment and test software. Traceable record maintenance shall be a requirement as evidence of control. The calibration records shall be made available to customers or their representatives upon request. **(SAE AS9000, par. 4.11.1.1)**

MINFOR INCORPORATED	Iss dt. 1/1/99	Rev dt. 1/1/99	Pg.
QUALITY OPERATING PROCEDURE	Sign AG	Rev # 0	4/6
REQUIREMENT: QUALITY SYSTEM MANUAL SEC: 4.11			
SUBJECT: CONTROL OF INSPECTION, MEASURING, AND TEST EQUIPMENT			
QOP 007 DOCUMENT REQ'D: This procedure and Form A-020			

3.01   Controlling the Identification of Gages

The Calibration Technician shall:

1. Uniquely identify (label) all inspection, measuring, and test equipment. Do this either on the instrument itself, or when that's not possible, on the container holding the instrument. Use discretion when metal stamping or other form of identification is applied so as not to damage the internal mechanism of the instrument;

2. Uniquely identify the employees owned measuring and test equipment which they use to verify and accept the Company's product(s);

3. Control the inventory of the measuring and test equipment which are under loan agreement or on borrowed basis. Identify them on Form A-020 (Calibration Status Record) only. Don't calibrate these as the owners do the scheduling of calibration. Verify their accuracy and repeatability by similarity comparison and record the results on Form A-020, under "Status," as required;

4. Ensure that the ID number assigned to each measuring and test equipment is always maintained on the specific equipment (calibration sticker) and on all the calibration related documents even if that equipment is removed from service;

5. Don't reassign the once issued ID number to any other equipment (new or refurbished) unless the old calibration record clearly states the replacement requirement and can substantiate the "was" versus the "is" calibration accuracy and repeatability status;

6. Identify all the equipment no longer suitable for measuring and test requirements as "No Longer in Service." Stamp, or sign and date the label. Or scrap the equipment.

3.02   Controlling the Handling, Preservation, and Storage of Gages

The Calibration Technician shall:

1. Maintain all the measuring and test equipment and comparative reference masters separately from the commonly assigned equipment;

2. Maintain the environmental conditions in order to provide a suitable area for calibration, preservation, and storage for the measuring and test equipment;

3. Safeguard the measuring and test equipment from tempering and pilferage;

4. Maintain "Sign-in and Return Log" in accordance with house rules;

5. Provide adequate instructions (verbal or documented, in-house or off-site) to those employees showing lack of skill in the handling and application of the measuring and test equipment;

MINFOR INCORPORATED	Iss dt. 1/1/99	Rev dt. 1/1/99	Pg.
QUALITY OPERATING PROCEDURE	Sign AG	Rev # 0	5/6
REQUIREMENT: QUALITY SYSTEM MANUAL SEC: 4.11			
SUBJECT: CONTROL OF INSPECTION, MEASURING, AND TEST EQUIPMENT			
QOP 007	DOCUMENT REQ'D: This procedure and Form A-020		

6. Keep the Training Log, Form A-016 (pg 471), up to date whenever training instructions to the employees have been given, especially for new employees;

7. Maintain surveillance throughout the processing areas or where the measuring and test equipment is being used to ensure safe containment and handling of the equipment. Ensure on spot corrective action through supervision.

3.03   Controlling the Calibration of Gages

The Calibration Technician shall:

1. Plan the calibration assignment of the measuring and test equipment in line with the requirement of production scheduling in order to ensure an uninterrupted processing environment;

2. Review and follow the "Gage Calibration Technique Sheets" according to the requirement of the type of measuring and test equipment being calibrated. Do the calibration. Record the calibration results onto Form A-020 simultaneously, as practical;

3. Calibrate the measuring and test equipment routinely as indicated by the cycle time (due date) posted on the Calibration Status Report (A-020); (or on the "Employee Gage Calibration Cycle Time" sheet shown under the Training Metrics QOP 007, Basic Information, pg 184)

4. Calibrate the measuring and test equipment at other times as follows: when
   - It's new,
   - It has been repaired,
   - It has been recalled,
   - The accuracy is in question,
   - Failure occurs (damage, etc.);

5. Ensure, when calibrating the measuring and test equipment, that the previously recorded calibration results are reviewed, case by case, in order to assess the progressive deterioration to reclassify the equipment or to determine what is wrong in order to rework/repair it;

6. Lubricate the mechanical equipment as recommended by the manufacturer;

7. Maintain a safe work area and keep the place clean and organized.

3.04   Controlling the Documentation of Gages

The Calibration Technician shall:

1. Maintain for each identified measuring and test equipment accurate documentation on Form A-020; (for the employee gages, see the "Employee Gage Calibration Record" shown under the Training Metrics QOP 007, pg 185).

MINFOR INCORPORATED	Iss dt. 1/1/99	Rev dt. 1/1/99	Pg.
QUALITY OPERATING PROCEDURE	Sign AG	Rev # 0	6/6
REQUIREMENT: QUALITY SYSTEM MANUAL SEC: 4.11			
SUBJECT: CONTROL OF INSPECTION, MEASURING, AND TEST EQUIPMENT			
QOP 007 DOCUMENT REQ'D: This procedure and Form A-020			

2. Place a stamped, or signed and dated Calibration Sticker on each identified measuring and test equipment according to the calibration and due date recorded on Form A-020. (For the employee gages, refer to the "Employee Gage Calibration Cycle Time" sheet shown in the Training Metrics QOP 007 under Basic Information, pg 184.) Observe the following:

   a) don't place calibration stickers on individual pin (plug) gages belonging to a set. Place the sticker on the box containing the pins. The same thing applies to size blocks and other gage sets. The "Calibration Certificate" is your "Proof of Record." Describe it as such on Form A-020, under the description "Status." Attach the "Cert" to the back of the applicable Form A-020 and tuck both documents in a "Sheet Protector." Then, stack these "Sheet Protectors" in a binder;

   b) Some shops have work reference plug gages and size blocks that are not in the calibration system anymore due to downgrading. That's OK. But put a note on the box containing the gages, "For Reference Only – Verify Before Use – Not For Product Acceptance";

   c) Some shops have granite surface plates and other equipment (placed in the most obvious area) that are not exactly to the "Standard" but are perfectly useable for reference applications. Put the following note on them, "For Reference Only – Not For Product Acceptance."

3.05   Controlling the Calibration Records

The Calibration Technician shall:

1. Group the calibration records into three categories:
   a) company owned measuring and test equipment;
   b) employee owned measuring and test equipment;
   c) customer and/or loaned measuring and test equipment.
   Note: Subcontractors' calibration control is regulated and satisfied under the survey protocol records. (See QOP 003, Sec One, par 3.04 on page 339)

2. Arrange the categorized calibration records in a binder according to your "paper" organizational skills in order to be able to locate them promptly; (Use "Sheet Protectors" as appropriate)

3. Maintain document order within your established filing system at all times.

3.06   Inspection, Measuring, and Test Equipment Release
Release any measuring and test equipment as required by the process owners.

OVERVIEW TO QOP 008
(CONTROL OF INSPECTION AND TEST STATUS)

The control of inspection and test status means a lot more than just keeping track of inspection stamps and passwords. The meaning behind it has everything to do with the authorized use or application of them. The authorized use of this media signifies that a product has been verified, accepted, and released to go on to the next determined step in the product's finishing cycle. Without it, we would not know whether any phase operation was checked or not. We would be going on and on finishing a product only to find out at the end that from the first piece on to the last piece everything was rejected. Nobody indicated along the way that any processing step was controlled. What a way to learn a lesson! So, those who are authorized to release products from operation to operation bear a very significant responsibility in the application of inspection stamps and passwords. QOP 008 will define for you how to go about controlling the use of signatures, stamps, and passwords.

MINFOR INCORPORATED	Iss dt. 1/1/99	Rev dt. 1/1/99		Pg.
QUALITY OPERATING PROCEDURE	Sign	Rev #	0	1/5
REQUIREMENT: QUALITY SYSTEM MANUAL SEC: 4.12				
SUBJECT: CONRTOL INSPECTION AND TEST STATUS				
QOP 008 DOCUMENT REQ'D: This procedure and Forms A-003, A-022, A-032				

QUALITY OPERATING PROCEDURE 008

CONTROL OF INSPECTION AND TEST STATUS

CONTENTS   **Page**

Form A-003 Rev 0 1999

MINFOR INCORPORATED	Iss dt. 1/1/99	Rev dt. 1/1/99		Pg.
QUALITY OPERATING PROCEDURE	Sign	Rev #	0	2/5
REQUIREMENT: QUALITY SYSTEM MANUAL SEC: 4.12				
SUBJECT: CONTROL OF INSPECTION AND TEST STATUS				
QOP 008 DOCUMENT REQ'D: This procedure and Forms A-003, A-022, A-032				

REVISION HISTORY

Rev Date	Rev No	Description	Approval

MINFOR INCORPORATED	Iss dt. 1/1/99	Rev dt. 1/1/99	Pg.
QUALITY OPERATING PROCEDURE	Sign	Rev # 0	3/5
REQUIREMENT: QUALITY SYSTEM MANUAL SEC: 4.12			
SUBJECT: CONTROL OF INSPECTION AND TEST STATUS			
QOP 008｜DOCUMENT REQ'D: This procedure and Forms A-003, A-022, A-032			

1.0   PURPOSE

Maintain control over inspection and test status indication, and computer access codes (passwords).

2.0   APPLICATION

This procedure shall apply to the application of inspection stamps and electronic passwords.**(see SAE AS9000, par. 4.12.1)**

3.0   PROCEDURE

General Requirement

Inspection and test status indication shall be applied to product acceptance and release during the entire processing cycle, including products that are on hold and/or waiting for disposition. The indication of inspection and test status may be executed on the product itself, as required, but not without backup documentation to prove it. All products stored for further processing or shipment to customers shall have the inspection and test status indicated (on the Job Traveler or A- 013).

Computerized (software based) document and data management for product related matters shall require access code (password) for all employees whose assignments affect product quality. Product acceptance and release indication executed through a software based documentation system shall be limited to designated personnel, responsible for the acceptance and release of the identified product(s). The issuance of access codes (passwords) shall be documented and controlled. Periodic monitoring by the issuing authority to ensure proper application of the access codes shall be enforced and adequately documented, as required.

3.01   The Control of Inspection Stamps

The Quality Engineer shall:
1. Determine the need as to the type(s) of inspection stamps required. Purchase adequate quantities of inspection stamps to ensure replacement coverage in case of wear, loss, damage, or application denial;
2. Determine who will be the stamp owners, authorized to use the stamps for specific applications, according to their assignments to accept and release products. Issue and distribute to the stamp owners a Quality Directive on the

MINFOR INCORPORATED	Iss dt. 1/1/99	Rev dt. 1/1/99	Pg.
QUALITY OPERATING PROCEDURE	Sign	Rev # 0	4/5
REQUIREMENT: QUALITY SYSTEM MANUAL SEC: 4.12			
SUBJECT: CONTROL OF INSPECTION AND TEST STATUS			
QOP 008	DOCUMENT REQ'D: This procedure and Forms A-003, A-022, A-032		

application of stamps and the associated responsibility. Use Form A-003 (blank, pg 457) on which to issue this directive, as appropriate;

3. Assign the stamps according to documentation requirement stated on Form A-032 (pg 491);

4. Monitor the application of the issued stamps, as required, to prevent improper use. Record the monitoring results onto Form A-022 (pg 478). When finished, sign and date this form and hand it over to the Management Representative. Take corrective action, as required. In severe cases, resort to the NCMR process (QOP 009) for Corrective Action implementation, as appropriate;

5. Maintain security over the stamp inventory;

6. Manage additional documentation requirement on Form A-032 according to personnel changes;

7. Retain Forms A-003 (Quality Directive) and A-032 in one file folder. When issuing stamps, issue the Quality Directive at the same time.

Standard: The stamp owner's signature on Form A-032 represents the acceptance of responsibility in the proper application of the issued stamp(s). The stamp(s) owner may apply his/her signature on processing related documents instead of using the authorized stamp(s) in the acceptance and release of product(s).

3.02 The Control of Electronic Passwords

The Communications Manager shall:

1. Determine the need as to methods used in the issuance, limitation, and application of electronic access codes (passwords). Accordingly, issue and distribute a Communication Directive, using Form A-003, as appropriate, informing the employees about basic requirements and conduct, regarding password ownership, application, and responsibility;

2. Maintain internal registry on Form A-032 on all password owners, including access limitations. Keep this registry secured. Provide access to it by upper management in case of your absence or termination;

3. Monitor the application of the issued passwords, as required, to prevent improper use. Record the monitoring results onto Form A-022. When finished, sign and date this form and send it (E-mail) over to the Management Representative. Take corrective action, as required. In severe cases resort to the NCMR process (QOP 009) for Corrective Action implementation, as appropriate;

4. Manage additional electronic documentation requirements on Form A-032 according to personnel changes.

MINFOR INCORPORATED	Iss dt. 1/1/99	Rev dt. 1/1/99	Pg.
QUALITY OPERATING PROCEDURE	Sign	Rev # 0	5/5
REQUIREMENT: QUALITY SYSTEM MANUAL SEC: 4.12			
SUBJECT: CONTROL OF INSPECTION AND TEST STATUS			
QOP 008 | DOCUMENT REQ'D: This procedure and Forms A-003, A-022, A-032			

5. Provide adequate system backup to ensure normal operations. Provide virus protection for all the equipment engaged in product related assignments, affecting product quality.

OVERVIEW TO QOP 009
(CONTROL OF NON-CONFORMING PRODUCTS)

This procedure is about controlling non-conforming products and it is the third longest procedure by design. Don't be surprised to learn that in a process-approach quality management system, the control of non-conforming products also means controlling deficiencies in documents, processes, and methods. We cannot have an effectively run business by controlling hardware problems alone when the contributing causes of the non-conformance originated from deficiencies in documents, processes, and methods. Controlling deficiency in every process from planning to the delivery of the product is the quintessential driver in a process-approach system to effect on time continuous improvement to the operating management system. The lack of this type of control in businesses of all types is the hornet's nest from which everybody runs away instead of facing it and eliminating it through effective corrective action. This QOP has been linked to other QOPs in the quality management system, and the other QOPs in turn have been linked to this QOP, in order to bring about solving the festering problems in all the core departmental activities, contributing directly or indirectly to the deficiencies in documents, processes, and methods. Once a non-conformance has been identified and written up on Form A-006, MRB is required to call for corrective action in line with the significance of the problem. The implementation of corrective action cannot be shoved aside, cannot be ignored, and cannot be turned over to Management Reviews as long as there are documented procedures in force to cure it. The Management Representative cannot close out any Form A-006 (Non-conforming Material Report) until he/she has followed up the effective implementation of corrective/preventive action. Under this system, the Management Representative is being independently audited for compliance. This is the significance of the stated provisions in QOP 009.

MINFOR INCORPORATED	Iss dt. 1/1/99	Rev dt. 1/1/99	Pg.
QUALITY OPERATING PROCEDURE	Sign	Rev # 0	1/17

REQUIREMENT: QUALITY SYSTEM MANUAL SEC: 4.13
SUBJECT: CONTROL OF NON-CONFORMING PRODUCTS
QOP 009|DOCUMENT REQ'D: This Quality Operating Procedure

QUALITY OPERATING PROCEDURE 009

CONTROL OF NON-CONFORMING PRODUCTS

MINFOR INCORPORATED	Iss dt. 1/1/99	Rev dt. 1/1/99		Pg.
QUALITY OPERATING PROCEDURE	Sign	Rev #	0	2/17
REQUIREMENT: QUALITY SYSTEM MANUAL SEC: 4.13				
SUBJECT: CONTROL OF NON-CONFORMING PRODUCTS				
QOP 009	DOCUMENT REQ'D: This Quality Operating Procedure			

QUALITY OPERATING PROCEDURE 009

CONTROL OF NON-CONFORMING PRODUCTS

Form A-003 Rev 0 1999

MINFOR INCORPORATED	Iss dt. 1/1/99	Rev dt. 1/1/99	Pg.
QUALITY OPERATING PROCEDURE	Sign	Rev # 0	3/ 17
REQUIREMENT: QUALITY SYSTEM MANUAL SEC: 4.13			
SUBJECT: CONTROL OF NON-CONFORMING PRODUCTS			
QOP 009	DOCUMENT REQ'D: This Quality Operating Procedure			

REVISION HISTORY

Rev Date	Rev No	Description	Approval

MINFOR INCORPORATED	Iss dt. 1/1/99	Rev dt. 1/1/99		Pg.
QUALITY OPERATING PROCEDURE	Sign	Rev #	0	4/17
REQUIREMENT: QUALITY SYSTEM MANUAL SEC: 4.13				
SUBJECT: CONTROL OF NON-CONFORMING PRODUCTS				
QOP 009	DOCUMENT REQ'D: Section One, Two, and Three of this procedure			

1.0 STANDARD

Maintain control over non-conforming products, documents, and actions affecting product quality.

2.0 APPLICATION

This procedure shall apply to the handling, reporting, and documentation of non-conforming products, documents, and actions affecting product quality.

3.0 Definitions

1. **Deviation** – a specific customer authorization issued prior to the manufacture of a product to allow departure from a defined design requirement for a specific number of units for a specific duration.
2. **Product non-conformance** – is any condition that violates the requirement of a specification, process, or procedure.
3. **Repair** – a non-conforming product which cannot be further processed to meet a specified requirement without written approval from the customer.
4. **Rework** – a non-conforming product which can be reprocessed under defined conditions to meet a specified drawing requirement.
5. **Waiver** – a written request sent to a customer to disposition a non-conforming product.

4.0 General Requirements

Non-conformance in product(s) and documented procedures shall be verified and documented in order to be processed as a reported non-conformance. Otherwise, it would lack the documented evidence to prove the indication of non-conformance (Forms A-018 and A-009).

Decisions made by the reviewing authority (MRB) shall be unanimous. In case of conflict, the management shall make the final decision.

A minimum of two (2) members of the reviewing authority shall be required to make a decision binding. One of the members shall always be a quality representative. The customer has final authority over non-reworkable products.

MINFOR INCORPORATED	Iss dt. 1/1/99	Rev dt. 1/1/99		Pg.
QUALITY OPERATING PROCEDURE	Sign	Rev #	0	5/17
REQUIREMENT: QUALITY SYSTEM MANUAL SEC: 4.13				
SUBJECT: CONTROL OF NON-CONFORMING PRODUCTS				
QOP 009|DOCUMENT REQ'D: Section One, Two, and Three of this procedure				

SECTION ONE

1.0   PURPOSE

Maintain control over the reporting of non-conformance.

2.0   APPLICATION

This procedural section shall apply to the process of reporting non-conformance:
1. internally
2. to customers
3. to subcontractors

3.0   PROCEDURE

3.01   Internal Reporting (item "1" under par. 2.0 above)

The Reporting Authority shall:
1. Determine whether the non-conformance is:
   – error in documented procedures (may involve the processes they control)

   – rejected material(s) (used collectively)

   – both of the above (through the NCMR process)

   – other (actions);
   a) Correct the error in documented procedures in accordance with QOP 002, Section Two, paragraph 3.03.1(pg 318) by the application of Form A-009 (Amendment to Procedures, pg 464);
   b) Process the rejected material(s) in accordance with this QOP 009, Section Two (pg 410) or Three (pg 417) by the application of Form A-006 (Non-conforming Material Report, pg 461, as applicable);

   Note: Notification of non-conformance to the affected parties shall be a requirement. **(see SAE AS9000, par. 4.13.1, Note 1 and paragraph 4.13.2.4)**

   c) When both error in documented procedures and rejected material(s) are present, process it in accordance with step "**b**" above;
   d) When corrective action is required due to lack of compliance with written procedures, or with audit results implementation, or with any other major violation, process it in accordance with this QOP 009, Section Two (pg 410) by the application of Form A-006 (NCMR).

MINFOR INCORPORATED	Iss dt. 1/1/99	Rev dt. 1/1/99		Pg.
QUALITY OPERATING PROCEDURE	Sign	Rev #	0	6/17
REQUIREMENT: QUALITY SYSTEM MANUAL SEC: 4.13				
SUBJECT: CONTROL OF NON-CONFORMING PRODUCTS				
QOP 009	DOCUMENT REQ'D: Section One, Two, and Three of this procedure			

3.02 **Customer Supplied Products** (item "2" under paragraph 2.0 above)

The Reporting Authority shall:

Follow the instructions stated in QOP 004, paragraph 3.03, pg 349 (Product Rejection), regarding the reporting and handling of customer supplied rejects.

3.03 **Reporting Non-conformance to Subcontractors** (item "3" under par. 2.0 above)

The Reporting Authority shall:

Follow the instructions stated in this QOP 009, Section Two, paragraph 3.01.4.4 (pg 413), regarding subcontractor rejects (RTV).

3.04 **NCMR Issuance Control**

The Reporting Authority shall:

Record (enter) the sequentially maintained NCMR serial number on Form A-007 (pg 462) before issuing an NCMR. (The NCMR Register (A-007) is located in front of the latest NCMR collection held in a binder.) Follow instructions in this QOP 009, Section Two, paragraph 3.01.1 (pg 410) regarding the processing of NCMRs.

MINFOR INCORPORATED	Iss dt. 1/1/99	Rev dt. 1/1/99		Pg.
QUALITY OPERATING PROCEDURE	Sign	Rev #	0	7/17
REQUIREMENT: QUALITY SYSTEM MANUAL SEC: 4.13				
SUBJECT: CONTROL OF NON-CONFORMING PRODUCTS				
QOP 009	DOCUMENT REQ'D: Forms A-003, A-006, A-007, A-009, A-012, A-018, A-023, A-024, A-033, A-037, A-039			

SECTION TWO

1.0   PURPOSE

Maintain control over the process of reporting, determining the cause, disposition, corrective/preventive action, and follow-up of non-conforming products, processes, and work assignments.

2.0   APPLICATION

The provisions in this procedure shall apply to the processing of Non-conforming Material Report(s) (A-006) for the purpose of determining the cause, disposition, corrective/preventive action, and follow-up in respect to non-conforming products, processes, and work assignments.

3.0   PROCEDURE

3.01   Performance Requirement

3.01.1   Processing Non-conforming Material Report(s) Through Form A-006

The Reporting Authority shall:
1. Obtain a blank form of A-006 (pg 461). Observe that the form is divided into 6 sections, each requiring specific entry;
2. Assign from the NCMR Register (A-007, pg 462) onto Form A-006 (top right) the next sequential serial number. Locate the NCMR Register in front of the latest collection of NCMRs in a binder. Use additional blanks of A-006 as continuation sheet under the same serial number, as required;
3. Transfer the product-related data from the Inspection Report (A-018, pg 473) onto the NCMR in section one and two. Post the NCMR S/N onto Form A-018 and other related documents, as required. Sign and date the NCMR under "Inspector" and turn it over to the supervisor for his review and approval;

3.01.2   Cause Determination (in section three of Form A-006)

The Supervisor shall: (rely on appropriate assistance as needed)
1. Review the reported non-conformance and determine the underlying reason(s) (the root cause) as to why this problem has happened;

MINFOR INCORPORATED	Iss dt. 1/1/99	Rev dt. 1/1/99	Pg.
QUALITY OPERATING PROCEDURE	Sign	Rev # 0	8/17
REQUIREMENT: QUALITY SYSTEM MANUAL SEC: 4.13			
SUBJECT: CONTROL OF NON-CONFORMING PRODUCTS			
QOP 009	DOCUMENT REQ'D: Forms A-003, A-006, A-007, A-009, A-012, A-018, A-023, A-024, A-033, A-037,A-039		

2. Describe the contributing reasons for the non-conformance in the third block on Form A-006. (Avoid explaining the end results of the problem. We all know that from the inspection report (A-018));

3. Sign and date under "Supervisor" as indicated and send this NCMR, together with all related documents (A-018, etc.) and the product, to the Material Review Board (MBR), as practical.

3.01.3   MRB Disposition (in section four of Form A-006)

The Material Review Board shall: (Quality and Engineering)

1. Review and evaluate the inspection results (from A-018) and the cause (from A-006) of the nonconformance in order to determine the appropriate action to take in despositioning the subject non-conformance;

2. Make the disposition and indicate under "MRB Disposition" the appropriate action required. Do this by encircling the activity indicator(s) (RWK, RPR, ACC, RTV, SCR, Waiver) and indicating the affected item(s) (from section two of A-006). Next, add under "Instruction/Comment," in line with the affected item, any relevant information required to be followed by those who will be implementing your disposition(s);

3. Determine whether any corrective/preventive action is required and enter this requirement under "C/A." Then, sign and date as shown, authorizing enforcement;

4. Describe the "Corrective Action" specifically (in section five of form A-006) as to who, where, and what to enforce. Then, indicate the "Required By" date as shown in order to know the implementation time deadline. Sign and date as shown to authorize the C/A implementation.

5. Distribute copies of the executed NCMR to the affected parties, as required. (This step is solely the Quality Engineer's responsibility)
   a) make as many copies of the dispositioned NCMR as required. Then,
   b) file the original NCMR behind the Register (A-007) in the NCMR binder;
   c) keep one copy on the product, attached to the other documents, as appropriate;
   d) send a copy to the responsible department or individual;
   e) notify the customer, as required. See page 408, par 3.01b "Note." **(AS9000, par. 4.13.2.4);**
   f) send a copy to the Management Representative to enable follow-up on corrective action implementation as per paragraph 3.01.6 below, and close out of the NCMR as per paragraph 3.01.7 below in this section;

MINFOR INCORPORATED	Iss dt. 1/1/99	Rev dt. 1/1/99		Pg.
QUALITY OPERATING PROCEDURE	Sign	Rev #	0	9/17
REQUIREMENT: QUALITY SYSTEM MANUAL SEC: 4.13				
SUBJECT: CONTROL OF NON-CONFORMING PRODUCTS				
QOP 009	DOCUMENT REQ'D: Forms A-003, A-006, A-007, A-009, A-012, A-018, A-023, A-024, A-033, A-037, A-039			

g) maintain continuously both the Inspection Product Folder and the NCMR Folder (the binder, containing the NCMRs), as required.

3.01.4   Compliance Activities Regarding Disposition(s)

3.01.4.1   Rework (RWK)

The Responsible Department or Individual shall:
1. Rework the product, as required:
   a) if processing still in progress, use the authority of the Job Traveler (no rework instruction (A-023) is required from Engineering or the Customer). Observe QOP 006, Section One, paragraph 3.02 rules (pg 365), regarding rework verification requirement;
   b) if the non-conforming product is an isolated unit (batch) that may or may not be part of a lot and requires a new setup, rework instruction on Form A-023 is required from Engineering. Observe QOP 006, Section One, paragraph 3.02 rules, regarding rework verification requirement;
   c) if the rework is a document change requirement, carry it out in accordance with QOP 002, Sec Two, par 3.03.2.2 (pg 319).

3.01.4.2   Repair (RPR) (Provisions of AS9000, par. 4.13.2.1 apply)

3.01.4.2.1 The Quality Engineer shall:
1. Issue a Waiver Request on Form A-024 (pg 480) to the customer, relying on Engineering technical support. Interact with customer's Quality to get a reply. Wait for the reply. Then,
2. Interact with all the departments needed to carry out the customer's instructions as indicated in the Waiver Request reply:
   a) if the disposition by the customer is to repair the product, interact with the Process Engineer to issue the Rework/Repair Procedure on Form A-023 (pg 479);
   b) if the disposition by the customer is to send the product for his evaluation, follow the instructions as stated in QOP 006, Section Four, paragraph 3.03 (pg 382), as appropriate. Observe "Note" under step 6 in paragraph 3.02 of the same procedure, regarding tagging;
   c) if the disposition by the customer is to scrap the product, follow the instructions as stated below in paragraph 3.01.4.5 in this section.

MINFOR INCORPORATED	Iss dt. 1/1/99	Rev dt. 1/1/99	Pg.
QUALITY OPERATING PROCEDURE	Sign	Rev # 0	10/17
REQUIREMENT: QUALITY SYSTEM MANUAL SEC: 4.13			
SUBJECT: CONTROL OF NON-CONFORMING PRODUCTS			
QOP 009	DOCUMENT REQ'D: Forms A-003, A-006, A-007, A-009, A-012, A-018, A-023, A-024, A-033, A-037, A-039		

3.01.4.2.2 The Process Engineer shall:

1. Issue the Rework/Repair Procedure on Form A-023 according to instructions given by the customer in the Waiver Request reply. Interact with the customer's Process Engineer in order to define the repair instructions, as appropriate;
2. Monitor the rework/repair progress and ensure adequate technical support.

3.01.4.2.3 The Production Supervisor shall:

1. Assign the rework/repair job to the appropriate process owner who will carry out the rework/repair according to the work instruction issued by the Process Engineer on Form A-023;
2. Monitor the process owner's work performance. Observe product verification requirements as stated under QOP 006, Section One, paragraph 3.02 (pg 365).

3.01.4.2.4 The Quality Engineer shall:

1. Monitor the completion of the reworked/repaired product through the final release verification by the Quality Inspector;
2. Work together with the Quality Inspector in line with instructions given under QOP 006, Section One, paragraph 3.02 (pg 365), as appropriate.

Note: This product, which has been repaired, shall be kept tagged with Form A-012 through all other processes. Ship the product tagged (A-012). Attach the customer's concession agreement (A-024) to the shipping documents.

3.01.4.3   Accept (ACC – Use-as-is condition)

The Process and Quality Engineers shall:
Observe that acceptance of non-conforming product(s) except to rework it is not permitted on customers' POs. Issue rework instructions on Form A-023, as required, or resort to the "Repair" process as outlined above under paragraph 3.01.4.2; **(AS9000, par. 4.13.2.1 apply)**

3.01.4.4   Return to Vendor (RTV)

The Quality Engineer shall:

1. Notify the Purchasing and Planning Departments, as required. (A copy of the dispositioned NCMR will do it. See par 3.01.3, step 5, above);

MINFOR INCORPORATED	Iss dt. 1/1/99	Rev dt. 1/1/99		Pg.
QUALITY OPERATING PROCEDURE	Sign	Rev #	0	11/17
REQUIREMENT: QUALITY SYSTEM MANUAL SEC: 4.13				
SUBJECT: CONTROL OF NON-CONFORMING PRODUCTS				
QOP 009	DOCUMENT REQ'D: Forms A-003, A-006, A-007, A-009, A-012, A-018, A-023, A-024, A-033, A-037, A-039			

2. Request Packing Slip (Form A-028, pg 487);
3. Attach to the Packing Slip the Inspection Report (Form A-018), a copy of the dispositioned NCMR (Form A-006), and the completed Supplier Corrective Action Request (SCAR, Form A-037, pg 496);
4. Release the product to the Shipping Department to return it to the subcontractor;
5. Maintain the Supplier and Inspection Product Folders regarding document retention, as required;
6. Follow up with the subcontractor the "SCAR" response requirement. Interact with the Management Representative regarding corrective action responses. (See paragraph 3.01.5 below)

3.01.4.5 Scrap (SCR)

The Quality and Process Engineers shall:
1. Observe that scrap status may be voluntarily declared; however, the customer will give direction on how to dispose of the product either under the quality provisions of the contract or on the submitted waiver request itself (Form A-024);
2. Observe that when raw material is a Company purchased product, scrap status may not require reporting to the customer. Disposal of the scrapped product, however, may be controllable by the customer under the quality provisions of the contract;
3. Dispose of the scrapped product, as required, **(AS9000 par. 4.13.2.3 and 4.13.2.4)**. (See QSM on page 286)

3.01.4.6 Waiver

The Quality Engineer shall:
Issue the "Waiver Request, Form A-024" to the customer and follow the instructions given above under paragraph 3.01.4.2.1

3.01.4.7 Regrading Material

The Quality Engineer shall:
1. Observe that regrading is a customer imposed requirement resulting from his/her dispositioning of a non-conforming product, which will be identified either on the Waiver Request or on the customer's processed NCMR (on his own form);

MINFOR INCORPORATED	Iss dt. 1/1/99	Rev dt. 1/1/99	Pg.
QUALITY OPERATING PROCEDURE	Sign	Rev # 0	12/17
REQUIREMENT: QUALITY SYSTEM MANUAL SEC: 4.13			
SUBJECT: CONTROL OF NON-CONFORMING PRODUCTS			
QOP 009	DOCUMENT REQ'D: Forms A-003, A-006, A-007, A-009, A-012, A-018, A-023, A-024, A-033, A-037,A-039		

2. Observe that product identification, traceability, and documentation requirements to handle regrading may be imposed by the customer. Carry this out according to the customer's requirements and in line with the internal documentation requirement as described in QOP 005, paragraph 3.02 (pg 355), as appropriate, **(AS9000, par. 4.13.2.2 shall be applicable, as required)**. See QSM on page 285.

3.01.5   Handling Corrective and Preventive Action

The Management Representative shall:
1. Determine from the dispositioned NCMR the extent of the corrective action requirement. Ensure timely response on SCARs from suppliers;
2. Interact with the responsible department head, or the individual, who has to implement the corrective action and work out the implementation plan with him/her in line with the required date noted on the specific NCMR;
3. Summarize the implementation plan on Form A-003 (blank) and issue it to the responsible agent. Maintain a copy of it for follow-up verification.
4. Observe that corrective action may involve change to the Quality System documents. Refer to QOP 002, Section Two (all paragraphs may apply) to implement it (pg 317).
5. After each corrective action, determine the type and level of training requirement and carry out the training, as required, from the corresponding Training Metrics. Evaluate the training results according to Form A-039 (pg 498). Less than 70% scoring requires retraining. Post the respective NCMR number on the "Training Evaluation Sheet" (Form A-039) and maintain it in the Training Log (pg 471).

3.01.6   Follow-up of Corrective/Preventive Action Implementation (last block on the NCMR)

The Management Representative shall:
1. Locate the appropriate and original NCMR according to your schedule to follow up the implementation of corrective actions (use your copy of the NCMR given to you above, under paragraph 3.01.3, step 5, to locate the original which is filed with the latest collection of NCMRs in the binder);
2. Review the C/A implementation summary previously issued by you on Form A-003 to the responsible department head, or individual;
3. Verify the effective implementation of the required C/A;

MINFOR INCORPORATED	Iss dt. 1/1/99	Rev dt. 1/1/99	Pg.
QUALITY OPERATING PROCEDURE	Sign	Rev # 0	13/17
REQUIREMENT: QUALITY SYSTEM MANUAL SEC: 4.13			
SUBJECT: CONTROL OF NON-CONFORMING PRODUCTS			
QOP 009	DOCUMENT REQ'D: Forms A-003, A-006, A-007, A-009, A-012, A-018, A-023, A-024, A-033, A-037,A-039		

4. Indicate any unsatisfactory implementation under "No" (shown in the last section (6) of the NCMR) and explain the action taken; otherwise,
5. Indicate "Yes" for satisfactory implementation. Sign and date the NCMR as shown;
6. Return the completed NCMR to its binder under its sequential S/N.

3.01.7   NCMR Close-out

The Management Representative shall:
1. Open the latest NCMR binder and locate in front of the collection the NCMR Register (A-007);
2. Observe the right side of the page "CLOSEOUT." (All NCMRs that were issued and registered under the sequential serial number listing on the left side of the page will have to be closed out on the right side of the page in order to demonstrate the effective compliance with the MRB disposition);
3. Determine the appropriate NCMR serial number and close it out after the corrective action implementation had been satisfactorily completed. Also, do the same thing where "No" corrective action was indicated;
Note: In certain cases corrective action implementation will take considerable time to be effectively implemented. Do not presume implementation completion. Follow it up, regardless of time lapse. Do not close out any NCMR without effective follow-up.
4. Answer any outstanding customer SCAR that was pending due to corrective action implementation, follow-up, and NCMR close-out.

3.01.8   Document Retention

The Management Representative and Quality Engineer shall:
Observe document retention in accordance with QOP 002, Section Three (pg 328).

MINFOR INCORPORATED	Iss dt. 1/1/99	Rev dt. 1/1/99		Pg.
QUALITY OPERATING PROCEDURE	Sign	Rev #	0	14/17
REQUIREMENT: QUALITY SYSTEM MANUAL SEC: 4.13				
SUBJECT:CUSTOMER COMPLAINTS, PRODUCT RETURNS, AND SATISFACTION REPORTS				
QOP 009 DOCUMENT REQ'D: Forms A-003, A-006, A-007, A-009, A-012, A-018, A-023, A-024, A-033, A-038				

SECTION THREE

1.0 PURPOSE

Maintain control over customer complaints, product returns, and customer satisfaction reporting.

2.0 APPLICATION

The provisions of this procedure shall apply to the processing of customer complaints, product returns, and customer satisfaction reporting.

3.0 PROCEDURE

3.01 Customer Complaints

3.01.1 Reporting, Documentation, and Resolution

General Requirement
All written customer complaints shall be directed to the attention of the Quality Assurance Manager or his/her designate;

Personnel receiving customer complaints via telephone or personal contact shall be required to refer the reporter of the complaint to the Quality Assurance Manager or his/her designate.

3.01.1.1 The Quality Manager or His/Her Designate shall: (see Form A-038, pg 497)

1. Take the complaint and record the customer and product related information in the first section of Form A-038. (Observe that Form A-038 is divided into two main sections. The first section is for documenting the customer complaint. The second is for reporting the investigation and action taken);
2. Review the complaint and launch your investigation:
   a) review the Inspection Product Folder in order to determine processing and final inspection history. Review the NCMR and Supplier Product Folders as may be applicable;
   b) determine the course of action to take, including product recall, in order to resolve the complaint, and act accordingly. Use the quality provisions of the established procedures stated under this QOP 009, Section One

MINFOR INCORPORATED	Iss dt. 1/1/99	Rev dt. 1/1/99		Pg.
QUALITY OPERATING PROCEDURE	Sign	Rev #	0	15/17
REQUIREMENT: QUALITY SYSTEM MANUAL SEC: 4.13				
SUBJECT:CUSTOMER COMPLAINTS, PRODUCT RETURNS, AND SATISFACTION REPORTS				
QOP 009	DOCUMENT REQ'D: Forms A-003, A-006, A-007, A-009, A-012, A-018, A-023, A-024, A-033, A-038			

and Two. The NCMR process equally applies to non-conforming products identified in the field, to voluntarily recalled products, or to customer returned products;

3. Complete the second part of Form A-038, indicating "Cause of the problem" and "Corrective action taken." Sign and date Form A-038. Send a copy of Form A-038 to the customer (to the person who reported the complaint).

4. Maintain the complaint related documents under the "Customer Complaint Folder(s)."

3.02   Product Return

3.02.1   Receiving

The Receiver shall:

1. Sign the shipping document to accept the returned product, as required;
2. Fill out Form A-012 (Rejected Material Ticket, pg 467) and attach it to the customer returned product;
3. Take the returned product related documents and hand them over to the Quality Department;
4. Place the returned product in a safe area, pending Receiving Inspection.

3.02.2   Identification, Verification, and Resolution

The Quality Engineer shall:

1. Review the customer supplied documents (customer's NCMR);
2. Identify the returned product through the final inspection records;
3. Visually inspect the product for handling and packaging damage;
4. Record the findings onto Form A-018 (Inspection Report);
5. Submit the product to re-verification, if required, in order to determine the validity of the rejection. Record the findings onto Form A-018;
6. Evaluate the verification results against the customer's Rejection Report;
7. Determine the course of action to take and act accordingly. Use the quality provisions of the established procedures stated under this QOP 009, Section One and Two, as required, in order to effect any remedial action including corrective action;

Note: The NCMR process equally applies to non-conforming products identified in the field, to voluntarily recalled products, or to customer returned products. Follow these procedures in order to resolve the product's rejec-

MINFOR INCORPORATED	Iss dt. 1/1/99	Rev dt. 1/1/99	Pg.
QUALITY OPERATING PROCEDURE	Sign	Rev # 0	16/17
REQUIREMENT: QUALITY SYSTEM MANUAL SEC: 4.13			
SUBJECT: CUSTOMER COMPLAINTS, PRODUCT RETURNS, AND SATISFACTION REPORTS			
QOP 009 DOCUMENT REQ'D: Forms A-003, A-006, A-007, A-009, A-012, A-018, A-023, A-024, A-033, A-038			

tion status to the level of customer satisfaction.

8. Return or replace the product: Follow the "Product Delivery" instructions as stated in QOP 006, Sec Four, par 3.03 (pg 382), as applicable.

9. Maintain control over the retention of the various documents involved in this process in line with QOP 002, Section Three (Document Retention, pg 328), as applicable.

3.03   Cause/Corrective Action Request by the Customer (SCAR)

The Quality Engineer shall:

1. Review the Cause/Corrective Action Request (SCAR) sent by the customer and determine the course of action to take in complying with it:
   a) go to the Inspection Product Folder. Identify the customer's folder by the product's part number and pull it out;
   b) review the folder's contents – Job Traveler (A-008) and the Final Inspection Report (A-018) – and look for any posted NCMR number. Review the NCMR also, if there is one, in order to establish relevancy to the case identified in the SCAR.

2. Answer the SCAR according to what you have found in the records, or if the records indicate no violation, accept the fact as stated in the SCAR and respond to it accordingly, or recall the product to substantiate the "purported" deficiency; or (go to step 3)

Note: Do not respond to any SCAR prematurely. Wait until you have done all the internal cause/corrective action as required by this QOP 009, Section Two.

3. Determine the necessary corrective action, if required, in line with the answer you have given to the customer in the SCAR under "Corrective Action Taken." Implement meaningful corrective action in accordance with the established procedures stated in this QOP 009, Section Two, paragraphs 3.01.5 and 3.01.6 (pg 415). Issue the NCMR, as appropriate;

4. Mail or fax the completed SCAR to the customer, as requested;

5. Maintain control over the documents that relate to this case in accordance with QOP 002, Section Three, under A-037, par 3.05w on page 333.

3.04   Customer Satisfaction Reporting

General Requirement

Product acceptance by the customers through the "Customer Complaint and Evaluation Report" (Form A-038) is not a contractual requirement unless it is specifically stated in the customer's Purchase Order. Product deficiencies

MINFOR INCORPORATED	Iss dt. 1/1/99	Rev dt. 1/1/99		Pg.
QUALITY OPERATING PROCEDURE	Sign	Rev #	0	17/17
REQUIREMENT: QUALITY SYSTEM MANUAL SEC: 4.13				
SUBJECT:CUSTOMER COMPLAINTS, PRODUCT RETURNS, AND SATISFACTION REPORTS				
QOP 009|DOCUMENT REQ'D: Forms A-003, A-006, A-007, A-009, A-012, A-018, A-023, A-024, A-033, A-038				

between the customer and the supplier's organization in contractual relationships are handled through the Quality Provisions of the customer's Purchase Orders (SCAR provisions) and the Company's documented procedures, the "Control of Non-conforming Products" stipulated in QOP 009. Therefore, the application of Form A-038 is a self-imposed requirement to document and implement those areas of product concerns that fall outside the established documented system. The purpose of the application of Form A-038 is to ensure ultimate customer satisfaction in product quality beyond those provisions stipulated in customers' Purchase Orders. Failure to define by the customers the Quality Provisions in their Purchase Orders shall not be construed by the customers as accepting responsibility by the supplier's organization through the application of Form A-038.

Form A-038 is a multipurpose document: 1. designed to handle customer complaints reported outside of the SCAR provisions of the customer's Purchase Order. (See paragraph 3.01 above) 2. designed to handle the unsolicited product evaluation responses that customers may wish to express when they receive a delivered product. (Covered under this paragraph 3.04)

3.04.1   Handling and Controlling Customer Evaluation Reports (A-038)

3.04.1.1   The Final Inspector shall:
1. After product release from final inspection, take a blank form of A-038, located in the Forms File;
2. Enclose this form with the shipping documents in the "Documents Enclosed" envelop of the deliverable product to the customer.

3.04.1.2   The Management Representative shall:
1. Review each customer responded Evaluation Report (A-038);
2. Determine the product quality impact;
3. Implement corrective action through the provisions of this QOP 009, Section Two, paragraphs 3.01.5 and 3.01.6 (pg 415), as appropriate;
4. Report the "Customer Satisfaction Measurement" at the next Management Review, as required;
5. Maintain document control in the applicable Customer Complaint Folder.

OVERVIEW TO QOP 010
(INTERNAL QUALITY AUDIT)

The quality system procedures in this process-approach system don't resemble the standard operating procedures (SOP) in conventional quality systems. The quality procedures in this system have been designed to be user-friendly in all areas as to the identification of requirement, the definition of responsibility, work assignment, accountability, and process flow. For an auditor to spend more time on compiling the checklist than on doing the audit itself is absolutely unacceptable in economic terms. This, of course, is not the auditor's problem, but how the quality system's procedures have been laid out and integrated lend a user-friendly flow process in which what you need is right in front of you. Since internal audits are mostly done by in-house employees, it makes a big difference economically whether the auditor spends a few hours or a few days in accomplishing internal audits. I paid serious attention to this also when I planned the layout of my procedures, for I used to do audits at many companies in my forty years service and could not easily find where and how responsibilities were linked to work assignments in the conventionally written SOPs. Every quality operating procedure in this process-approach system is also a checklist, not only for the auditor, but also for the process owner and his supervisor as well. Note also the dramatic shift from conventional methods as to how we maintain continuous improvement in training employees whenever corrective action requirement points to the operators' lack of understanding of their work assignment. The details of the audit process are the subject of this QOP.

MINFOR INCORPORATED	Iss dt. 1/1/99	Rev dt. 1/1/99	Pg.
QUALITY OPERATING PROCEDURE	Sign	Rev # 0	1/6
REQUIREMENT: QUALITY SYSTEM MANUAL SEC: 4.17			
SUBJECT: INTERNAL QUALITY AUDITS (System Audit)			
QOP 010|DOCUMENT REQ'D: Forms A-006, A-007, A-021, A-022			

QUALITY OPERATING PROCEDURE 010

INTERNAL QUALITY AUDITS

Form A-003 Rev 0 1999

MINFOR INCORPORATED	Iss dt. 1/1/99	Rev dt. 1/1/99	Pg.
QUALITY OPERATING PROCEDURE	Sign	Rev # 0	2/6
REQUIREMENT: QUALITY SYSTEM MANUAL SEC: 4.17			
SUBJECT: INTERNAL QUALITY AUDITS (System Audit)			
QOP 010 | DOCUMENT REQ'D: Forms A-006, A-007, A-021, A-022			

REVISION HISTORY

Rev Date	Rev No	Description	Approval

MINFOR INCORPORATED	Iss dt. 1/1/99	Rev dt. 1/1/99	Pg.
QUALITY OPERATING PROCEDURE	Sign	Rev # 0	3/6
REQUIREMENT: QUALITY SYSTEM MANUAL SEC: 4.17			
SUBJECT: INTERNAL QUALITY AUDITS (System Audit)			
QOP 010|DOCUMENT REQ'D: Forms A-006, A-007, A-021, A-022			

1.0   PURPOSE

Maintain control over the process of carrying out internal quality system audits.

2.0   APPLICATION

The provisions of this operating procedure shall apply to: determining internal audit requirements, carrying out internal quality system audits and documentation of those activities.

3.0   PROCEDURE

3.01   Definition

Audit – a planned, independent review and verification process to determine compliance with the approved procedures of the Quality Management System.

3.02   General Requirement

The manual defines in policy statements the quality system's requirements and gives linkage by reference or inclusion to the applicable process instructions, known as Quality Operating Procedures (QOPs). The manual follows the layout of ISO 9002/1994 and within these provisions incorporates the requirements of ISO 9001/2000. It reflects every clause of the Standard according to its numerical layout, except clauses 4.4 and 4.19. **The provisions of AS9000 are complementary to the basic ISO requirements.**

Auditors must first understand the specific paragraph requirement in the manual and have familiarity with the referenced or included process instructions (QOPs) before undertaking audits. This is a prerequisite for auditors to enable them to carry out any planned audit(s) effectively. Auditors shall not impose their own set of requirements to contradict the provisions of the established and implemented Quality Management System.

3.03   Performance Requirement

3.03.1 Checklist Guidance
The Auditor shall:
1. Observe: that the Quality Operating Procedures (QOPs) outline the specific work compliance requirements stated in the Quality Manual: that the QOP pages have been page plated to give guidance on linkage and document requirement;

MINFOR INCORPORATED	Iss dt. 1/1/99	Rev dt. 1/1/99		Pg.
QUALITY OPERATING PROCEDURE	Sign	Rev # 0		4/6
REQUIREMENT: QUALITY SYSTEM MANUAL SEC: 4.17				
SUBJECT: INTERNAL QUALITY AUDITS (System Audit)				
QOP 010	DOCUMENT REQ'D: Forms A-006, A-007, A-021, A-022			

2. Observe: that the process instruction procedures embodied in QOPs guide you forward to understand the task at hand and at the same time refer you to the specific forms used by the process owners to record work results; that each QOP gives indication of linkage to other operating procedures, enabling you to find the exact enforcement requirement for the identified task;

3. Observe: that each operating procedure is a stand-alone document that may contain several interrelated work assignments, step by step, requiring work performance accomplishment and documentation of work results by the process owners;

    Example: QOP 006 (Inspection and Test Control, pg 357) contains seven (7) interrelated work assignments – first piece inspection; first article inspection; process control; final inspection; receiving inspection; customer source inspection; and source inspection at subcontractors. Others follow the same pattern;

4. Observe: that when auditing is carried out, the QOP sections and the related paragraphs should be identified and documented in the Audit Plan (Reporting) Form A-022 (pg 478), to indicate the specific audit effort (the form is also the checklist);

5. Observe: that in order to determine conformance or nonconformance to the specified requirements, the various applicable procedures and forms will have to be reviewed and understood ahead of time;

6. Observe: that the Management Representative may perform the quality system audit(s) provided that he/she is not the direct process owner of the work assignment(s) being audited;

7. Observe: that when the Management Representative performs a scheduled (A-021, pg 477) quality system audit, his/her audit performance shall be verified (audited) by any other independent agent to establish definite audit compliance. (Use the same QOP 010, including Audit Plan, A-022)

3.03.2 Doing the Audit

The Auditor shall:

1. Obtain from the Management Representative the latest (approved and released) Quality Manual (pg 267), the Quality Operating Procedures Manual (pg 291), and the Audit Schedule (form A-021). Obtain the Audit Plan (Reporting) form (A-022) from the Forms File (many will be needed);

2. Take the Audit Schedule (A-021, pg 477) and determine which clause paragraph is scheduled to be audited for the specified month. Find that clause paragraph in the Quality Manual;

MINFOR INCORPORATED	Iss dt. 1/1/99	Rev dt. 1/1/99	Pg.
QUALITY OPERATING PROCEDURE	Sign	Rev # 0	5/6
REQUIREMENT: QUALITY SYSTEM MANUAL SEC: 4.17			
SUBJECT: INTERNAL QUALITY AUDITS (System Audit)			
QOP 010|DOCUMENT REQ'D: Forms A-006, A-007, A-021, A-022			

3. Review all the requirements given under the clause paragraph in the manual. Then, review the work performance requirement, given either under the clause paragraph in the manual, or linked to a Quality Operating Procedure. Review the applicable form(s) assigned either in the manual, or in the Quality Operating Procedure;

4. Prepare (fill out) the left side of the Audit Plan (Reporting) form (A-022). Write down in the form what will be audited. Show, step by step, the applicable QOP, the Section in the QOP, and the paragraphs;

5. Carry out the audit in accordance with the process instructions given in the applicable sections of the Quality Operating Procedure, or the Quality Manual. Verify, as required, the work result documentation, recorded by the process owner(s) in the applicable forms. Determine compliance;

6. Record on the right side of the Audit Plan (Reporting) form (A-022), in line with the identified paragraph, the non-compliance found during your audit. If none found, report "None." Report the facts and not what you would like to be done. Follow the procedure. If the procedure is wrong or misleading, report that also. You are not allowed to make judgments which would impact product quality or the process owners' job performance;

7. Write down your observations (the facts) and not your recommendations under "Comments" in the Audit Plan (Reporting) form;

8. Sign and date the Audit Plan form as shown. Turn in the Audit Plan form, the Quality Manual and the Quality Operating Procedure to the Management Representative. Put the remainder blanks (A-022) back to the Forms File under the correct serial number location.

3.03.3 Reviewing the Audit Plan (Reporting) (A-022)

The Management Representative shall:
1. Receive the Audit Plan (Reporting) (one or several) from the Auditor;
2. Review what has been audited and evaluate any non-conformance written down by the Auditor. Evaluate "Comments" made, as appropriate;
3. Determine any negative impact to product quality, to processes, and to procedures;
4. Implement corrective action in accordance with the following:
   a) if performance has not been carried out and documented as required by the work instruction, implement corrective action through supervision in accordance with QOP 006, Sec Three (pg 374), as appropriate, in line with the identified supervisory responsibility. If NCMR enforcement is required, do it in accordance with QOP 009, Sec Two (pg 410);

MINFOR INCORPORATED	Iss dt. 1/1/99	Rev dt. 1/1/99	Pg.
QUALITY OPERATING PROCEDURE	Sign	Rev # 0	6/6
REQUIREMENT: QUALITY SYSTEM MANUAL SEC: 4.17			
SUBJECT: INTERNAL QUALITY AUDITS (System Audit)			
QOP 010│DOCUMENT REQ'D: Forms A-006, A-007, A-021, A-022			

b) if product related work procedures (Job Traveler, Operation Sheet, etc.) are in error or exhibit unauthorized changes, implement corrective action in accordance with QOP 002, Sec Two, par 3.03.1, pg 318 (through the application of form A-009). If NCMR enforcement is required, do it in accordance with QOP 009, Section Two (pg 410);

c) if the Quality System's procedures are in error, implement corrective action in accordance with QOP 002, Section Two, paragraphs 4.0 (pg 320) and 5.0 (pg 324), as appropriate;

d) if product quality had been affected due to system audit findings, implement corrective action in accordance with QOP 009 Sections One (pg 408), Two (pg 410), and Three (pg 417), as applicable;

e) determine training requirement in conjunction with any corrective action implementation and carry it out in accordance with QOP 009, Section Two, paragraph 3.01.5, step 5, (pg 415);

f) maintain document retention in all the above actions in line with QOP 002, Section Three (pg 328).

5. Close out the Audit Schedule (A-021) whenever a scheduled system audit has been completed. Do this by signing and dating under "AUDIT CLOSEOUT" in Form A-021;

6. Maintain the system's audit-related documents in their respective folders as per QOP 002, Section Three (pg 328).

3.04   Reporting Audit Results to Upper Management

The Management Representative shall:

1. Assess periodically the quality system audit results as to the overall impact (positive or negative) to product quality, to processes, and to procedures in respect to internal (organizational) and external (customer satisfaction) effectiveness;

2. Issue the assessment results on Form A-003 (blank, pg 457) to upper management. State the level of effectiveness and efficiency of the Quality Management System in respect to overall operations. Do this (quarterly) in line with the reporting by management to the stakeholders on "the state of business";

3. Present any significant problems and recommend solutions to upper management during "Management Review" to improve the effectiveness and efficiency of the Quality Management System. (see QOP 011 on page 428)

OVERVIEW TO QOP 011
(MANAGEMENT REVIEW)

Management Review in a process-approach system is not the forum to present a stupendous pile of papers to impress upper management that lots of things were done but there are still numerous unmanageable problems outstanding, making the operating system ineffective. This sort of thing happens when companies implement ISO compliant systems according to the way they always ran their businesses. They have documented all their weaknesses and not much has been improved in product quality. They have now little more to show except a mountain of records and lots of bills, awaiting review. You see, under a process-approach system, the responsibilities for process-owners are defined from the beginning to the end in the product's processing cycle and work assignment accountability to requirement is demonstrated at every process in the chain of events. The product is only an object, moving through the cycle, controlled by the process and not by the problems. Every process owner checks his/her work, because nobody else can build quality into the product. At the end you are ready to ship that product to the customer. This is how customer satisfaction is also built into the processes.

Controlling non-conformance in documents, processes, methods, and in products is also managed as a process. The problems are not pushed aside to be forgotten until they strike back again. So, management is given all the good news on how problems were solved as part of process control and not as pending management review to look for improvements where improvements would do nothing more than fan the wind – too late to add value to the process. Yet, management reviews are very important to assess successes and through them set new business plans for growth in an effectively managed production environment. In a process-approach system, management makes improvements on the basis of successes – a very significant change from the ISO philosophy.

MINFOR INCORPORATED	Iss dt. 1/1/99	Rev dt. 1/1/99		Pg.
QUALITY OPERATING PROCEDURE	Sign	Rev # 0		1/5
REQUIREMENT: QUALITY SYSTEM MANUAL SEC: 4.1.3				
SUBJECT: MANAGEMENT REVIEW				
QOP 011|DOCUMENT REQ'D: Forms A-003, A-006, A-007, A-034				

QUALITY OPERATING PROCEDURE 011

MANAGEMENT REVIEW

CONTENTS   **Page**

Form A-003 Rev 0 1999

MINFOR INCORPORATED		Iss dt. 1/1/99	Rev dt. 1/1/99		Pg.
QUALITY OPERATING PROCEDURE		Sign	Rev #	0	2/5
REQUIREMENT: QUALITY SYSTEM MANUAL SEC: 4.1.3					
SUBJECT: MANAGEMENT REVIEW					
QOP 011	DOCUMENT REQ'D: Forms A-003, A-006, A-007, A-034				

REVISION HISTORY

Rev Date	Rev No	Description	Approval

MINFOR INCORPORATED	Iss dt. 1/1/99	Rev dt. 1/1/99		Pg.
QUALITY OPERATING PROCEDURE	Sign	Rev # 0		3/5
REQUIREMENT: QUALITY SYSTEM MANUAL SEC: 4.1.3				
SUBJECT: MANAGEMENT REVIEW				
QOP 011 DOCUMENT REQ'D: Forms A-003, A-006, A-007, A-034				

1.0   PURPOSE

Define the Company's management review requirements in order to assess, as often as necessary, the suitability and effectiveness of the quality management system.

2.0   APPLICATION

The provisions of this procedure shall apply to the preparation, presentation, conclusion, and follow-up of management reviews.

3.0   PROCEDURE

3.01   General Requirement

Management Reviews shall be based on the results of effective assessment of the internal audits of the Quality Management System. Effectiveness of performance shall be measured against targeted limits (threshold) established in line with the internal (organizational) and external (customer satisfaction) business interest (risk limits) of the Company. Reason: Without being able to demonstrate compliance versus noncompliance (ratios) against practically established threshold limits, the goals to accomplish the set objectives of the quality system would miss the reason for additional improvements beyond the routine improvements accomplished through the corrective action provisions of QOP 009 (pg 415).

The Management Representative shall be responsible to establish reasonable threshold limits (maximum allowable non-compliance figures) in all the areas where this can be objectively implemented in order to know when to trigger (plan, do, check, act) corrective measures to remove the causes that negatively impacted the goals to achieve the business objectives of the Company. (See par. 3.02, step 2, below.) When the corrective measures, identified through assessment, cannot be implemented through the provisions of QOP 009, Sections Two and Three, the Management Representative shall set the agenda for Management Review.

Management shall not be called to session to review issues for which the answers to solve them are given in the documented procedures, for that would be dereliction (shirking one's duty). Conversely, when the lack of enforcement on the part of the delegated authority (supervision) impacts compliance with the provisions of the documented system, management shall be called to session to implement corrective action.

MINFOR INCORPORATED	Iss dt. 1/1/99	Rev dt. 1/1/99	Pg.
QUALITY OPERATING PROCEDURE	Sign	Rev # 0	4/5
REQUIREMENT: QUALITY SYSTEM MANUAL SEC: 4.1.3			
SUBJECT: MANAGEMENT REVIEW			
QOP 011 DOCUMENT REQ'D: Forms A-003, A-006, A-007, A-034			

3.02   Preparation for Management Review

The Management Representative shall:

1. Review the statistical results of the internal audit assessment. Determine the levels of noncompliance. Determine whether the causes have been cured through the C/A provisions of QOP 009, Sec Two (pg 410) and Three (pg 417). If they have not been cured, set the agenda for Management Review to resolve them. Determine new objectives, recommendations, and solutions;

2. Gather the supporting data to justify Management Review – supporting data to demonstrate to management both the positive and the negative areas of the internal audit assessment. Show the effectiveness versus the lack of effectiveness in respect to the overall performance objectives. Prepare proposal on how to resolve the lack of effectiveness. Describe these on Form A-003 in order to support your discussion (persuasion) points during Management Review. Issue it to all;

3. Select the participants to the Management Review. Ensure that those against whom corrective action may be issued are present. Management wants to hear their input for solutions;

4. Compile the supporting data in order of presentation. Make enough copies of these documents so that each participant can have a copy ahead of time;

5. Draw up the agenda for the Management Review under step 1 and 2 of Form A-034 (pg 493). Complete the "Attendees, Place and Date" sections. Distribute Form A-034 to all participants.

6. Prepare the participants so that the management review doesn't turn into a one man show, for the whole quality management system is dependent on teamwork to operate effectively. That's why you gave each participant a copy of the presentation material under step 4 above;

7. Hold the Management Review as scheduled.

Note: Management Reviews in small business don't require the elaborate setup big companies do. There aren't enough people to fill every position to carry out assignments. One manager may be responsible easily for a half a dozen assignments. It often happens that the Management Representative may hold the Management Review alone with the General Manager anywhere, even standing in the middle of the shop. There is nothing wrong with this type of setup for it's not the setting that counts but the outcome for solutions and improvements. Nevertheless, the documentation end of the process still requires the formalities associated with record keeping. That's why Form A-034 was designed for quick management reviews in small business.

MINFOR INCORPORATED	Iss dt. 1/1/99	Rev dt. 1/1/99		Pg.
QUALITY OPERATING PROCEDURE	Sign	Rev # 0		5/5
REQUIREMENT: QUALITY SYSTEM MANUAL SEC: 4.1.3				
SUBJECT: MANAGEMENT REVIEW				
QOP 011 DOCUMENT REQ'D: Forms A-003, A-006, A-007, A-034				

3.03   Conclusion of Management Review

The Management Representative shall:

1. Summarize the corrective action decisions made during Management Review under section 3 of Form A-034;
2. Describe the "Corrective/Preventive Action Summary" under section 4, as appropriate, and post the next NCMR serial number onto Form A-034. Sign and date Form A-034;
3. Issue Form A-034 to all participants.

3.04   Implementation of Corrective/Preventive Action

The Management Representative shall:

1. Take the blank NCMR (A-006) assigned in paragraph 3.03, step 2, above and fill out all sections, as appropriate. Observe that this is not a product-related NCMR but a process related matter. Accordingly, enter the relevant information as required. Put "N/A" in the blocks not applicable to Management Review. After you have described the details in the "Cause," "MRB Disposition," and "Corrective Action" blocks, you and the General Manager sign and date under each indicator. Distribute the NCMR, as required;
2. Interact with the responsible department head, or the individual, that has to implement the corrective action and work out the implementation plan with him/her in line with the required date noted on the NCMR, as agreed;
3. Summarize the implementation plan on Form A-003 (blank) and issue it to the responsible agent. Maintain a copy of it for follow-up review. Note: corrective action may involve change to the quality system's documents, resulting from Management Review. When this happens, carry it out in accordance with QOP 002, Sec Two, pg 317 (all paragraphs may apply) to implement it;
4. Maintain control over the retention of documents involved in this process in line with QOP 002, Sec Three (pg 328).

3.05   Follow up the Implementation of Corrective Action

The Management Representative shall:

1. Follow up implementation of corrective action in accordance with QOP 009, Sec Two, par 3.01.6 (pg 415).
2. Close out subject NCMR in accordance with QOP 009, Sec Two, par 3.01.7 (pg 416,) as applicable.

OVERVIEW TO QOP 012
(PERFORMANCE STANDARD)

QOP 012 is designed to give you a quick review of linkage as to how a process-approach system integrates assignments to procedures in handling two types of product categories, – conforming and non-conforming products.

Both must be simultaneously dealt with, right through a product's realization processes. If we don't, the conforming product eventually will be shipped, but the non-conforming product will die on the vine in many cases. At some time it will become a quickly fixed product, without corrective action. We don't prevent problems this way. We repeat them. This is not the *continual improvement* opportunity ISO 9001/2000 has in mind. Continuous improvement in a process-approach quality management system primarily takes place by implementing corrective action through process control, hence preventing problems from flowing down, process to process. It is for a very good reason then, that we must also control non-conforming products as we do conforming products.

Section One and Two of this QOP communicate the put through mechanism, first for the conforming products, second for the non-conforming products. These two procedures impact directly or indirectly every other Quality Operating Procedure in the quality system, simply because they effectively handle the two product categories mentioned above. They are performance standards, for every identified process has a performance requirement linked to it, which requires work result documentation and end result measurement to evaluate achievement in both category types.

Section Three is a completely different procedure, dealing with developmental projects to qualify processing methods and from it, formalizing the project related work instructions for an ongoing production effort often required with new customer purchase orders.

MINFOR INCORPORATED	Iss dt.7/16/99	Rev dt. 7/16/99		Pg.
QUALITY OPERATING PROCEDURE	Sign AG	Rev # 0		1/17
REQUIREMENT: QUALITY SYSTEM MANUAL SEC: 4.9				
SUBJECT: PERFORMANCE STANDARD, PROCESSING CONTROL				
QOP 012|DOCUMENT REQ'D: Form A-008 (Job Traveler), A-001				

QUALITY OPERATING PROCEDURE 012

PROCESSING CONTROL (Put Through)

CONTENTS   Page

Form A-003 Rev 0 1999

MINFOR INCORPORATED	Iss dt.7/16/99	Rev dt. 7/16/99	Pg.
QUALITY OPERATING PROCEDURE	Sign AG	Rev # 0	2/17
REQUIREMENT: QUALITY SYSTEM MANUAL SEC: 4.9			
SUBJECT: PERFORMANCE STANDARD, PROCESSING CONTROL			
QOP 012|DOCUMENT REQ'D: Form A-008 (Job Traveler), A-001			

QUALITY OPERATING PROCEDURE 012

PROCESSING CONTROL (Put Through)

CONTENTS (continued)   Page

Form A-003 Rev 0 1999

MINFOR INCORPORATED	Iss dt.7/16/99	Rev dt 7/16/99		Pg.
QUALITY OPERATING PROCEDURE	Sign AG	Rev # 0		3/17
REQUIREMENT: QUALITY SYSTEM MANUAL SEC: 4.9				
SUBJECT: Performance Standard, Processing Control (put through)				
QOP 012| DOCUMENT REQ'D: Form A-008 (Job Traveler), A-001				

REVISION HISTORY

Rev Date	Rev No	Description	Approval

MINFOR INCORPORATED	Iss dt.7/16/99	Rev dt. 7/16/99	Pg.
QUALITY OPERATING PROCEDURE	Sign AG	Rev # 0	4/17
REQUIREMENT: QUALITY SYSTEM MANUAL SEC: 4.9			
SUBJECT: Performance Standard, Processing Control (put through)			
QOP 012|DOCUMENT REQ'D: Form A-008 (Job Traveler), A-001			

SECTION ONE

1.0   PURPOSE

Standardize work assignments in order to demonstrate the process-approach put through system in an actual work environment.

2.0   APPLICATION

Subject procedure applies to contractually defined products going through the phases of planning, processing, verification, release and delivery to the customer.

3.0   PROCEDURE

General

A Job Traveler is that controlled document which defines all the required process steps that regulate the movement of a product from the beginning of a process to the end of a process in a way that process owners can determine from it, step by step, the identification of a product and its related documents, the equipment and tooling, the work instruction on what to do and what not to do in order to process that product to meet the scheduled work process requirement and ultimately, the customer's requirement, without unnecessary interruptions.

In order for a Job Traveler to become a useful, valid, and an approved process control document for all those who are required to follow it, the issuing authority (Process Engineer) must interact with Contracts, Production Planning, Purchasing, Manufacturing, and Quality to integrate the processing cycle evolution of the product, equipment, people, and methods. This would have to be done in a manner so that not one of the members of the operation's team will have to wait to fulfill its responsibilities to carry out his/her work assignment in the product's realization processes, due to poor planning in the beginning phases of the layout processes. After all, the Job Traveler is the road map of the processing cycle, which requires continuous input and follow-up by those in charge of it, so that when it is released to the traffic (the work realization processes), the people (process owners) will not have to be rerouted to take alternate paths to reach their targets (production goals), because somebody failed to do his part to make the road map (Job Traveler) an accurately reflected schedule-execute-verify-control driven directive (Plan-Do-Check-Act).

MINFOR INCORPORATED	Iss dt.7/16/99	Rev dt. 7/16/99		Pg.
QUALITY OPERATING PROCEDURE	Sign AG	Rev # 0		5/17
REQUIREMENT: QUALITY SYSTEM MANUAL SEC: 4.9				
SUBJECT: Performance Standard, Processing Control (put through)				
QOP 012│DOCUMENT REQ'D: Form A-008 (Job Traveler), A-001				

This process-approach quality management system planned the Job Traveler (including Split Travelers) to be the single most important document to control the product's entire routing requirements in a manner so as to ensure that accurate direction is given, in sequential order, to the process owners. This will enable them to carry out their work assignments so that the product can be processed through the cycle the first time in a controlled manner, ensuring that quality is continuously built into the product. Of course, to make this a reality, everybody will have to be entrusted to take charge of his/her work assignment in order to help each other to work as a team. This brings us then to integrating the work assignments within the process approach system so the above can be realized, ensuring along the way that each process owner verifies his/her work results and documents it, as required, to prevent transferring mistakes, regardless of who made them, from process to process.

3.01   Planning the Product Realization Processes

3.01.1   The Contract Administrative Secretary does:
Sends copy of customer's Purchase Order and Product Specifications to Engineering, Planning, and Quality.
Requirement: QOP 001, Section Two, paragraph 3.01 (pg 301).

3.01.1.1 The Quality Engineer does:
Reviews the customer's Purchase Order and Product Specification to determine the product quality requirements for the whole production cycle. Summarizes the results on Form A-001 (Purchase Order Review Sheet). Distributes it.
Requirement: QOP 001, Section Two, paragraph 3.02 (pg 302).

3.01.1.2 The Process Engineer does:
Reviews the customer's Purchase Order, the Product Specifications and Form A-001 (sent by the Quality Engineer) in order to plan the production layout, technical and equipment requirements. Incorporates the inspection steps identified by the Quality Engineer on A-001 into the production layout.
Requirement: QOP 001, Section Two, paragraph 3.03 (pg 302).

3.01.1.3 The Contract Administrator does:
Holds contract review meeting in executive session with Production Planning, Engineering, Purchasing, Manufacturing, and Quality in order to determine process evolution and direction. Issues summary of commitments, action items, and target dates on Form A-003. Distributes it to the attendees.

MINFOR INCORPORATED	Iss dt.7/16/99	Rev dt. 7/16/99	Pg.
QUALITY OPERATING PROCEDURE	Sign AG	Rev # 0	6/17
REQUIREMENT: QUALITY SYSTEM MANUAL SEC: 4.9			
SUBJECT: Performance Standard, Processing Control (put through)			
QOP 012	DOCUMENT REQ'D: Form A-008 (Job Traveler), A-001		

Requirement: QOP 001, Section Two, paragraph 3.04 (pg 303).

3.02   Enacting the Planning Provisions for the Product Realization Processes

3.02.1   The Production Planner does:
Releases the Master Production Scheduling (MPS) to all affected departments.
Requirement: QOP 001, Section Two, paragraph 3.05, step 1, (pg 303).

3.02.2   The Quality Engineer does:
Sets up the Inspection Product Folder and plans the inspection controls for the product's entire processing cycle.
Requirement: QOP 001, Sec Two, par 3.02, steps 4 and 5, (pg 302).

3.02.3   The Process Engineer does:
Prepares, reviews, approves, and releases the Production Folder to the Production Planner.
Requirement: QOP 001, Sec Two, par 3.03, steps 3,4,5, and 6, (pg 302).

3.02.4   The Purchasing Agent does:
Carries out the procurement activities, as required.
Requirement: QOP 003, Sections One (pg 338) and Two (pg 341).

3.02.5   The Production Planner does:
Interacts regularly with all departments to maintain schedule.
Requirement: QOP 001, Section Two, paragraph 3.05, step 2, (pg 303).

3.03   The Product Realization Process

3.03.1   The Production Planner does:
Issues the Production Folder, as required.
Requirement: QOP 001, Section Two, paragraph 3.05, step 3, (pg 303).

3.03.2   The Production Supervisor does:
Reviews the Production Folder (the Job Traveler) and production scheduling (MPS) and carries out process controls before setting up operations (pg 375).

MINFOR INCORPORATED	Iss dt.7/16/99	Rev dt. 7/16/99		Pg.
QUALITY OPERATING PROCEDURE	Sign AG	Rev # 0		7/17
REQUIREMENT: QUALITY SYSTEM MANUAL SEC: 4.9				
SUBJECT: Performance Standard, Processing Control (put through)				
QOP 012	DOCUMENT REQ'D: Form A-008 (Job Traveler), A-001			

Requirements:

a) material identification as per QOP 006, Section Five, paragraph 3.02.2, step 5, (pg 384);

b) tooling, fixture, special gages as per QOP 001, Section Two, paragraph 3.03, step 6, (pg 303);

c) equipment maintenance on Form A-002 as per QSM, par 4.9g, (pg 281);

d) availability of standard gages as per QOP 007, par 3.06, (pg 396).

3.03.3   The Process Owner (operator) does:
Sets job up, as required. Makes first production piece.
Requirement: QOP 006, Sec One, par 3.01, steps 1 through 6, (pg 364).

3.03.4   The Production Supervisor does:
Determines production continuation. Reviews product completion on Job Traveler and signs Form A-033 under "Audited."
Requirement: QOP 006, Section One, paragraph 3.01, step 4, (pg 364).

3.03.5   The Quality Inspector does:
Verifies and approves the first production piece.
Requirement: QOP 006, Sec One, par 3.01, steps 1 through 9, (pg 365).

3.04   Final Inspection, Product Release and Delivery

3.04.1   The Quality Inspector does:
Carries out final inspection and controls product release and delivery, as required.
Requirement: 1. QOP 006, Sec Four (pg 380), all paragraphs, as required;
              2. QOP 009, Sec Two (pg 410), all paragraphs, as required.

MINFOR INCORPORATED	Iss dt.7/16/99	Rev dt. 7/16/99		Pg.
QUALITY OPERATING PROCEDURE	Sign AG	Rev # 0		8/17
REQUIREMENT: QUALITY SYSTEM MANUAL SEC: 4.13				
SUBJECT: Handling Non-conformance (put through)				
QOP 012	DOCUMENT REQ'D: Form A-006 (Non-conforming Material Report)			

SECTION TWO

1.0   PURPOSE

Integrate the handling and control of non-conformities as part of process control during the product's overall processing cycle.

2.0   APPLICATION

This procedure provides specific instructions on how to handle and control non-conformities identified at any point in the processing cycle of the product, including customer complaints and product returns.

3.0   PROCEDURE

General

Since a non-conforming state could contribute to process deficiencies anywhere in the processing cycle of a product, timely resolution of the constraint, created by it, must be implemented to ensure effective process continuation. A nonconforming state does not only mean hardware in a process-approach system, but it also includes errors in documents, in processes, and in the methods. These may eventually affect production and product quality. To stop the transfer of deficiencies creeping from process to process, we have implemented the controls to isolate them and cure them when and where they occur. Not every nonconformity leads to hardware deficiency and when it doesn't, we have implemented other methods in this process-approach system to cure them. Carefulness must be exercised therefore between the two distinct varieties – hardware vs. non-hardware – for the cost difference in handling is substantial. Applying the NCMR (MRB) process is primarily hardware related. It should also be used when no other documented procedure exists to handle corrective action. The four main categories in which we may discover non-conformities that we must control are:

1. **hardware related (handled by the application of the NCMR (MRB) process);**
2. **production related documents (handled by the application of Form A-009,** pg 464**);**
3. **customer related documents (handled through QOP 001, Section Four);**
4. **documents making up the Quality System manual and the Quality Operating Procedures (handled by draft revisions).**

MINFOR INCORPORATED	Iss dt.7/16/99	Rev dt. 7/16/99		Pg.
QUALITY OPERATING PROCEDURE	Sign AG	Rev # 0		9/17
REQUIREMENT: QUALITY SYSTEM MANUAL SEC: 4.13				
SUBJECT: Handling Non-conformance (put through)				
QOP 012	DOCUMENT REQ'D: Form A-006 (Non-conforming Material Report)			

3.01   Hardware related non-conforming products identified during production

3.01.1   The Process Owners (operators) do:
Notify supervision when you have verified that a product is non-conforming.
Requirement: QOP 006, Section One, paragraph 3.01, step 6, (pg 364).

3.01.2   The Supervisor does:
Review that the process owner properly verified the product non-conformance. Take the necessary corrective action to continue production. Process NCMR (for MRB) through the Quality Inspector, as required.
Requirement: QOP 006, Sec One, par 3.01, steps 1 through 5, (pg 364).

3.01.3   The Quality Inspector does:
a)   verify the product non-conformance and document it on Form A-018;
b)   process Non-conforming Material Report (Form A-006)
Requirement: a) QOP 006, Section One, paragraph 3.01, step 8, (pg 365).
               b) QOP 009, Sec Two, par 3.01.1, steps 1 through 3, (pg 410).
Note: Further processing of NCMR is picked up under paragraph 3.02.2 below.

3.02   Hardware related non-conforming products identified at any point during the processing cycle of a product

3.02.1   The Quality Inspector does:
a)   verify the non-conforming product and document the results onto Form A-018. Tag the non-conforming product applying Form A-012 (pg 467). Segregate it, as practical;
b)   obtain a blank form of A-006 (pg 461) and fill out, as instructed under:
Requirement: QOP 009, Sec Two, par 3.01.1, steps 1 through 3, (pg 410).

3.02.2   The Inspection Supervisor or the Quality Engineer does:
Reviews the NCMR for completeness and accuracy. Signs and dates it. Submits hardware with documents to MRB action, as appropriate.
Requirement: QOP 009, Sec Two, par 3.01.1, step 3 (last sentence), (pg 410).

3.02.3   The Production Supervisor does:
Determines the cause of the non-conformance as instructed under:
Requirement: QOP 009, Sec Two, par 3.01.2, steps 1 through 3, (pg 410).

MINFOR INCORPORATED	Iss dt.7/16/99	Rev dt. 7/16/99		Pg.
QUALITY OPERATING PROCEDURE	Sign AG	Rev # 0		10/17
REQUIREMENT: QUALITY SYSTEM MANUAL SEC: 4.13				
SUBJECT: Handling Non-conformance (put through)				
QOP 012	DOCUMENT REQ'D: Form A-006 (Non-conforming Material Report)			

3.02.4 The Material Review Board (MRB) does:
Carries out the disposition activity as instructed under:
Requirement: QOP 009, Sec Two, par 3.01.3, all steps apply, (pg 411).

3.02.5 Executing the MRB disposition
The Quality Engineer is in charge of interacting with all departments to ensure timely execution and follow-up on MRB activities to *move* people to make timely resolution, as instructed under:
Requirement:
a) **Rework:** QOP 009, Sec Two, par 3.01.4.1, all steps apply, (pg 412);
b) **Repair:** QOP 009, Sec Two, par 3.01.4.2, all subparagraphs and steps apply under each identified responsibility, (pg 412);
c) **Accept:** QOP 009, Section Two, paragraph 3.01.4.3, (pg 413);
d) **Return to Vendor:** QOP 009, Section Two, paragraph 3.01.4.4, all steps apply, (pg 413);
e) **Scrap:** QOP 009, Section Two, paragraph 3.01.4.5, all steps apply, (pg 414);
f) **Waiver:** QOP 009, Section Two, paragraph 3.01.4.6, (pg 414 and 412);
g) **Regrading Material:** QOP 009, Section Two, paragraph 3.01.4.7, all steps apply, (pg 414).

3.03 Handling Corrective Action

3.03.1 The Management Representative does:
Carries out corrective/preventive action planning and determines the implementation process together with the responsible department head or individual. Documents the planning results. Implements training requirement and its recording.
Requirement: QOP 009, Sec Two, par 3.01.5, all steps apply, (pg 415).

3.03.2 The Management Representative does:
Follows up the corrective/preventive action's effective implementation as stated under:
Requirement: QOP 009, Sec Two, par 3.01.6, all steps apply, (pg 415).

3.03.3 The Management Representative does:
Closes out the NCMP (MRB) process and answers any outstanding SCAR, as required. Maintains document retention.
Requirement: QOP 009, Section Two, paragraphs 3.01.7 and 3.01.8, all steps apply, (pg 416).

MINFOR INCORPORATED	Iss dt.7/16/99	Rev dt. 7/16/99		Pg.
QUALITY OPERATING PROCEDURE	Sign AG	Rev # 0		11/17
REQUIREMENT: QUALITY SYSTEM MANUAL SEC: 4.13				
SUBJECT: Handling Non-conformance (put through)				
QOP 012 DOCUMENT REQ'D: Form A-006 (Non-conforming Material Report)				

3.04 Production related document-change requirement identified by process owners, anywhere in the product's life-cycle

3.04.1 The Process Owners do:

Any process owner who discovers errors, mistakes, or misleading instructions in any document which they are required to use during the process of carrying out their work assignments must report the problem to their immediate supervisor.
Requirement: QOP 002, Section Two, paragraph 3.0, (pg 317), under "General Requirement."

3.04.2 The Supervisor does:

Determine the nature of the document anomaly. If product quality is impacted, stop production. If not, continue production. In either case, call the Process and Quality Engineers to the site and point out to them the document problem.
Requirement: QOP 002, Section Two, paragraph 3.0, (pg 317), under "General Requirement."

3.04.3 The Process and Quality Engineers do:
a) carry out the "Immediate and Provisional" change requirements when the process and/or product is impacted as stated under:
   Requirement: QOP 002, Sec Two, par 3.03.1, all steps apply, (pg 318);
b) carry out the "Standard Document Changes" when the process and/or product is not immediately impacted as stated under:
   Requirement: QOP 002, Sec Two, par 3.03.2.1, all steps apply, (pg 319);
c) carry out "NCMR (MRB)" affected document changes as stated under:
   Requirement: QOP 002, Sec Two, par 3.03.2.2, all steps apply, (pg 319);
d) carry out "Software Changes" as stated under:
   Requirement: QOP 002, Section Two, paragraph 3.03.2.3, (pg 320).

3.05 Document-change requirements affecting the Quality System's Manual and the Quality Operating Procedures

3.05.1 The Management Representative does:
a) review the type of change requirements as stated under:
   Requirement: QOP 002, Section Two, paragraph 4.0, the "General" requirement and all "Notes" apply, (pg 320);
b) carry out the indirect changes to the procedures as stated under:
   Requirement: QOP 002, Sec Two, par 4.01.1, all steps apply, (pg 321);

MINFOR INCORPORATED	Iss dt.7/16/99	Rev dt. 7/16/99		Pg.
QUALITY OPERATING PROCEDURE	Sign AG	Rev # 0		12/17
REQUIREMENT: QUALITY SYSTEM MANUAL SEC: 4.13				
SUBJECT: Handling Non-conformance (put through)				
QOP 012|DOCUMENT REQ'D: Form A-006 (Non-conforming Material Report)				

    c) carry out the direct changes to the procedures as stated under:
      Requirement: 1. QOP 002, Sec Two, par 4.02, "General," (pg 322);
               2. QOP 002, Section Two, paragraph 4.02.1, specific requirement, all steps apply, (pg 322);

    d) carry out the revision indication changes in the Quality System Manual as stated under:
      Requirement: 1. QOP 002, Section Two, paragraph 5.0, "General," including "Notes," (pg 324);
               2. QOP 002, Section Two, paragraph 5.01, specific requirement, all steps apply, (pg 324);

    e) carry out the revision indication changes in the Quality Operating Procedures as stated under:
      Requirement: QOP 002, Sec Two, par 5.02, all steps apply, (pg 325);

    f) carry out the revision indication changes to the Quality System Forms as stated under:
      Requirement: QOP 002, Sec Two, par 5.03, all steps apply, (pg 326).

3.06 Handling non-conforming products resulting from document changes

3.06.1 The Quality Engineer does:
Handle the non-conforming material resulted from imposing document changes as stated under:
Requirement: 1. QOP 002, Sec Two, par 6.0, "General" requirement, (pg 327).
               2. QOP 002, Section Two, paragraph 6.01, specific requirement, all steps apply, (pg 327).

3.07 Handling customer-related document changes

General Information
Customer-related document changes involve two separately initiated efforts. The first one is initiated by the customer through the provisions of Purchase Order Amendments. The second one is initiated internally, due to deficiencies discovered in the customer's documents. The customer-initiated changes involve extensive review process due to cost factors, and therefore they are handled much like a quotation process (RFQ). It is covered under QOP 001, Section Four. The internally initiated changes involve the application of a "Waiver Request" through the provisions of QOP 009, Section Two, paragraph 3.01.4.2.1, step 1, under "Repair." (pg 412).

MINFOR INCORPORATED	Iss dt.7/16/99	Rev dt. 7/16/99		Pg.
QUALITY OPERATING PROCEDURE	Sign AG	Rev # 0		13/17
REQUIREMENT: QUALITY SYSTEM MANUAL SEC: 4.13				
SUBJECT: Handling Non-conformance (put through)				
QOP 012	DOCUMENT REQ'D: Form A-006 (Non-conforming Material Report)			

3.07.1   Customer initiated contract (PO) amendments

3.07.1.1 The Contract Administrative Secretary does:

Handles the amendment-related documents to maintain administrative control over them as stated under:

Requirement: QOP 001, Section Four, paragraphs 3.01, 3.03, 3.05, and 3.06, all steps apply, (pg 307, 308);

3.07.1.2 The Contract Administrator does:

Determines what impact the customer's amendment has on the current contract and initiates action accordingly as stated under:

Requirement: QOP 001, Section Four, paragraphs 3.02, 3.04, 3.10, all steps apply, (pg 307, 308, 309);

3.07.1.3 The Process Engineer does:

Implements internally the processing related document changes through Form A-009 as stated under:

Requirement: QOP 001, Sec Four, par 3.08, all steps apply, (pg 309);

3.07.1.4 The Process and Quality Engineers do:

Work together to implement the customer's amendment as stated under:

Requirement: QOP 001, Section Four, paragraphs 3.09 and 3.11, all steps apply, (pg 309).

3.08   Customer complaints, product returns, and customer evaluation (satisfaction) reporting

3.08.1 Customer complaints

3.08.1.1 The Complaint Receiver does:

Directs the complaint reporter to the Quality Manager as stated under:

Requirement: QOP 009, Section Three, paragraph 3.01.1 "General Requirement," (pg 417).

3.08.1.2 The Quality Manager or his/her designate does:

Investigates and interacts to resolve the complaint as stated under:

Requirement: QOP 009, Section Three, paragraph 3.01.1.1, all steps apply, (pg 417).

MINFOR INCORPORATED	Iss dt.7/16/99	Rev dt. 7/16/99	Pg.
QUALITY OPERATING PROCEDURE	Sign AG	Rev # 0	14/17
REQUIREMENT: QUALITY SYSTEM MANUAL SEC: 4.13			
SUBJECT: Handling Non-conformance (put through)			
QOP 012|DOCUMENT REQ'D: Form A-006 (Non-conforming Material Report)			

3.08.2   Product Return

3.08.2.1 The Receiver of the product does:
Receives the customer returned product and initiates action as stated under:
Requirement: QOP 009, Sec Three, par 3.02.1, all steps apply, (pg 418);

3.08.2.2 The Quality Engineer does:
Investigates background, subjects the returned product to verification, cause determination, and corrective action as stated under:
Requirement: QOP 009, Section Three, paragraphs 3.02.2 and 3.03, all steps apply, (pg 418, 419).

3.08.3   Customer satisfaction reporting

3.08.3.1 All personnel do:
Understand the nature of customer satisfaction reporting as stated under:
Requirement: QOP 009, Section Three, paragraph 3.04 "General Requirement," (pg 419);

3.08.3.2 The Final Inspector does:
Initiates the customer evaluation and satisfaction report (Form A-038) as stated under:
Requirement: QOP 009, Sec Three, par 3.04.1.1, all steps apply, (pg 420);

3.08.3.3 The Management Representative does:
Handles the replied customer evaluation and satisfaction report as stated under:
Requirement: QOP 009, Sec Three, par 3.04.1.2, all steps apply, (pg 420).

MINFOR INCORPORATED	Iss dt.7/16/99	Rev dt. 7/16/99		Pg.
QUALITY OPERATING PROCEDURE	Sign AG	Rev # 0		15/17
REQUIREMENT: QUALITY SYSTEM MANUAL SEC: 4.9				
SUBJECT: Developmental Process Control (put through)				
QOP 012	DOCUMENT REQ'D: Forms A-008, A-001, A-018			

SECTION THREE

1.0   PURPOSE

Maintain control over those activities that involve the product's processing quali-fication when no documented procedures exist on how to do it, as with new cus-tomer Purchase Orders.

2.0   APPLICATION

This procedure applies to the planning and implementation of qualification pro-cesses on new jobs where the capability as to how to do it have not been previous-ly documented and demonstrated.

3.0   PROCEDURE

General Information

Processing qualification of new products is often regarded by many customers as nothing more than setting up a new job in an otherwise similar production envi-ronment. While the equipment, tools, and operator's skill may be the same, the processing steps and work methods involved can be completely different. This type of product cannot be regarded as identical with those products for which the processing capability has already been demonstrated and documented. Under uncontrolled conditions, the entire order could become scrap. A process-approach quality management system must provide the procedural mechanism under which the processing qualification can be orderly carried out and the product released to the customer according to his/her PO requirement. Since we have already provid-ed controlling procedures for First Article Production Approval (ESA) under QOP 006, Section Two, we will be applying certain provisions of the same (text) proce-dures to implement basic requirements for the processing qualification of new jobs. The applicable steps are marked with a star "✰."

3.01   Planning

3.01.1   The Process Engineer does:

Reviews the customer's PO and product specification in order to determine the planning and processing layout for developing the procedures on how to imple-ment the qualification steps on new jobs, which procedures when formalized would become the standard work instruction for the process owners.
Requirement: QOP 006, Sec Two, par 2.0, steps 1, 2, 4, and 5, (pg 368);

MINFOR INCORPORATED	Iss dt.7/16/99	Rev dt. 7/16/99		Pg.
QUALITY OPERATING PROCEDURE	Sign AG	Rev # 0		16/17
REQUIREMENT: QUALITY SYSTEM MANUAL SEC: 4.9				
SUBJECT: Developmental Process Control (put through)				
QOP 012 DOCUMENT REQ'D: Forms A-008, A-001, A-018				

QOP 006, Section Two, paragraph 3.0, step 2, (pg 368).

3.01.2 The Quality Engineer does:

Reviews the customer's PO and product specification in order to determine the planning of product quality verification requirements for the processing cycle as stated under:

Requirement: QOP 006, Sec Two, par 2.0, steps 1,2,3, 4, and 5, (pg 368);
QOP 006, Sec Two, par 3.0, steps 1, 3, and 4, (pg 368).

3.02 Issuance of Developmental Procedures

3.02.1 The Quality Engineer does:

Issues the Purchase Order Review Sheet (form A-001) in order to identify the product quality verification requirements for the processing cycle as stated under:
Requirement: QOP 006, Sec Two, par 4.0, steps 1, 2, 3, 4, and 5, (pg 368).

3.02.2 The Process Engineer does:

Determines, documents, reviews, approves, and issues the developmental Production Folder as stated under:
Requirement: QOP 006, Sec Two, par 5.0, steps 1, 2, 3, 4, 5, and 6, (pg 369).

3.03 Process Control

3.03.1 The Quality Engineer does:

Maintains process control, as required, as stated under:
Requirement: QOP 006, Sec Two, par 6.0, steps 1, 2, 3, 4, 5, 6, and 7, (pg 370).
(Apply these requirements as suitable for the developmental processing effort.)

3.03.2 The Quality Engineer does:

Handles developmental non-conforming materials as follows:

1. If the contract (PO) has material (parts) allowance for developmental purposes, no accountability for the non-conformance is required. Correct the non-conformance as part of processing development. Note that this type of allowance many times is not stated in the contract. What may be stated, however, is that you are allowed a certain percentage of parts for attrition (scrap). Note also that physical scrapping may still be controlled under certain contract provisions **(as in the case of AS9000)**;

MINFOR INCORPORATED	Iss dt.7/16/99	Rev dt. 7/16/99		Pg.
QUALITY OPERATING PROCEDURE	Sign AG	Rev # 0		17/17
REQUIREMENT: QUALITY SYSTEM MANUAL SEC: 4.9				
SUBJECT: Developmental Process Control (put through)				
QOP 012 DOCUMENT REQ'D: Forms A-008, A-001, A-018				

2. If there is no allowance for developmental purposes given and the material (parts) were supplied by the customer, handle the non-conforming product as stated under:

   Requirement: QOP 006, Section Two, paragraph 7.0, steps 1 through 8, as required, (pg 370);

3. If the material (parts) is company purchased, any fallout due to processing development should not be controlled as non-conforming material. Document it on Form A-006 (NCMR) as part of development and destroy the product, as required.

3.04   Closure of Developmental Procedures

3.04.1   The Process Engineer does:

After the developmental processes are over, you are required to formalize all procedures to become a standard operating work instruction issued in Production Folders, as stated under:

Requirement: QOP 006, Section Two, paragraph 5.0, step 7, (pg 370).

3.04.2   The Process and Quality Engineers do:

Maintain retention of the Developmental Procedures as stated under:

Requirement: QOP 006, Section Two, paragraph 11.0, all provisions apply as required, (pg 373).

3.05   Release of Product Completed Under Development

3.05.1   The Final Inspector does:

Carry out final inspection on products made under Developmental Process the same way as products made under standard operating conditions. The customer's product specification (drawing) is the controlling document. If the developmental product meets all the specification requirements, the product can't be rejected. For background information, rely on documents contained in the First Article Inspection Product Folder now marked as "Development." Carry out the final inspection, release, and delivery requirements as stated under:

Requirement: QOP 006, Sec Four, all provisions apply as required, (pg 380).
(Make up a standard Inspection Product Folder at this time in order to retain the final inspection records separately from the "Developmental Records." Interact with the Quality Engineer to work out document handling details.)

BOOK SECTION FOUR

FORMS MANUAL

OVERVIEW TO THE FORMS MANUAL

The forms in this manual represent actual and typical forms used in manufacturing shops to record work results. Without work results demonstration, there is no way to convince anybody that there is a controlled quality system in operation. The application of these forms has been linked to the specific work assignments elaborated in the Quality Operating Procedures. In a sense they are part of the operating procedures, but for the purpose of maintenance and control, they have been split off from the operating procedures. The experienced quality practitioner will immediately recognize that a number of these forms are typically used in metal cutting shops and, therefore, need no additional customizing. Other shops may want to customize them according to their product lines. Remember though that when you do need to customize a form, the product quality requirement, customer satisfaction, continuous improvement objectives, how minimal they are, will not change. Only the way you want to record the work results to prove accomplishment will change.

Because the forms have a controlled structure insofar as their title, form number, and revision control are concerned, any change in the body of the forms will not alter this controlled structure. But if you add or remove a form, the controlled structure will be impacted. You will have to align any form addition or removal, first in the Master List, second in the applicable operating procedure and third in the form itself. Each form has a note on the left bottom of the page, indicating where the application of it is described. Also, QOP 002, Section Three, gives you details on the application of forms in addition to retention requirements. The form number is the control number and that is how each form should be remembered and referenced in communication.

There is a strict retention requirement for most forms because they demonstrate product quality enforcement and work results documentation. For this reason, clear instruction has been defined in QOP 002, Section Three, as to where the forms should be filed. This becomes a very important aid in filing control as you need to refer back to the records when traceability or audit is carried out.

Please refer to the "Guidance" section in front of the book should you need information on how to carry out changes to the forms.

Remember, ISO 9001/2000 requires a documented procedure to control records (forms). You may follow the controls I have already implemented.

MINFOR INCORPORATED

SUMMARY OF FORMS (Master List)

Form #	Title	Rev.	Dt.	Status
A-001	Purchase Order Review Sheet	0	1998	New
A-002	Maintenance Record	0	1998	New
A-003	Quality Operating Procedure Blank	0	1998	New
A-004	RFQ Worksheet (alternate provided)	0	1998	New
A-005	RFQ, PO, and Amendment Log	0	1998	New
A-006	Nonconforming Material Report (NCMR)	0	1998	New
A-007	Nonconforming Material Report Register	0	1998	New
A-008	Job Traveler	0	1998	New
A-009	Amendment to Procedures	0	1998	New
A-010	Document Issuance Control (QSM, QOP)	0	1998	New
A-011	Control of Customer Documents	0	1998	New
A-012	Rejected Material Ticket	0	1998	New
A-013	Acceptance Material Ticket	0	1998	New
A-014	Hold Ticket	0	1998	New
A-015	Record of Received Materials	0	1998	New
A-016	Training Log	0	1998	New
A-017	Purchase Order (internal)	0	1998	New
A-018	Inspection Report	0	1998	New
A-019	Inspection Report Continuation Sheet	0	1998	New
A-020	Calibration and Status Record	0	1998	New
A-021	Audit Schedule	0	1998	New
A-022	Audit Plan (Reporting)	0	1998	New
A-023	Rework/Repair Record	0	1998	New
A-024	Waiver Request	0	1998	New
A-025	Shipping Log	0	1998	New
A-026	Supplier Survey Questionnaire	0	1998	New
A-027	Material Identification Tag	0	1998	New
A-028	Packing Slip	0	1998	New
A-029	Certificate of Conformance	0	1998	New
A-030	Receiving Log	0	1998	New
A-031	Issue and Traceability of Customer Drawing	0	1998	New
A-032	Password and Inspection Stamp Control	0	1998	New
A-033	Operator Product Verification Record	0	1998	New
A-034	Management Review Status Record	0	1998	New
A-035	Supplier Survey Form (short form)	0	1998	New
A-036	Source Inspected Product Approval	0	1998	New
A-037	Supplier Corrective Action Request	0	1999	New
A-038	Customer Complaint and Evaluation Report	0	7/16/99	New
A-039	Training Evaluation Sheet	0	1999	New

MINFOR INCORPORATED		Iss dt. 1/1/99	Rev dt. 7/16/99	Pg.
QUALITY OPERATING PROCEDURE	Sign		Rev # 1	N/A
REQUIREMENT: QUALITY SYSTEM MANUAL SEC: 4.1 – 4.20				
SUBJECT: MASTER LIST – Forms				
QOP All |DOCUMENT REQ'D: QOP 001 through QOP 012				

REVISION HISTORY

Rev Date	Rev No	Description	Approval
7/16/99	1	Added Form A-038	AG

A-001	**MINFOR INCORPORATED**

PURCHASE ORDER REVIEW SHEET   PO Date _____

PO No_____ Rev_____ Formal ☐   Verbal ☐   Amendment ☐

Cust_____ Ph_____ Fax _____

Cust Qual Rep_____   Ph_____   Fax_____

Part No _____ Rev _____ Matrl _____ Qty _____

Job No _____   New ☐   Repeat ☐   Other ☐

PO Qual Prov: | | | | | | | |

PO Item Qty: 1 ___ 2 ___ 3 ___ 4 ___ 5 ___ 6 ___ 7 ___ 8 ___

Del Date: 1 ___ 2 ___ 3 ___ 4 ___ 5 ___ 6 ___ 7 ___ 8 ___

List Contracted Rqmts ____ Incl Subs and Special Processes

1	5
2	6
3	7
4	8

Inspection Requirements:

SPC Requirement _____ ☐	ESA Requirement_____ ☐
Material Control _____ ☐	Vendor Source Insp_____ ☐
Marking and Ident_____ ☐	Special Gaging_____ ☐
First Pc Insp_____ ☐	Special Process_____ ☐
First Article Insp_____ ☐	Special Workmanship_____ ☐
Final Insp_____ ☐	Subcontracting_____ ☐
Cust Source Insp _____ ☐	Traceability_____ ☐
In-process Control_____ ☐	Special Packaging_____ ☐

Shipping Documents: Insp Rprt ___ C of C ___ Test Rprt ___ Deviations___
Delivery Cond: Cust Pickup___ Air ___ Land ___ Other _____

Additional Requirements:

Record of Shipments: 1 ___ 2 ___ 3 ___ 4 ___ 5 ___ 6 ___ 7 ___ 8 ___

Signature_____ Date _____ Audited by: _____ Date:_____

QOP 001, Sec. One and Two

Form   A-001   Rev 0   1998

A-002

MINFOR INCORPORATED

MAINTENANCE RECORD

Equipment No _____   Location _____   Dept._____

Frequency Date Requirement	JAN	FEB	MAR	APR	MAY	JUN	JUL	AUG	SEP	OCT	NOV	DEC
1												
2												
3												
4												
5												
6												
7												
8												
9												
10												
11												
12												
13												
Signature												
Date done												
Audited By												
Date Audited												

Notes:

QSM par. 4.9f   Form A-002   Rev 0 1998

A-003 (Blank Document for Recording and Internal Communication)

MINFOR INCORPORATED	Iss dt.	Rev dt.	Pg.
QUALITY OPERATING PROCEDURE	Sign	Rev #	

REQUIREMENT:

SUBJECT:

QOP	DOCUMENT REQ'D:

A-004 (ALTERNATE)	MINFOR INCORPORATED RFQ WORK SHEET		

RFQ #	PART NUMBER	CUSTOMER	DUE DATE

PROCESS	QTY:	QTY:	QTY:	QTY:	QTY:
Total Hours					

SUBCONTRACTS					

COSTS US $					
SUBCONTRACTS					
TOTAL PROCESS					
TOOLS, FIXTURE					
NON-RECURRING					
DELIVERY					
FINAL PRICE					

T E R M S: FOB: _____ NET: _____ DATED: _____

NOTES: _____
_____

Audited by: _____ Date: _____

QOP 001, Section One (alternate form)   Form A-004 Rev. 0 1999

MINFOR INCORPORATED   RFQ WORK SHEET

(COST BREAKDOWN)

A-004

DATE	COMPANY	RFQ#	PART NUMBER	ENGINEER

Raw Mat'l	Setup & Tooling	Machining 1	Machining 2	Machining 3	Machining 4	Machining 5	Secondary Op.	Handling	Prod Supp.	Subcontract	Inspection	Gaging (Tool Room)	Qual. Support	Documentation	NOTE:
1	2	3	4	5	6	7	8	9	10	11	12	13	14	15	1. Computation papers are allowed separately.
														x	2. Select items as req'd.
													x		3. Take notes as needed.
				N	O	T	E	S				x			4. Add additional cost on bottom.
											x				
										x					
									x						
								x							
							x								
						x									
					x										
				x											
			x												
		x													
	x														
x															

Cost Per

1 pc.

10 pc. / ea.

50 pc. / ea.

100 pc. / ea.

200 pc. / ea.

Subtotal

Additional Cost

Final Cost

QOP 001 Section One   Audited by: _____ Date: _____   Form A-004 Rev 0 1998

460	A-005	MINFOR INCORPORATED

RFQ, PO, AND AMENDMENT LOG Indicate Type:_____

LOG-IN			LOG-OUT	
COMPANY	Document #	DATE	STATUS	DATE

QOP 001, Section 1, 2, 3, and 4   **Audited by:** _____ **Date**: _____ Form A-005 Rev 0 1998

A-006   **MINFOR INCORPORATED**   No _____

① **NONCONFORMING MATERIAL REPORT**

Doc or P/N		Rev	P/Name		Mat'l		Job No

Op No/SN		Qty Insp	Qty Acc	Qty Rej		Insp By	Date

PO No		Cust/Vend	Operator	Buyer		RM or Shipper No	

Item	Rqmt	Description
1	②	
2		
3		
4		
5		

Inspector _____ Date _____ Supervisor _____ Date _____

CAUSE:

③

Supervisor _____ Date _____

MRB DISPOSITION		Instruction/Comment
RWK	Item#	
RPR	Item#	④
ACC	Item#	
RTV	Item#	
SCR	Item#	
Waiver	Item# ——→	*Regrade is determined by Customer on the submitted waiver!*

C/A Yes ☐ No ☐   Signature: Eng _____ Date _____Qual _____ Date _____

Corrective Action: ⑤   Req'd By _____

Engineering_____ Date_____Quality_____ Date _____

Follow up Audit By: Name_____ Date _____

C/A Completed Yes ☐ No ☐ N/A ☐   ⑥
If "No" action taken. Explain:_____

Audited by: _____ **Date:** _____

Mgmt Rep _____ Date _____

QOP 009 (all sections)   Form A-006 Rev 0   1998

462	A-007	MINFOR INCORPORATED			

MRB REGISTER

ISSUANCE			CLOSEOUT		
NCMR S/N	JOB No	DATE	NOTES	BY	DATE

MINFOR INCORPORATED

JOB TRAVELER

Issuer_____ Date _____ Page _____

JOB No _____
Part No _____ Revision _____ ← ① Req'd Date _____ Mat'l _____
Part Name_____   Scheduled_____ Qty _____
Prod Qty_____ S/N _____ Sched Due _____ HT Lot _____

CUSTOMER _____

Final Assembly	DESCRIPTION	Prod Notes
Part No _____ Rev_____ Qty _____	PO No _____ NCMR No _____ PROCESS REV _____ ← ② APPROVED BY ____ Date ____	

OPERATIONAL PROCESSES

Seq	W/C	Operation	Process Description	
10	Insp	Release	Inspected Yes☐ No ☐ Sign _____ Date _____	**Performance Status**
20	Shop	Turn	*(Indicate what to do here)* First pc. Insp. Yes☐ No☐ Sign_____ Date_____ Operator Acc. Sign _____ Date _____	*(Indicate measurement requirements here)*

③

QOP 012, Sec. One   **Audited by:**_____ **Date:**_____

MINFOR INCORPORATED

AMENDMENT TO PROCEDURES

Doc/Proc #	Name	P/N	Page /
Present Rev	Proposed Rev	Effective Date	
Process Eng	Date	NCMR No	

1. Describe reason for amendment: _____

_____

_____

_____

2. Describe what is present requirement: _____

_____

_____

_____

3. Describe what is new requirement: _____

_____

_____

_____

Reviewed By_____ Date_____ Accepted By _____ Date_____

Action By Quality:

4. This document has been changed on: Date _____

5. This document has been issued on: Date _____

6. List all affected documents within the Quality System _____

_____

_____

7. The affected documents have been updated on: Date:_____

8. Amendment is N/A to the quality documents ☐

Notes: _____

_____

_____

Signed By _____   Date _____ Audited by _____ Dt. _____

Audited by:_____ Date: _____

QOP 002, Section Two, par. 3.02   Form A-009 Rev 0 1998

ISSUANCE CONTROL OF QSM & QOP

QUALITY SYSTEM MANUAL				QUALITY OPERATING PROCEDURE			
#	Issued To	Issued By	Date	#	Issued to	Issued by	Date
1				1			
2				2			
3				3			
4				4			
5				5			
6				6			
7				7			
8				8			
9				9			
10				10			
#	Reissued To	Reissued By	Date	#	Reissued To	Reissued By	Date

QOP 002, Sec. One, par. 3.01 & 3.02   **Audited by:** _____   **Date:** _____   Form A-010 Rev 0 1998

MINFOR INCORPORATED

CONTROL OF CUSTOMER DOCUMENTS

CUSTOMER:_____ PURCHASE ORDER:_____DATE_____

Documents	Document No	Rev	Amendment	Rev	Effective Dt.	Authorized By	Date
Contract							
Attachment							
Blueprint 1	1of _____						
	2of _____						
	3of _____						
	4of _____						
Blueprint 2	1of _____						
	2of _____						
	3of _____						
	4of _____						
Blueprint 3	1of _____						
	2of _____						
	3of _____						
	4of _____						
Blueprint 4	1of _____						
	2of _____						
	3of _____						
	4of _____						
Blueprint 5	1of _____						
	2of _____						
	3of _____						
	4of _____						
Mat'l Spec 1							
Mat'l Spec 2							
Spl Proc 1							
Spl Proc 2							
Fixture 1							
Fixture 2							
Spl Tooling 1							
Spl Gaging 1							
Other							

Audited by _____ Date _____

REJECTED MATERIAL

REJECTED MATERIAL

PO # _____   **Job #** _____

Part No _____   **S/N** _____

○ **Quantity:** _____   **Date:** _____

                         NCMR # _____

Reason for Rejection: _____

_____

_____

☐

Insp. Stamp   **Signature:** _____

QOP 002, Sec. Three, par. 3.05h   Form A-012 Rev 0 1998

ACCEPTED MATERIAL

ACCEPTED MATERIAL

PO# ——————— JOB# ———————
Part No ——————— S/N ———————
Quantity ——————— Date ———————

NCMR # ———————————

Insp. Stamp ☐ **Signature** ———————

QOP 006, Sec. One through Seven   Form A-013 Rev 0 1998

HOLD MATERIAL

HOLD MATERIAL

PO #_____ JOB # _____

Part No _____ S/N _____

○ Quantity_____ Date _____

NCMR # _____

Insp. Stamp ☐ Signature _____

QOP 006, Sec. One through Seven   Form A-014 Rev 0 1998

MINFOR INCORPORATED

RECORD OF RECEIVED MATERIAL

PRODUCT RELATED MATERIALS

NOTE TO RECIEVER:

ATTACH THIS FORM TO ALL OTHER RECEIVED DOCUMENTS AND PASS IT ON TO THE QUALITY DEPARTMENT AFTER YOU HAVE RECORDED THE NEEDED INFORMATION. <u>EACH PART NUMBER MUST HAVE SEPARATE REPORT.</u>

CUSTOMER NAME _____

SHIPPER'S NAME _____

PACKING SLIP No _____

P/N OF MATERIAL _____

TOTAL QTY OF MATERIAL _____ COUNT EACH PIECE IN EACH HEAT LOT!

S/N'S OF MATERIAL	HEAT LOT	HEAT LOT	HEAT LOT	HEAT LOT

Do not rely on supplied Packing Slip for information. Confirm the whole shipment against the Packing Slip by counting and recording.

RECEIVER: _____ **Date:** _____ **Inspector:** _____ **Date:** _____

AUDITED: _____ **Date:** _____

QOP 006, Sec. Five, par. 3.02.1, step 3

Form A-015 Rev 0 1998

TRAINING LOG

EMPLOYEE NAME	DEPT	SUBJECT OF TRAINING	START DT	FINISH DT	BY

AUDITED BY: Date:

MINFOR INCORPORATED

PURCHASE ORDER

No _____ Rev _____

Page _____ /

FROM: _____

_____

_____

Order Date: _____

Due Date: _____

FOB _____

Phone: _____

Ship Via: _____

Fax: _____

Terms: _____

Tax Exempt No: _____

TO: _____

_____

_____

BUYER _____

Phone: _____

Fax: _____

Line Item	Qty	Part Description	Tax	Unit price	U/M	Extended
1						

Purchase Quality Requirements: _____

☐ **PRODUCT RELATED**

☐ **NOT PRODUCT RELATED**

CONFORMANCE CERTIFICATION REQUIRED

AUDITED BY: **Date:**

_____   _____

QUALITY MANAGER   **DATE**

QOP 003, Section One

Form A-017 Rev 0 1998

MINFOR INCORPORATED

INSPECTION REPORT	**Indicate Type of Inspection:** _____ **1 of**

P/N	REV.	JOB#	OP#	CUST.	HT	DT

Verify all items as required by the P O Review Sheet (A-001). PO Line Item: _____ **Qty:**

Job Traveler _____ Verified ☐   N/A ☐   Identification Marking _____ Verified ☐   N/A ☐
Material Certification _____ ☐   ☐   Overall Documentation _____ ☐   ☐
Special Process Certification _____ ☐   ☐   Shipping Documentation _____ ☐   ☐
First Piece Inspection _____ ☐   ☐   Special Handling & Packaging _____ ☐   ☐
First Article Inspection _____ ☐   ☐   Source Inspection _____ ☐   ☐
In-process Inspection _____ ☐   ☐   NCMR Activity _____ ☐   ☐
Receiving Inspection _____ ☐   ☐   Other (1) _____ ☐   ☐
Final Inspection _____ ☐   ☐   Other (2) _____ ☐   ☐
Amendments Incorporated _____ ☐   ☐   Other (3) _____ ☐   ☐

Comments: (S/N's) _____ **NCMR#** _____
AQL _____ **Sample Size** _____ **NCMR#** _____

DIMENSIONAL, WORKMANSHIP & VISUAL INSPECTION RESULTS: | TOTAL QTY:

SEQ	B/P dim & tol.	Method	S/N	S/N	S/N	S/N	S/N
1							
2							
3							
4							
5							
6							
7							
8							
9							
10							
11							
12							
13							
14							
15							
16							
17							

Qty. Accepted	Qty. Rejected	Waiver: Yes ☐ No ☐

Notes:

Inspector Sign _____ **Date** _____ **Audited By** _____ **Date** _____

Methods Legend

1. Micrometer	6. CMM	11. Layout Equip	16. Surf. Comparator	21. Ultrasonic	26. Plating Thickness Analyser
2. Vernier Caliper	7. PI Tape	12. Hardness Tester	17. CNC Mach. Ck.	22. FPI	27. Demagnetiser
3. Dial Caliper	8. Scale	13. Shadowgraph	18. Visual (Optical)	23. MPI	28. Hipot
4. Depth Gage	9. Boroscope	14. Profilometer	19. Swing gage	24. Chem. Analysis	29. Volt Meter
5. Gage Pin	10. Rad. Gage	15. Height Gage	20. X-Ray	25. Feeler Gage	30. Other

QOP 006 All Sections

Form A-018 Rev. 0 1998

| INSPECTION REPORT | Indicate Type of Inspection ——————————— 2 of |

| P/N | | Rev. | Job # | Op # | Cust. | | HT # | Date |

Seq	B/P dim & tol.	Method	S/N	S/N	S/N	S/N	S/N
18							
19							
20							
21							
22							
23							
24							
25							
26							
27							
28							
29							
30							
31							
32							
33							
34							
35							
36							
37							
38							
39							
40							
41							
42							
43							
44							
45							
46							
47							
48							
49							
50							
51	Visual Inspection		Acc☐ Rej ☐	Acc☐ Rej ☐	Acc☐ Rej ☐	Acc☐ Rej ☐	Acc☐Rej ☐
52	Workmanship		Acc☐ Rej ☐	Acc☐ Rej ☐	Acc☐ Rej ☐	Acc☐ Rej ☐	Acc☐Rej ☐
53	Verify all B/P notes		Acc☐ Rej ☐	Acc☐ Rej ☐	Acc☐ Rej ☐	Acc☐ Rej ☐	Acc☐Rej ☐

Inspector Sign _____ Date _____ Audited By _____ Date _____

NOTE: If needed, use continuation sheet Form A-019.

P/N		Rev.	Job #	OP #	Cust.		HT #	DT
Seq	B/P dim & tol.	Method	S/N	S/N	S/N	S/N	S/N	S/N

Notes: _____

_____

Inspector Sign _____ **Date** _____ **Audited By** _____ **Date**_____

QOP 006 All Sections

Form A-019 Rev 0 1998

476	A-020	MINFOR INCORPORATED				

GAGE CALIBRATION AND STATUS RECORD OWNER:

S/N	Description	Cycle	Tolerance	Use/Class	Mfr's S/N	Nom. Units of.

Date	Due Dt.	Status	Env Cor Fact'r	Funct. Status	Cal. By	I.D. of Std

Audited by _____ **Date**_____

QOP 007 par. 3. 01, 3.03, and 3.04

Form A-020 Rev 0 1998

MINFOR INCORPORATED

A-021													AUDIT CLOSEOUT	
PAR	AUDIT SCHEDULE												SIGNATURE/DATE	
ISO 9001	JAN	FEB	MAR	APR	MAY	JUN	JUL	AUG	SEP	OCT	NOV	DEC		
4.1	□												/	/
4.2		□											/	/
4.3			□										/	/
4.5				□									/	/
4.6					□								/	/
4.7					□								/	/
4.8									□				/	/
4.9						□						□	/	/
4.10	□			□			□						/	/
4.11										□			/	/
4.12								□					/	/
4.13			□						□				/	/
4.14					□						□		/	/
4.15		□										□	/	/
4.16							□						/	/
4.17								□		□			/	/
4.18													/	/
4.20									□				/	/

Re: QOP 010

Form A-021 Rev. 0 1999

478	A-022	MINFOR INCORPORATED

Checklist: ↓   <u>**AUDIT PLAN**</u>   **Reporting:** ↓

Paragraph Requirement to be Audited	Non-conformance Found
Quality Manual Section ———— **Quality O.P. Number** ———— **Section** ———— **Paragraph** ————	
Quality Manual Section ———— **Quality O.P Number** ———— **Section** ———— **Paragraph** ————	
Quality Manual Section ———— **Quality O.P. Number** ———— **Section** ———— **Paragraph** ————	
Quality Manual Section ———— **Quality O.P. Number** ———— **Section** ———— **Paragraph** ————	
Quality Manual Section ———— **Quality O.P. Number** ———— **Section** ———— **Paragraph** ————	

Comments:

NCMR ————————

Auditor ——————— **Date** ——— **Mgmt Rep** ——————— **Date** ———

QOP 010   Form A-022 Rev 0 1998

A-023	MINFOR INCRPORATED			

REWORK/REPAIR PROCEDURE

Job No	Part No	Customer	NCMR No	Date

REWORK INSTRUCTION Engineer: Date:

Description	Requirement	Result

Operator Acc: Date: | Insp Acc: Date:

REPAIR INSTRUCTION Engineer: Date:

Description	Requirement	Result

Operator Acc: Date: | Insp Acc: Date:

Attach all required drawings:

Audited by _____ Date _____

QOP 009, Sec. Two, par. 3.01.4.1 & 3.01.4.2   Form A-023 Rev 0 1998

WAIVER REQUEST

Date _____ Customer _____ NCMR No _____

Job Number: _____   Purchase Order No _____
Raw Material: _____   Shipping Memo: _____
Part Number: _____ Rev. _____ Part Name: _____ S/N _____
Operation: _____ QTY: _____ Dept. Resp. _____

NONCONFORMANCE

Description: _____
_____
_____
_____

Cause: _____
_____
_____

Corrective/preventive action: _____
_____
_____
_____
_____

Quality Manager _____

CUSTOMER DISPOSITION   Date: _____

		Special Instruction
Accept	☐	
Rework	☐	_____
Repair	☐	_____
Scrap	☐	_____
Regrade	☐	_____
Other	☐	_____

Authorized Authority _____

Audited by _____ Date _____

A-025			MINFOR INCORPORATED			481

SHIPPING LOG

Part No	Qty	Date	Customer	Memo No	PO #	Comments By

Audited by_____ **Date** _____

SUPPLIER SURVEY QUESTIONNAIRE

Company Name: _____

Address (street): _____ City: _____

State: _____ Country: _____ Zip Code: _____

Phone: _____ Fax: _____

List Your Product and/or Service: _____

Number of employees: Total:_____ Production:_____ Inspection: _____

Your Quality System conforms to the following standard(s):

ISO 9001 ☐   ISO 9002 ☐   ISO 9003 ☐   QS 9000 ☐

AS 9000 ☐   MIL-Q-9858 ☐   MIL-I-45208 ☐   FAR ☐

Other (Specify): _____

List third party which has registered your Quality System: _____

ORGANIZATION:

Title	Name
President/Gen. Mgr.: _____	
Senior Quality Position: _____	
Senior Eng'g Position: _____	
Senior Mkt'g Position: _____	
Mfrg. Mgr.: _____	
Comments: _____	

MANAGEMENT RESPONSIBILITY

		YES	NO
1	Do you have a company Quality Policy?	☐	☐
2	Do procedures describe the authority of those responsible for managing, performing, and verifying work affecting quality?	☐	☐
3	Are resources for inspection, test, and audits of the quality system provided?	☐	☐
4	Is the Quality System reviewed at least annually by executive management?	☐	☐

QUALITY SYSTEM

5	Do you have a company approved Quality Manual?	☐	☐
	If so, specify current revision level and date: Rev: _____ Date: _____		
6	Are quality plans written, approved and used as required?	☐	☐

CONTRACT REVIEW

7	Do procedures include contract review activities and are records maintained?	☐	☐

DESIGN CONTROL

8	Do procedures include the control and verification of product design?	☐	☐
9	Are plans written and approved for each design and development activity?	☐	☐
10	Do the plans include verification that design output meets input requirements?	☐	☐
11	Are all design changes and modifications documented, reviewed, approved, and maintained?	☐	☐

DOCUMENTATION CONTROL

12	Do procedures include the control of external and internal documents identified within the Quality System?	☐	☐
13	Are the documents available and controlled where they are needed?	☐	☐
14	Are obsolete documents promptly removed from all points of issue or use?	☐	☐
15	Are document changes reviewed and approved by those who issued them?	☐	☐
16	Where appropriate, are document changes identified or referenced?	☐	☐

PURCHASING

17	Are your suppliers controlled to meet your purchase quality requirements and are your records maintained to prove this?	☐	☐
18	Do purchase orders correctly define the product ordered?	☐	☐
19	Does Quality review and approve product related purchase orders?	☐	☐

CUSTOMER SUPPLIED PRODUCTS

20	Are procedures written and maintained for the control of customer supplied products with adequate feedback on lost, damaged or unacceptable conditions?	☐	☐

PRODUCT IDENTIFICATION AND TRACEABILITY

21	Where applicable, does documentation cover product identification during all stages of production, installation, and delivery?	☐	☐
22	Where traceability is a requirement, do you have documented control for it?	☐	☐

PROCESS CONTROL

23	Do you have written instructions for planned product quality verification during the various processing stages before product release?	☐	☐

	YES	NO

PROCESS CONTROL continued

24 Do you perform special processes? If so, please list them: ☐ ☐

_____

25 Are records maintained for qualified processes, equipment, and personnel? ☐ ☐

INSPECTION AND TESTING

26 Do procedures ensure that incoming material is identified, documented, verified and controlled to ensure product recall in case of in-process rejection. ☐ ☐

27 Are in-process inspections and tests performed and documented for first piece acceptance according to quality plans? ☐ ☐

28 Do final inspection procedures ensure that all previous inspections have been reviewed and correctly documented including MRB activities? ☐ ☐

29 Do final inspection procedures ensure that all the customer's purchase order requirements have been verified and documented prior to product release? ☐ ☐

30 Are records maintained which provide evidence that the final product has passed all the required inspections and tests prior to product release? ☐ ☐

31 Are procedures clearly defined document control for product released for shipment? ☐ ☐

INSPECTION, MEASURING AND TEST EQUIPMENT

32 Is measuring and test equipment which is used for product conformance verification identified, calibrated and procedurally controlled? ☐ ☐

33 Is all measuring and test equipment calibrated at prescribed intervals or prior to use? ☐ ☐

34 Are calibration masters traceable to national or international Standards? ☐ ☐

35 Are calibration technique sheets used and are they present and up to date? ☐ ☐

36 Is calibration, measuring and test equipment identified with suitable indicators to show calibration status? Is it traceable to individual equipment? ☐ ☐

37 Are environmental conditions suitable for the calibration of equipment? ☐ ☐

38 Is test hardware (such as jigs and fixtures) including software checked to ensure it is capable of verifying the product? ☐ ☐

INSPECTION AND TEST STATUS

39 Is inspection and test status of products identified by using markings (such as controlled stamps, tags and/or labels) throughout production and installation? ☐ ☐

40 Do inspection and test records identify the inspection authority responsible for the release of conforming product? ☐ ☐

CONTROL OF NON-CONFORMING PRODUCT

41 Do documented procedures prevent the inadvertent use or installation of non-conforming product? ☐ ☐

42 Is non-conforming product identified, documented, segregated (when practical). Is notification given to concerned parties? ☐ ☐

43 Do documented procedures describe the MRB activities? ☐ ☐

	YES	NO

CONTROL OF NON-CONFORMING PRODUCT continued

44 Do you have a waiver procedure in place to request customer disposition on non-conforming product? Is reworked and repaired product re-inspected? ☐ ☐

CORRECTIVE/ PREVENTIVE ACTION

45 Do procedures describe how corrective/preventive action is implemented? ☐ ☐
46 Do procedures describe how follow-up of corrective/preventive action implementation is carried out and how documentation is closed out? ☐ ☐

HANDLING, STORAGE, PACKAGING AND DELIVERY

47 Do you have procedures written for handling, storage, packaging and delivery of the conforming product? Are they adequate to protect product quality? ☐ ☐

CONTROL OF QUALITY RECORDS

48 Do procedures describe the retention and control of quality records, the process of retention, period of retention, retrieval and accessibility? ☐ ☐
49 Do procedures describe the safe-keeping of records and the accessibility to them by customers? ☐ ☐

INTERNAL QUALITY AUDITS

50 Are internal quality audits at least annually scheduled, performed and documented in accordance with written procedures? Are follow-up actions scheduled and documented to ensure any corrective action implementation? ☐ ☐

TRAINING

51 Do you maintain records of training activities for all personnel? ☐ ☐
52 Are personnel who perform special processes certified to do so? ☐ ☐

SERVICING

53 When servicing is specified by contract, do procedures include that verification servicing meets the specified requirements? ☐ ☐

STATISTICAL TECHNIQUES

54 Where appropriate, do procedures adequately define the application of statistical techniques for verifying and controlling process capability and product characteristics? ☐ ☐

COMPLETED BY: _____ SIGNATURE: _____

TITLE: _____ DATE: _____

Please mail or fax to:

Audited by _____ Date _____ Approved by _____ Date _____

QOP 003, Section Two

Form A-026 Rev 0 1998

MATERIAL IDENTIFICATION TAG

<u>**MATERIAL IDENTIFICATION TAG**</u>

DATE RECEIVED: ——————————————

CUSTOMER: ——————————————

PART NUMBER: ——————————————

PO NUMBER: ——————————————

○ LOCATION: ——————————————

COMMENTS: ——————————————

——————————————

SIGNATURE OF RECEIVER: ——————————

QOP 006, Sec. Five, par. 3.02.1, step 4   Form A-027 Rev 0 1998

Pg 1

Packing Slip #

From: Minfor Incorporated

_____

_____

Phone:_____

Fax:_____

Bill To: _____

_____

_____

Ship To:_____

_____

_____

Ship Date:_____
Your Order: _____
FOB   _____
Ship Via:_____
Terms: _____
Our Order: _____

Sales Person: _____

Order Qty	Ship Qty	Part Name/description	U/M Each

Released By: _____ Date: _____

Received By: _____ Date: _____

QOP 006, Sec. Four, par. 3.02, step 9   Form A-028 Rev 0 1998

No _____

CERTFICATE OF COMFORMANCE

Purchase Order _____ Packing Slip _____ Shipping dt. _____

Part No: _____ Rev: _____ Part Name: _____ Qty: _____

This is to certify that the below listed item(s) correspond to the requirements of blueprint, standard, specification or clauses stipulated in the Purchase Order and which are supported by the appropriate records on file available on request.

Document No	Rev.	Process Description	Manufacturer	Quality Standard

Supplier Name and Address

_____

_____

_____

Signature of Quality Manager   Date: _____

RECEIVING LOG

Quantity Received	Date Received	Received From	Part No or Description	Material Belongs to	Heat Code	Condition Received	Rec'd By

Optional usage as required. Not a procedural requirement of this QMS   Form A-030 Rev 0 1998

ISSUE AND TRACEABILITY OF DRAWINGS

Customer: Part No: Job No:

Document Number ↓	Rev.	No of Copies Issued Qty / Area	Date Issued	Issued By	Obsolete Date	Destroyed Date	Returned Date	Returned By
		Engineering						
		Production						
		Inspection						
		Tool Room						

FILE NOTES: _____

_____

QOP 001, Sec. Two, par. 3.01, step 6   **Audited by** _____ **Date** _____   Form A-031 Rev 0 1998

PASSWORD, INSPECTION STAMP, AND SIGNATURE CONTROL

The following passwords and/or inspection stamps have been assigned or inactivated to the indicated personnel and are to be controlled as defined in QOP 008, paragraph 3.01 and 3.02.

Stamp Type and Stamp #	Password (encoded)	Issued to	Signature (of owner)	Issued By	Date Issued	Inactivation Date and/or Limitation Notes

Audited by _____   **Date** _____

QOP 008, par. 3.01 and 3.02   Form A-032 Rev 0 1998

492	A-O33	MINFOR INCORPORATED			

OPERATOR PRODUCT VERIFICATION RECORD			Sht 1 of		

Part Number: _____ OP # _____ Part Name: _____

Name of Empl					
Date: ⟶					
S/N ⟶					
B/P Dimension						
Insp Acc. ⟶					
NCMR # ⟶					

| A-034 | MINFOR INCORPORATED | 493 |

MANAGEMENT REVIEW
QUALITY SYSTEM REVIEW RECORD

| SYSTEM STANDARD ISO 9001 | MANAGEMENT REPRESENTATIVE _____ |

ATTENDEES: _____

PLACE OF MEETING:

DATE OF MEETING:

1. Agenda to review present state of the Quality System:

2. Agenda for proposals of new goals and objectives:

3. Summary of Management Review Decisions:

4. Corrective/Preventive Action Summary: NCMR No _____

Management Representative _____ Date _____

SUPPLIER SURVEY FORM (Short Form)

Supplier Name:_____ **Date:** _____

Address: (street) _____ **City:** _____ **Zip Code** _____

Phone: _____ **Fax:** _____

Contact Person: _____ **Title:** _____

Survey Performed By: _____ **Title:** _____

Reason for Survey: _____

No. of Employees: Production: _____ **Quality:** _____ **Other:** _____ **Total:** _____

Quality System Used: ISO 9001 ☐   **AS9000** ☐   **QS9000** ☐   **Other:** _____

Survey results reported for the following requirements:

	YES	NO
INSPECTION AND TESTING		
Do procedures ensure that incoming product is inspected and documented?	☐	☐
Are in-process inspections and tests performed and documented to quality plans?	☐	☐
Do final inspection records indicate a complete review and documentation of Identify Purchase Order imposed requirements, including workmanship?	☐	☐
CONTROL OF NON-CONFORMING PRODUCTS		
Is non-conforming product identified, documented, and segregated (when practical)?	☐	☐
Is there a corrective action system?	☐	☐
Is reworked and repaired product re-inspected and controlled to written procedures?	☐	☐
CONTROL OF INSPECTION, MEASURINGS AND TEST EQUIPMENT		
Is measuring and test equipment controlled and calibrated?	☐	☐
Are all measuring and test equipment identified and calibrated at determined intervals or prior to use and are the records traceable to individual equipment?	☐	☐
Are environmental conditions suitable? Are the test masters traceable to standards?	☐	☐
Do records indicate a recall system is in place? Can the stickers prove this out?	☐	☐
DOCUMENTATION CONTROL		
Are documents traceable and controlled according to written procedures in purchasing, engineering, production, and inspection for revisions, retrieval, and retention?	☐	☐

Summarize deficiencies found: _____

_____

_____

Signature: _____   **Approved By** _____ **Date:** _____
Audited by _____ **Date** _____
QOP 003, Sec Two, par. 3.02.1 & 3.02.2

SOURCE INSPECTED PRODUCT APPROVAL

Purchase Order No: _____ **Part No:** _____ **Rev:** _____

Part Name: _____ **Job Number:** _____

Supplier Name: _____ **Location:** _____

Total Qty: _____ **Total Submitted:** _____ **Total Accepted:** _____

S/N of accepted product: _____

_____

S/N of rejected product: _____

NCMR No: _____

Comments: _____

_____

_____

_____

Source Authority: _____ **Date:** _____ **Stamp:** [   ]

NOTE: Must be attached to shipment   **Audited** _____ **Date** _____

QOP 006, Sec. Seven, par.3.02, step 6   Form A-036 Rev 0 1998

| 496 | A-037 | **MINFOR INCORPORATED** |

MINFOR INCORPORATED
SUPPLIER CORRECTIVE ACTION REQUEST

NCMR No: _____ Date _____

Issued to: _____ Part No: _____
Name: _____ Part Name: _____
Address: _____ PO No: _____
Location: _____ Packing Slip No: _____
Tel: _____ Qty. Rec'd: _____ Qty. Rej: _____
Fax: _____ S/N: _____

DESCRIPTION OF DISCREPANCY: _____
_____
_____
_____

DISPOSITION:
☐ Returned for Evaluation   ☐ Returned for Rework

☐ Use As Is   ☐ Returned At Your Expense

☐ Other (Specify)

Authorized Authority: _____ Purchasing: _____

SUPPLIER TO COMPLETE THE FOLLOWING AND RETURN IT BY: _____

CAUSE OF DISCREPANCY: _____
_____
_____

CORRECTIVE ACTION TAKEN TO ELIMINATE CAUSE: _____
_____
_____
_____

SUPPLIER SIGNATURE: _____ TITLE: _____ DATE: _____

☐ CORRECTIVE ACTION ACCEPTED   ☐ CORRECTIVE ACTION REJECTED

COMMENTS: _____
_____

SIGNATURE: _____ Date: _____ Audited _____ Date: _____

QOP 009, Sec Two, par. 3.01.4.4, step 3 & 6   Form A-037 Rev 0 1999

CUSTOMER COMPLAINT AND EVALUATION REPORT

NOTE: Return the completed form to the Quality Manager without delay.   *Put return address in this block or put N/A*

CUSTOMER RELATED INFORMATION:

Company: _____ Location: _____

Name of customer representative who is reporting the complaint or evaluation:

Name: _____ Title: _____ Tel: _____ Fax: _____

Product Information:

Purchase Order No: _____ Packing Slip No: _____
Part No: _____ Part Name: _____
Qty: _____ Serial Number: _____

What is the complaint or satisfaction evaluation: _____
_____
_____
_____

Taken by: _____ Date: _____

INTERNAL INVESTIGATION AND ACTION TAKEN: NCMR: _____

Cause of the problem: _____
_____
_____
_____
_____

Corrective action taken: _____
_____
_____
_____

Followed up by: _____ Title: _____ Date: _____

Instruction to the Quality Department:
After the cause has been investigated and the corrective action implemented, this report must be sent back to the customer as official reply and a copy of it retained in the Customer Complaint File. Customer satisfaction evaluations must be tabulated and reported to upper management.

QOP 009, Section Three, par. 3.04   **Audited by:** _____ **Date:** _____   Form A-038   Rev. 0   1999

A-039 **MINFOR INCORPORATED**

TRAINING EVALUATION SHEET
ORAL EXAMINATION

DATE OF TEST: _____ NCMR # _____

NAME OF STUDENT: _____ DEPT: _____

TRAINING METRICS QOP # _____

LIST PARAGRAPH NUMBERS AND TOTAL STEPS IN EACH PARAGRAPH ON WHICH TRAINING WAS GIVEN:

Par #					
Steps					

TOTAL QUESTIONS ASKED FROM STUDENT: _____(10)_____ (b)

TOATAL QUESTIONS ANSWERED CORRECTLY: ____(7)_____ (a)

DIVIDE THE NUMBER OF TOTAL QUESTIONS ASKED INTO THE TOTAL NUMBER OF CORRECT ANSWERS GIVEN BY THE STUDENT:

FORMULA: 7 (a) $\div$ 10 (b) = .7 (c)   (a/b = c)
(Take your calculator, press 7, press division sign, press 10, press % button and your answer is given in percentage. That's your scoring)

INDICATE SCORE: _____

Note: The figures given in the parentheses are only examples.
       Base your questions on the subject you have taught.

INSTRUCTION:
1. Less than 70% scoring requires retraining the student in the wrong answers.
2. Maintain this test record in the Training Log (Binder).

 Mgmt Rep Signature: _____

NOTES: _____

_____

_____

_____

QOP 009, Section Two, par. 3.01.5, step 5   Form A-039 Rev. 0 1999

APPENDIX ONE

This matrix represents the alignment compatibility of the QSM to ISO 9001/2000

QUALITY SYSTEM MANUAL	ISO 9001/2000
1.0 Scope	5.1
1.1 Mission	5.2
1.2 Objectives	5.4.1
2.0 Exclusion	1.2
3.0 Definition	
3.1 References	
3.2 Revision History	5.5.6
3.3 Annual Review	5.3
4.0 Quality System Requirements	4.1
4.1 Management Responsibility	5, 5.1
4.1.1 Quality Policy	5.3
4.1.2 Organization Chart	5.1, 5.5, 5.5.1
4.1.2.1 Responsibility and Authority	5.5.2
4.1.2.2 Resources	6, 6.1, 6.2.1, 6.3,6.4
4.1.2.3 Management Representative	5.5.3
4.1.3 Management Review	5.6, 5.5.7, 5.6.1, 5.6.2, 5.6.3
4.2 Quality System	4.1, 5.5.4, 5.5.5
4.2.2 Quality System Procedures	4.2
4.2.3 Quality Planning	5.4, 5.4, 5.4.2
4.3 Contract Review	5.2, 5.5.7, 7, 7.1, 7.2, 7.2.1, 7.2.2, 7.2.3
4.4 Design Control	N/A
4.5 Document and Data Control	5.5.6, 5.5.7
4.6 Purchasing	7.4, 7.4.1, 7.4.2, 7.4.3, 5.5.7, 8.4d
4.7 Control of Customer Supplied Product	7.5.3
4.8 Product Identification and Traceability	5.5.7, 7.5.2
4.9 Process Control	5.5.7, 7.5.1, 7.5.5, 8, 8.2.3
4.10 Inspection and Testing	5.5.7, 7.5.1, 8, 8.1, 8.4, 8.2.3, 8.2.4
4.11 Control of Inspection, Measuring and Test Equipment	5.5.7, 7.6
4.12 Inspection and Test Status	7.5.1
4.13 Control of Non-conforming Product	5.5.7, 7.2.3, 8, 8.1, 8.2.4, 8.3, 8.4
4.14 Corrective and Preventive Action	5.5.7, 8.3, 8.5, 8.5.1, 8.5.2, 8.5.3
4.15 Handling, Storage, Packaging, Preservation, and Delivery	7.5.4
4.16 Control of Quality Records	5.5.7
4.17 Internal Quality Audits	5.5.7, 8, 8.1, 8.2.2, 8.2.3, 8.4
4.18 · Training	5.5.7, 6.2.2, 8.1
4.19 Servicing	N/A
4.20 Statistical Techniques	7.5.1, 8.1, 8.4

APPENDIX TWO

BOOK RECOMMENDATIONS
(to advance the process-approach quality management to higher levels within your organization)

1. Critical Shift: The future of Quality in Organizational Performance
   By Lori L. Silverman with Annabeth L. Propst
2. Total Quality Management Handbook
   By John L. Hradesky
3. Juran On Quality By Design: The New Steps for Planning Quality Into Goods and Services
   By J.M. Juran
4. Implementing Six Sigma
   By Forrest W. Breyfogle III (John Wiley & Sons, Inc.)
5. Mapping Work Processes
   By Dianne Galloway
6. Gemba Kaizen
   By Masaaki Imai
7. Statistical Thinking
   By Galen C. Britz, Donald W. Emerling, Lynne B. Hare, Roger W. Hoerl, Stuart J. Janis, Janice E. Shade
8. The Northbound Train: Finding the Purpose, Setting the Direction, Shaping the Destiny of Your Organization (New York: Amacom,1994)
   By K. Albrecht

Note: All the above books, except items 4 and 8, are available from
      ASQ Quality Press

APPENDIX THREE

The Suggestion Path

There are many different ways improvements to an operating quality management system can be made. We have provided in this book, basically, the documented ways to achieve them.

In every shop where teamwork and open-mindedness are nurtured and sincerely recognized by management, improvements in just about anything process owners do can be realized in conjunction with the documented system. In fact, a documented system can be the biggest prompter for improvements, because of its one-sidedness – the routine structure. Given freedom to do so, employees, including new employees bringing with them the innovations from resourceful companies, can turn a "routine structure" into marvelous simplicity for the better, just for the asking.

Behind this page, I have provided a simple form that can be added to the Production Folder for the employees to use in challenging the "routine structure" in their work methods, or in anything else where they see that things can be improved for the better. Believe me, this would be the best performing "bank account" you have ever opened. Please read up on this great idea from Gemba Kaizen by Masaaki Imai, available from Quality Press.

HISTORY OF PROBLEMS – *LESSONS LEARNED*	
(Your idea can make the difference for the better. Share it with others)	
Part Number: **Operation No:** **Date:**	
Describe the Problem	**Your Solution**
1	1
2	2
3	3
Uncontrolled Form	**Thank you**

APPENDIX FOUR

Book Summary

In this process-approach quality management system, I have designed the structural framework to support all the required processes and procedures that embrace the documented quality requirements serving contract manufacturing operations. The quality management system is a <u>throughput system</u> functioning in line with the manufacturing flow cycle of a contracted product from the Request for Quote phase (RFQ) to the delivery of a product. This is the shortest distance between input from the customer to the output from the supplier (organization) and back to the customer in terms of product quality management. I am managing only the involved processes to ensure product quality, customer satisfaction and continuous improvement. It is a Total Quality Management (TQM) system for small business operations in contract manufacturing. Several quality management techniques have been integrated into the documented system to achieve operational cohesion and consistent product quality maintenance. These are listed below:

1. Individual responsibility in respect to work assignments has been identified for all the processes in the product's life-cycle in order to recognize accountability and process ownership to enable the organization to <u>measure performance and output evaluation.</u>
2. The product related quality objectives have been identified and defined in documented procedures to ensure the effective control of <u>conforming as well as non-conforming products.</u>
3. The throughput quality system is fully documented, integrated, and implemented, <u>targeting product quality objectives</u> to result in demonstrable and verifiable output from process owners.
4. The documented procedures include process instructions to enable the <u>monitoring of performance</u> in order to prevent the migration of discrepancies from process to process, impacting the end item.
5. For ensuring the achievement of quality objectives, all the documented procedures have been <u>interactively structured to support "global" quality administration.</u>
6. The documented quality management system has <u>locked in the controlling features to ensure corrective/preventive action</u> when and where problems are identified in order to maintain effective continuous improvement.
7. The periodic Management Review has been <u>restructured to facilitate also on call participation of management</u> in the decision making processes to enable the organization timely implementation of corrective action in maintaining continuous improvement for customer satisfaction.

The process-approach quality management system in this book is based on the continuous application of the plan-do-check-act cycle in two major areas. First, it is applied to the documented objectives as a preventive quality tool to weed out variables impacting both the process and the product. Second, it is applied when non-conformities have been identified anywhere in the operating system. In the first cycle, the PDCA provisions are applied to prevent the flow down of errors and mistakes, migrating from process to process and impacting throughput and finally the end item. In the second cycle, the PDCA provisions are applied to correct the deficiencies identified anywhere in the throughput cycle. This explains why the plan-do-check-act provisions in a process-approach system are never really ending, for as one product has been delivered, other products are coming through the production cycle, each with its own Purchase Order quality requirements within the documented and controlled system. Each requires its own quality planning, work execution, conformity verification, and the implementation of task-related follow-ups in line with the specific customer's quality requirements.

Nowhere else in this book has the application of the above management techniques been explained. The reason for this should be easily understood from the evolving processes enfolding these management techniques through which this process-approach system was implemented. All seven management tools listed above were created to build into the quality management system forces that would impel operational effectiveness in maintaining product quality, cost control, continuous improvement and customer satisfaction. No quality system will survive long unless it has these objectives interwoven in the organization's operational fibers. Without paying careful attention to these objectives in the creation of a quality management system, the uncontrolled variables would erode the financial resources of the organization resulting in red ink. To prevent this from happening, the seven management techniques have been designed into the process-approach network of this quality management system by the application of the most commonly used statistical tools. First, the entire process-approach structure was laid out in line with *cause-and-effect* diagramming techniques. The *fishhead* was titled: *Quality System Variables*. The *ribs* attached to the *backbone* were labeled as the contributing core departments. The objective was to show interrelationships to cross-functional work cells charged with aligning customer requirements in work processes upstream. Each work cell in each core department was identified as a probable contributor to process variation in the chain of events. Once this was done, *tree diagrams* were created separately for each core department showing each work cell within the department and its workflow relationship within each other as well as to each cross-functional work cell. The next move was the application of *failure mode effect analysis* (FMEA) identifying in each work cell the *critical to customer (CTC), critical to quality (CTQ), and critical to process* (CTP) requirements. When this was done, *quality function deployment* (QFD) procedures were written for each work cell incorporating process ownership, responsibility, interaction, interrelationship, integration, process steps, performance measurement, corrective action alert as was required by each work cell. This ensured that *critical to customer, critical to quality, and critical to process* requirements were orderly flown down from each work cell until each process cycle was completed in the closed-loop system arrangement. Then, the various forms were designed and integrated with each work cell's requirement in order to record performance data as proof of work results measurement and demonstration of compliance. Finally, the quality manual was written to bring cohesion, linkage and system integration into focus in line with ISO 9001/2000 requirements. The *quality function deployment* (QFD) procedures have been renamed as *Quality Objective Procedures* (QOP) to bring about common usage of "quality objectives" applied throughout in the ISO Standard. For the same reason "documented procedures" have been renamed as *Quality Operating Procedures* (QOP). This process-approach quality management system is comprised of 12 Quality Operating Procedures and 209 Quality Objective Procedures. These procedures are shown in two separately but identically laid out matrices: Flowcharts in Book Section One and Text in Book Section Three.

It's no accident that this book is titled "Process-Approach Quality Management System." Let's break the title down in order to understand what's involved here.

System

"System" means that I have built into it the management commitments, responsibilities, and the quality objectives to direct the effective operation and maintenance of the quality system. To make the management system an interactive organizational network, I have created the quality objective procedures in order to communicate and implement the commitments and responsibilities to support the achievement of the network's quality objectives. Then, I interlinked them with the quality system's processes. I did this in order to flow down customer requirements to the relevant work assignments, focusing the activities within each work cell in the overall network in one direction, the simultaneous achievement of product quality, cost control, continuous improvement and customer satisfaction. First, I have expressed the quality objective requirements for the quality system in policy statements under the Quality Manual's processes. Second, I have created the quality objective procedures to allow the implementation, control, and maintenance of these objectives. Third, I have designed the forms to enable the employees to record work results to prove the achievement of these quality objectives. This is the triad within which all the quality system's objectives have been planned and documented to ensure customer satisfaction, internally and externally.

I have done something else that the reader might pass over without recognizing it. Every requirement identified in the Quality Manual and implemented through the Quality Objective Procedures has been painstakingly subjected to the plan-do-check-act provisions of TQM to prevent the migration of hidden variables from work cell to work cell.

Every business owner's goal should be that the objectives of his/her quality management system be so implemented that a contracted product could be put through the manufacturing cycle the first time without problems. That is also the primary goal of this process-approach system. Since every source of error can be identified as a variance that could, to a degree, negatively impact every activity, it makes a lot of sense to remove it at the initial stages of every planning process. Every quality objective procedure in this book contains provisions to observe and remove variables through monitoring and corrective action enforcement, hence ensuring continuous improvement.

We should realize as we read the quality objective procedures in this book that I am controlling the processes of the integrated system not in a here-and-there fashion, but constructively right from the beginning – the contract phase of a manufacturing business. This is the area where I have done some very serious initial planning to ensure that customer requirements are flown down to the responsible work cells. Had I not done it here, I would have been doomed later, for I would have lost cost control, the discipline to stay within the quoted figures of the job. It would not have been enough that I put in all kinds of procedures to maintain control over the defined processes and left out the requirement to tie them to the allocated cost figures of a project. It was in this area where the Plan-Do-Check-Act theory played a major role for me. Applying the PDCA theory ensured the removal of the deficiency contributors from the whole processing cycle that would otherwise have robbed me blind from being able to realize any profit that I had calculated making when I quoted the job. All of these requirements that have been carefully woven into the procedures of this process-approach system may be taken for granted by the reader at implementation time. Don't!

Procedures written to enable us to control the quality management system in this book are not the run-of-the-mill SOPs. The way traditional SOPs are written, they don't allow us to enforce the dynamic nature of an integrated process control system, for they lack the definition of continuity, process owners' accountability, system integration, linkage and flow-down description of requirements. Therefore, I have planned all these into this process-approach quality management system. This is how the dynamics of the presented system removes the common and hidden mistakes that plague conventional quality systems. What we have here are value added, targeted objectives to accomplish a schedule-based quality delivery system. This is a monitored system that sustains itself through the continuous and effective work discipline of the organization. How? By the clearly defined work assignments, process instructions, identified responsibilities, process ownership, and integrated flow processes. But, the system is also designed so that modification to it can be rather easily accomplished to suit one's circumstances without upsetting the basic document and control structure.

Management

Management in this process-approach quality system, regardless of its executive position, is also a process owner of all the business and quality objectives that it has established and documented for its employees to observe in the product's realization processes. Its attention, foremost, is focused to ensure that all the stated objectives for the business and quality are consistently maintained, as documented. Ownership in my case means having certain responsibilities assigned individually, or shared, or delegated. Whatever the case, I have identified and stated them in writing. Accountability, according to responsibility, I'm enforcing through process instructions everywhere. System cohesion plays a major roll in cementing the continuing processes together and focusing process owner's attention on product quality, cost control, and continuous improvement. This is how internal and external customer requirements have been imbedded in the continuing processes toward product realization within the documented system. This process-approach system is communicated within the organization by the observance of the documented quality objective procedures, tailored to manage core departmental operations in contract manufacturing environments. The completely integrated system requires management's full participation in order to realize its business goals.

Quality

The word 'Quality' is not used in this process approach-system in a limited sense – checking parts. Furthermore, it is not applied as an 'insert' to plug up gaps, here-and-there, created by system deficiencies. 'Quality' here is the controlling mechanism of management policies in a manufacturing environment, encompassing Total Quality Management (TQM) for directed and targeted objectives for the maintenance of product quality, cost control, continuous improvement and customer satisfaction. These system drivers are woven into the fibers of the process approach-system presented in this book. 'Quality' here is a definition of clearly defined policy objectives for all the identified process owners with identified responsibilities and assignments.

In this process-approach quality management system, I have documented the process control procedures so that the instructions in them unearth the hidden mistakes in a continuous fashion as the assignments flow from the beginning to the end of a process within the integrated system. This is where this process-approach system rips apart conventional systems relying on major system repair procedures to correct here-and-there the system deficiencies that I am correcting in the whole system through the application of process controls.

It's worth observing the totality of the presented process approach mechanism as to how, as a whole, I have interlinked the targeted core departmental objectives of a contract manufacturing operation to achieve product quality and continuous improvement through interactive processes within the framework of the quality management system. The key to how I have been able to accomplish this was predicated on interactive communication. Broadening the system to include additional business objectives of the organization would require similar interaction within the system. This process-approach system presents an accurate framework to do that.

Today, the application of the word <u>quality</u> is more advertised than it ever has been. It seems that there is a <u>quality face</u> to everything we do. And that's perfectly okay. But, there is a problem. Unless the acceptance boundaries for quality requirements are defined, we will always be pursuing subjective quality – everybody interprets it his/her own way. And this is essentially the major problem with conventional quality systems, where responsibilities to assignments are not clearly identified and defined to indicate acceptance boundaries of those assignments. It poses a critical problem of how to measure work performance. It was for this reason that I broke with the past and designed my Quality Objective Procedures.

So, 'Quality' in this process approach system also defines the acceptance boundaries of all the activities, expressed in the documented management system, whether it is to control the overall quality system performance or the increments thereof. What I was very careful of, however, was that I didn't impose tolerance boundaries without considering value to the processes I was controlling, for cost overrun would have resulted.

Process-Approach
Don't let reality escape your attention regarding the meaning and application of process-approach. Although the compound word 'process-approach' is totally new in quality systems, it has been practiced by some industries for a long time in a limited way without defining its overall implication. ISO 9001/2000 clearly defines its meaning without shedding any light on how to interface it within a controlled system. Without calling it 'process-approach,' civilizations practiced it ever since man had intelligence. Doing things correctly, systematically, consistently and in an integrated fashion, requires a lesson-learned process from each of us. This is the developmental process. After that, process-approach will become second nature to us as we repeat our actions. We will recognize from the practice that everything we do has a *natural* flow to it in the established order. It does not have to be an organization to prove this out, for every one of us applies process-approach to accomplish daily routines. But, because we got used to doing it every day, we forget about the process of doing it. On the other hand, try to teach your own kid correctly about something you are doing. You would most certainly think out the step by step plans (the process-approach) as to how to implement the process of teaching. Wouldn't this be an easier task for you to achieve, if you already had it on

paper? Organizations are the same way. The difference is that instead of one individual, they have many. Therefore, they need to <u>integrate</u> their process-approach to achieve acceptable results. This is where this book becomes essential.

Process-approach methods can be systematically identified and implemented, yet dismally ineffective from the lack of system integration and cohesion. I can tell you this from my own experience. The lack of a fully documented and integrated procedural quality system is the surest way to introduce hidden mistakes by anybody and at any point into a product's flow cycle. I used to detect these hidden mistakes at final inspection after the product was already completed. Many times, I traced these hidden defects to either engineering or to contracts, and even to the customers and subcontractors. Without a documented system to put corrective action into effect, I ended up sorting the product every time the order was repeated. Inadequate planning, human errors, negligence and disregard for process controls all played into the hand of creating roadblocks. To overcome these problems, I have integrated an interactive documented system to completely implement corrective action upstream/downstream in this quality management system. I have been able to accomplish this through the implementation of a locked-in corrective/preventive action system.

Many companies to some extent have process-approach resemblance in operation, because, realistically, that is the shortest distance between the beginning and end points of a process cycle. The trouble is that it is not practiced shop-wide and integrated with all the other product-related assignments through documented procedures. It is, therefore, lacking in organizational cohesion that would otherwise bind commitments to actions to consistently focus on achieving the many quality objectives of a process-approach throughput system. The overall interconnected flow process is a primary requirement in any type of effectively operated process-approach system, for where one process ends the other must logically follow until the cycle is completed. This was the primary reason why I have placed so much importance in the orderly layout and integration of the controlling system, preventing variations in processes by first employing statistical tools.

The benefits of a clearly defined, implemented, and maintained process-approach quality management system can only be realized when we compare it to what's lacking in traditional quality systems that rob dollars from the bottom line. Traditional quality systems are:

- Lacking in organizational focus on throughput process controls;
- Lacking in process ownership – the responsibility, accountability, and performance measurement;
- Lacking in upstream/downstream document control;
- Lacking in clearly defined product quality objectives right from the job's beginning;
- Lacking in communicating the overall product quality requirements for the throughput cycle;
- Lacking an integrated structure to maintain shop-wide corrective action to cure deficiencies, thus the continuous improvement tools;
- Lacking the documented structural system to maintain cohesion and interaction and integration in line with customer requirements from the beginning to the end in the product's flow-cycle.

This list could go on and on, but that's not what the point is. The point is that we should apply more the most commonly used statistical tools in the implementation of quality management systems in order to identify and control the cross-functionally hidden *common- and special-cause variables* that could negatively impact the achievement of every quality objective of the management system. In your hand is a model to follow that does not require six sigma calculations before you can implement it. I already did the preliminary work for you through the application of the above mentioned statistical tools.

APPENDIX FIVE

IDENTIFICATION OF NEW REQUIREMENTS

ISO 9001/2000

In part one, we find two types of amendments identified and presented, the completely new and the complementary requirements. All are add-on requirements to ISO 9001/1994. With the incorporation of these changes, the '94 edition became obsolete and the year 2000 edition took effect on December 13th, 2000. A three-year grace period is allowed by the new standard for the registered companies to change over from the old to the new standard.

The composition of ISO 9001/2000 is not at all similar to ISO 9001/1994. The new standard bases its requirements on Total Quality Management (TQM) and the application of P-D-C-A (Plan-Do-Check-Act) principles within the embodiment of process-approach quality management. The declarations of requirements are stark and repetitive and can be easily misunderstood by those lacking quality management experience. Therefore, we can expect different interpretations and implementations even within the same industry sector. There are no formal publications available on the market at this time, besides this book, that will explain and implement a quality management system based on process-approach quality management.

Quality documents that were designed to implement the '94 standard will need considerable conversion to meet the requirements of the new standard. Updating the quality manual will present very little conversion impediment, for master models can already be purchased as easily as the military standards were. The measurable objectives and the interactively integrated quality objective processes to fulfill internal and external customer requirements of the new standard will necessitate, however, the creation of quality objective procedures. Patching up the current and traditional Standard Operating Procedures will not truly accomplish the measurable quality objective criteria of the new standard. They don't contain the measurable process objective criteria, the linkage of interrelated processes and activities, and the organizational integration to allow process-based continual improvements to be effectively carried out. The reader, this time, has a choice. Follow the patchwork process already pushed by the consulting world, or follow the process-approach model presented in this book that ensures effective operation and control.

I have identified below eleven completely new changes for which there are no equivalent requirements in ISO 9001/1994 and eighteen complementary requirements. The completely new changes are *italicized*. The tabulation of clauses below does not include those clauses of the new standard where there are no basic changes encountered from the '94 standard, although the entire text is completely reworded in the year 2000 standard for each clause. Each change is not only explained below, but also implemented to indicate how compliance with the new standard is accomplishable through the use of this book.

Part two of this appendix will list what has been either reduced or contracted from ISO 9001/1994 and not made a requirement in ISO 9001/2000. The user can either operate his/her quality system with or without them. For the manufacturing industry, I do not recommend hasty abridgment of previously imposed requirements without examining the impact first, for it would impede the effective operation of the quality management system. With these reductions and/or contractions, the new standard went out of its way to accommodate the service industry at the expense of the manufacturing sector. The reverse was the case in the '94 standard.

PART ONE
NEW REQUIREMENTS

CLAUSE
4.2   DOCUMENTATION REQUIREMENTS

4.2.1   GENERAL

(1)   **Document the quality management system's processes to ensure effective planning, operation and control**

COMMENTS
There was no explicit requirement for "effective operation and control" imposed in ISO 9001/94 under clause 4.2.2

The effective operation and control of a quality management system means that quality management takes place under a documented procedural system. What we must be foremost concerned with, however, is that when creating a documented system, we don't stifle efficiency with it. We can be so effectively controlling everything through documented means that we short cut efficiency, thus driving up production cost. The new standard leaves all this out of the picture for the organization to worry about it. And we do. One way to do this is to write Quality Objective Procedures (QOPs) to control the processes (elements) of the quality management system through process instructions, less the work methods associated with it. Leave the work methods to be documented and controlled through the Job Travelers and Operation Sheets.

The work methods are dependent on the competency level of the process owners who carry out the tasks under a defined process. While the work methods may change from operator to operator performing the same task, the process instructions will remain the same for everybody. This way, the process instructions will control the achievement of the identified quality objectives' processes without interfering with the efficiency requirements of production. We will also keep the third party auditors at bay from interfering with the methods of operation. Effectiveness would now be demonstrated through process control in line with process instructions for the realization of each quality objective.

What to do:
One way to prove *effective operation and control* is through measuring the work results of the quality system's objectives. How to go about it must be defined in the quality manual. Please go to the quality manual now (book page 276) and observe under clauses 4.2.1 and 4.2.2 how this requirement has been integrated into the overall documented system.

4.2.2   QUALITY MANUAL

Any exclusion dealing with ISO requirements shall be identified and justified. Define interaction between processes.

COMMENTS
Clause 4.2.1 of ISO 9001/94 did not have these requirements.

What we have to remember here is that exclusions are limited only to the product realization processes. Clause 7 of ISO 9001/2000 defines what these processes are. Conformity to the standard should not be claimed if the exclusions exceed those that are permitted. See clause 1.2 in the new standard.

The definition of interaction between processes is nothing less than integrating the activities of the organization within the quality management system.

What to do:
The easiest way to claim exclusions is to justify them in the body of your current quality manual where the exclusions would normally apply (see book pages 278 and 289). I also made a general remark about it under clause 4.2.1 in the quality manual (pg 276). To describe the interaction requirement, I have used two words 'process-approach.' See book page number 270, clause 1.0 (Scope).

5   MANAGEMENT RESPONSIBILITY

5.1   MANAGEMENT COMMITMENT

Provide evidence of commitment to the development and improvement of the quality management system:

a) **communicating the importance of meeting customer, regulatory and statutory requirements**
b) **establishing the quality policy and**
c) **quality objectives** (measurable)
d) **conducting management reviews**
e) **ensuring resources**

COMMENTS

Most of these requirements were covered in ISO 9001/94 under clauses 4.1.1; 4.1.2.2; 4.1.3 except: communicating the importance of meeting customer, regulatory and statutory requirements; and establishing quality objectives

Evidence of Commitment:

Evidence of commitment will be demonstrated through the effectiveness of the implemented system, showing compliance with the imposed requirements through measuring work results and maintaining continual improvement. For this to take place, the quality objectives of the management system must be cohesively interconnected through the mechanics of written communication. This way, all employees will have the same understanding of what it means to fulfill customer, statutory and regulatory requirements. Interconnection means process integration, flow-down of requirements upstream/downstream within the documented system to ensure effective linkage between processes. We communicate this through written procedures to enable us to realize completion of the quality objectives within a closed-loop process arrangement. Communication under ISO 9001/2000 is the means to relay information to support the effective operation and control of the quality management processes. *Commitments* are achieved through the enforcement of written procedures.

What to do:

Amend your current quality manual in line with clause 4.1.1 (Quality policy) of ISO 9001/1994 to include the required commitments. Please go to book page 273 and read clause 4.1.1 to see how I have enfolded all these commitments in one paragraph. What we need to remember here is not that we have written these commitments down, but that we must live up to them in the overall scheme of the documented quality system. (You may want to read one more time 'Evidence of Commitment,' above.)

Quality Objectives:

In manufacturing the identification, documentation, implementation and maintenance of quality objectives are central to the framework of the quality management system. The various activities involved in complying with customer, statutory and regulatory requirements should be implemented through the quality objective definitions in the work assignments. This should include all those requirements we had in the '94 standard, plus the new requirements imposed by the year 2000 standard, allowing for justified exclusions (see clause 4.2.2a).

We can include all the functions and levels within the organization, or just limit the quality objective criteria to the core departmental activities as most of us did it under the '94 standard. The core departmental activities include the following departments: Contracts, Engineering, Purchasing, Production Planning, Manufacturing, and Quality. (This is also the centerpiece of clause 7 (Product Realization) of the year 2000 standard.) My recommendation is: first, stay with the core departmental responsibilities as this is the backbone of the organization. Manage your shop first because this is where most of the quality problems are converged. Add other functions later, as needed.

Identify all those quality objectives that are needed to cover all aspects that relate to the core departmental functions and levels. Aim for the maximum number of <u>standard</u> quality objectives conventional to your type of industry. This way, you will have procedures to cover the requirements of the high-tech industry in your field of operation as well. This is exactly what I did also.

What to do:
The best way I can help you in determining quality objectives is to ask you to read mine. I have identified twelve (12) <u>main</u> quality objectives in this book. These are titled: Quality Operating Procedures. To keep paperwork down to a minimum, I have combined other similar quality objectives under one Quality Operating Procedure (QOP). Please go to book page 357 and see how all the inspection related objectives are enfolded and integrated in QOP 006. Other QOPs followed the same layout as much as the subject matters allowed.

5.2   CUSTOMER FOCUS

Top management shall ensure that customer requirements are determined and are met to enhance customer satisfaction

COMMENTS
These requirements were covered in ISO 9001/94 under clause 4.3 (Contract Review), but without process and quality objective criteria. (See clause 5.4.1 and 5.4.2 of ISO 9001/2000.)
Note: Customer satisfaction is the <u>main theme</u> in ISO 9001/2000.

For a business to survive in today's competitive environment, satisfying customer needs and expectations must be the foremost commitment of top management. The year 2000 standard has given us the requirements around which to create an integrated process-approach quality management system in order to fulfill the customers' needs and expectations. In order to make this happen, the many processes (elements) of the quality management system dealing with this criteria must be interlinked and cohesively integrated to create the effective interaction between departmental work assignments and focus them in one direction, customer satisfaction. Defining the quality criteria for every process and product related work assignment, and identifying the process owner of each, will bring about fulfilling customer requirements through process instructions. Fulfilling customer satisfaction must hinge on communicating the necessary information to the internal customers first. Process instructions detailing the steps of the processes for every quality objective will do that. In this particular case, the internal customers are those employees handling the contract activities, the external customer's requirements. These requirements will impact eventually just about everybody in the chain of events, known as the product realization processes. This then is the first department where we must establish controls as to how the external customers' needs and expectations are determined and converted into requirements.

What to do:

Amend your current quality manual under clause 4.3 (Contract Review) and indicate that customer needs and expectations are fulfilled by the implementation of applicable Quality Objective Procedures. Give references in the manual to the applicable procedures. Please go now to book page 278 and observe there how I have handled this add-on requirement. Observe also the beginnings of the interactive processes coming to life by the tools of integration in Quality Operating Procedure 001, referenced under the same clause. Read it in its entirety, all four sections of QOP 001, beginning on book page 295. For ease of process flow and better understanding of what's going on here, please go to the equivalent diagrammatic procedure located on book page 38. This entire QOP sets the stage for integrating work responsibilities downstream. Without it, the whole organization would have a tough job understanding the external customer requirements as they impact the continuing processes in the product realization cycle.

5.3   QUALITY POLICY

(5) **The quality policy must include commitments to meeting requirements and to continual improvement (b)**

(6) **The quality policy must provide a framework for establishing and reviewing quality objectives (c)**

(7) **The quality policy is reviewed for continuing suitability (e)**

COMMENTS

All requirements of clause 4.1.1 Quality Policy of ISO 9001/94 are retained.

Plus, these three distinctly separate requirements have been added to clause 4.1.1 of the '94 standard and enfolded with the others under clause 5.3 of the new standard.

Continual Improvement: (5.3b)

The new standard implies that continual improvement is a process driven objective. How we are doing this strikes at the heart of how we are solving problems. Currently, we are backtracking from the end of the line to make improvements, instead of front-tracking to identify problems to eliminate variables. The backtracking methods indicate lack of process control while front-tracking indicates the opposite. The new standard requires process control by monitoring the work processes (see clause 4.1e) and thus, maintaining continual improvement. Doing continual improvement through monitoring the realization processes confronts us with the creation of interactive procedures. This is not an easy task to consistently fulfill, for enacting continual improvement practices when and wherever problems manifest themselves involves controlling changes also, while preserving the integrity of operations. Neither the backtracking nor the front-tracking methods work alone, but in combination. Continual improvement is always easier to implement after the problems have been found than before it. We must strive, therefore, to create procedures that can regulate the implementation of continual improvements for both the front-tracking as well as the backtracking activities, because they impact each other at unknown levels in the product's life-cycle.

What to do:

While the commitment for continual improvement should be stated in your current quality manual in line with clause 4.1.1 (Quality Policy) of the '94 standard, the implementation of it should take place in combination with process control procedures under clauses 4.10.3 (In-process Inspection and Testing) and 4.13 (Controlling Non-conforming Product). To illustrate how this has been implemented and integrated in a manufacturing environment, please go to book page 374 (Process Control, Section Three of QOP 006) and read there how corrective action is enforced to maintain continual improvement in the various realization processes (front-tracking). To maintain full documentation of corrective actions for system-wide continual improvements, please read QOP 009 (Control of Non-conforming Product on book page 403), all three sections apply (for both front- and backtracking).

Framework for Quality Objectives: (5.3c)

Quality objectives are identified from the various processes (elements) of the standard. We convert these to requirements and list them in sequential order under the various clauses in our quality manual. This is the first part of the framework. The second part is when we create the Quality Objective Procedures (QOPs) to enable us to fulfill these requirements. And the third part of the framework is when we design the appropriate forms to prove work results. This should create an interactive chain of events. When interlinked in a logical order, the elements of the framework should present an orderly alignment of the quality system's processes, which in turn, should identify the quality objectives, implemented through the application of the various quality operating procedures. By establishing this type of structural framework, not only the quality objectives, but also the interrelated process instructions can be consistently communicated between the various functions and levels within the organization. By creating this type of framework, expansion or contraction of quality requirements could be accomplished in an orderly manner without corrupting the integrity of the quality management system.

What to do:

The structural framework of a quality management system depends on how the creator of it determined the systematically interactive processes within the functions and levels of the organization. I have created mine in line with the same scenario given above for shop quality management in manufacturing. In order to understand what is involved here, please spend some time in reviewing the entire integrated system presented in this book. Start with the quality manual, proceed with the operating procedures, and conclude by understanding the application of the different forms. Guide yourself by observing linkage to interactions and controls. They are applied to ensure that each quality objective is achieved to satisfy organizational commitments for product quality, customer satisfaction, and continual improvement.

The first step is to generally state the "framework" requirement in the quality manual in line with clause 4.1.1 (Quality Policy) of the '94 standard. No specific procedure is required, for this is a declaration as to how the framework requirement will be developed later. For practical advice regarding construction methods, please go to page 16 in the 'Guidance Section' of this book and read clauses 2.0 through 9.0. Then go to book page 64 and read Section Two of QOP 002 (Training Metrics) to see how it is actually done.

Continuing Suitability: (5.3e)

Continuing suitability invokes the first question we should always ask ourselves in quality management. Do I have all the controlling procedures in place to satisfy product quality, customer satisfaction, and continual improvement requirements for my established customers. If the answer is "No," your quality system will not meet the ISO 9001/2000 requirements. It will not even meet the business goals of your company, regardless of ISO. Continuing suitability is to provide adequate quality support by which to enforce every quality objective deemed necessary in your type of business, taking into consideration that not every customer imposes the same quality requirement. In this light, you should have all the standard quality objectives documented and integrated into your system so that even the most stringent high-tech requirement would be covered in your type of business. Continuing suitability not only means just to fulfill the minimum product quality requirements, but also means that you advance your company's business aspirations to be able to fulfill the quality requirements of new business in your field of operation. As an example, take a good look at what I have included in my Quality Operating Procedures to cover just about every circumstance possible in shop-quality management, regarding product related quality enforcement. Do I need them all at the same time? Of course not. But when I do, I can satisfy my customers' needs and expectations at the moment of asking. I have in this book more than 209 quality objectives in documented procedures, both text and flowchart format, completely integrated for your immediate implementation, study, conversion or just reference. Continuing suitability should always be measured with futuristic outlook if you want to beat the competition.

What to do:

Likewise, add to your current quality manual under clause 4.1.1 (Quality Policy) that "continuing suitability" is a policy requirement. I have provided you with enough information to realize how important this requirement is for the future of your business. Aim for the maximum quality objectives possible in your business and be forward looking. There are no rules on the books against it.

5.4   PLANNING

5.4.1   QUALITY OBJECTIVES

(8) *Ensure that quality objectives are established at relevant functions and levels, and are measurable and consistent with the quality policy, including the need to meet the requirements of product*

COMMENTS

This is a completely new requirement. It belongs under clause 4.2.3 Quality Planning of 9001/94. The emphasis is on <u>measurable objectives</u> at relevant functions and levels of the organization.

Identifying Quality Objectives

Work assignments encompassing the activities of functions and levels within the organization are those through which we interactively fulfill the internal and external customer requirements. In each of these work assignments the possibility of error or mistake is real, and we transfer them from process to process during the chain of process realizations. In order to stop this from happening, we write procedures with clear process instructions and tell the process owners of these assignments that we want them to check their work results to ensure that they meet the following … quality requirements. Assignments with imposed quality requirements become the <u>quality objectives</u> of the management system. In a nutshell, they become quality objectives because there are quality requirements added to them.

The quality requirements ensure that mistakes are not transferred from process to process and further ensure that the work assignment is carried out to meet both the process and the quality requirement at the same time. The year 2000 standard requires that quality objectives are planned and established at relevant levels and functions within the organization, providing measurable results. So, if we integrate all the work assignments process after process, we have created an interactive system that should have built-in features to link commitments to quality objectives with measurable results from the beginning to the end in a product's life-cycle. This is essentially the meaning of this clause.

What to do:

Since this is a completely new requirement, add sub-clauses to your current quality manual in line with clause 4.2.3 (Quality Planning) of the '94 standard and state your reasoning in a similar fashion as I have discussed it above. Better yet, turn to book page 277 and look up how I have satisfied this seemingly enigmatic requirement. The implementation of what this clause wants us to do cannot be achieved through writing one procedure to take care of all its provisions. It requires that all the quality system's procedures cohesively integrate the intent declared in this clause. This requires serious quality planning which we will discuss under the next clause.

5.4.2   QUALITY MANAGEMENT SYSTEM PLANNING

Plan, identify, and document the processes and resources needed to achieve the quality objectives in line with clause 4.1

COMMENTS

Planning, identifying, and documenting to achieve the quality management system's processes (objectives), including continual improvement, were not an explicit requirement of ISO 9001/94 under clause 4.2.3, but we did it in some measure anyway, since a quality manual without planning to achieve the quality objectives would be a worthless document. This time around, the requirement is clearly stated under four separate clauses. This clause is about general planning, while the other three are tied to specific objectives.

7.1   Plan product realization, product verification and validation processes;
7.3.1   Plan design and/or development processes;
8.1   Plan the measurement and monitoring of products, processes, performance and customer satisfaction.

Regardless how the Standard dished out the planning requirements, Quality Planning should embrace the entire quality management system's processes, for these processes must work together within one system. Interconnected they must be for a common purpose, satisfying internal and external customer requirements. At the end, every quality objective must achieve its purpose in cohesion with other quality objectives to consistently achieve effective operations. This is really the purpose of a process-approach quality management system.

Although the year 2000 standard requires planning to be undertaken under four separate clauses, the job still boils down to one comprehensively coordinated planning effort to bring about one cohesively integrated and controlled quality system. Each of these different planning clauses deal with different objectives, but when taken together, they have one central purpose: the creation of the necessary controlling documents to ensure the effective operation of the quality management system. These different controlling documents serve as instructional materials to be followed by all the process owners at all functions and levels to ensure the fulfillment of each and every quality objective we have designed into the quality management system's processes.

Trying to address each of these planning provisions individually, without realizing their interactive nature within the whole quality management system, would give us one hell of a convoluted system without focus. We should remember, these planning declarations mean nothing, unless we can convert and reduce them to the shop floor and apply them sensibly and consistently in the product's realization processes.

What to do:

Amend and/or add sub-clauses to your current quality manual in line with clause 4.2.3 (Quality Planning, ISO 9001/94) to include the 'Quality Planning' objectives. Make references to the applicable procedures that implement and maintain these activities. Please go to book page 277 and observe there how I have brought together the planning activities under one clause. Then, look up any of the referenced procedures in order to understand how I have implemented them in the shop. The equivalent QOPs found under 'Training Metrics' will flowchart the implementation of every quality objective needed to manage shop quality practices in manufacturing.

Ensure that change (updating) is done in a controlled manner while the integrity of the quality management system at the same time is maintained

COMMENTS

The emphasis regarding change (updating) is that it is done in a controlled manner and that the <u>integrity</u> of the quality management system is maintained during this change. This requirement is a lot more than just document change control under clause 4.5.3 in ISO 9001/94. It applies to change (updating) made anywhere in the quality management system. This is flow-down control of changes upstream/downstream in the quality management system. This requirement embraces the control of customer related documents as well.

What to do:

Managing change control in the quality management system's documents requires special planning and documented procedures, for it could impact every aspect of the product realization processes, including the customer. For this reason, I have consolidated this effort under 'Document and Data Control,' clause 4.5 of the '94 standard and made reference to the controlling documents implementing change requirements. The complexity involved in managing change control cannot be expressed here in a few words. For this reason, I ask you to go directly to book page number 279 and see there what the quality manual says, and then go to the referenced procedures to study the extensive controlling requirements involved with managing changes.

5.5.2   MANAGEMENT REPRESENTATIVE

Ensure that the management representative promotes the awareness of customer requirements throughout the organization

COMMENTS

This requirement has been added to the management representative's other requirements under clause 4.1.3 of ISO 9001/94 and carried over to the new standard under the subject clause.

Promoting awareness of customer requirements, first of all, should happen through the availability of information at specified work cells, using the applicable procedures that we have created to fulfill customer requirements. Secondly, it should happen through the general training of new and established employees. Thirdly, it should happen when we implement continuous improvements resulting from corrective/preventive actions. Important to understand here is that the awareness information is communicated in a continuous and purposeful fashion, dependent on the specific and general needs to understand requirements, employee by employee. We don't promote awareness of customer requirements on a loud-speaker (intercom), but rather from the documented processes of the quality management system. Not every employee will have a need to know all the specifics of the whole quality management system. But they will have to know first, how they, individually, contribute to meeting customer requirements. Second, they will have to know the general workings of the quality management system.

What to do:

Add a statement in your current quality manual under the "Management Representative" responsibilities in line with clause 4.1.3 of ISO 9001/94 and define there how he/she will undertake promoting the awareness of customer requirements. Since this is a policy declaration, no specific documented procedure is needed. Please go to book page number 276 and see there how I have simplified the inclusion of this responsibility.

5.5.3   INTERNAL COMMUNICATION

Ensure internal communication and its effectiveness through-out the organization regarding the processes of the quality management system

COMMENTS

This new requirement is both oral and written dissemination of information, regarding the entire documented quality management system's policies, procedures, objectives, data, records, and instructions. It has everything to do with bringing the documented system together in a logical framework so that communication is targeted where needed. This is the integrating mechanism (referencing), the structural network, ensuring that the written stuff is effectively linked upstream/downstream within the quality management system and brought to the source where the information is needed by the appropriate personnel assigned to achieve quality objectives.

What to do:

Since this is a completely new requirement, add a sub-clause under Quality Planning in line with clause 4.2.3 of the ISO 9001/94 standard and state there how internal communication will be achieved. A separate documented procedure is not required by ISO 9001/2000. Please see book page number 278, clause 4.2.3k, and observe there how I have inserted this requirement. If you wish to become familiar with how to convey relevant data needed to guide people to targeted information, please see how I did it. The whole book is full of it. And if you like it, don't hesitate to follow it. This is more of an experience driven art than a college course, although both should be pursued to be effective in the skills of communication.

5.6   MANAGEMENT REVIEW

5.6.1   GENERAL

Assess the need for changes to the organization's quality management system, including policy and quality objectives, to ensure its continuing suitability, adequacy and effectiveness

COMMENTS

Clause 4.1.3 of ISO 9001/94 did have similar requirements for management reviews, but without the specifics for "Review Input and Review Output."

The evaluation of the quality management system is now focused on the documented and implemented system to determine whether it had achieved product quality requirements, problem prevention, continual improvement, and customer satisfaction (see clause 1.1). Now that the processes of the quality system have been implemented according to the lofty requirements of the new standard, reference to it no longer is made as it was the case under the '94 standard. (So goes the futuristic interpretation of the new standard.)

If we design our quality management system based on the process-approach definition of the year 2000 standard, we will be operating our quality system to the requirements of quality objective procedures. Since quality objectives shall be measurable (see clause 5.4.1), we will have to define in these procedures what to measure and record. This brings up a couple of questions. Who will do the measurement? When will the measurement be made? The new standard is mute about these. (Futuristic in its presumption)

I can most assuredly say from my experience that a process-approach quality management system has built-in controls to achieve product quality, problem prevention, continual improvement, and customer satisfaction without management reviews. Management review under a process-approach quality management system is handled as a quality objective, just like any other management meeting, part and parcel of the daily business process. You will discover that I have written my quality objective procedures based on shop reality, proactively and spontaneously responding to make changes to meet the requirements of the product realization processes. The greatest enemy against the periodic management reviews is the customer himself. He cannot wait. The product must move together with the decision making process. Changes resulting from the management review output must be concurrent with the product's realization processes. This is what experience tells us to do. In this light, management reviews are frequently done.

The next clause "Review input" (5.6.2) will add confirming weight to the above discussion as it lists the input requirements to management reviews. The Standard puts forth for the assessment of system performance the most relevant quality objectives that actually embody the attributes of success or failure of the quality management system's daily operations. This also confirms that we must write our quality objective procedures to support the achievement of product quality, problem prevention, continual improvement, and customer satisfaction as an ongoing part of the process realization cycle.

What to do:

Amend your current quality manual under "Management Review" in line with clause 4.1.3 of the '94 standard and define there how you will conduct management reviews in accordance with the requirements stated in the new standard. Although there is no requirement for a documented procedure to conduct management reviews, the subject should be treated as a quality objective and therefore, the whole review agenda should be formalized in a written procedure. Please see book page number 276 and observe there how I have stated the management review requirements in the quality manual. Then, see my quality objective procedure, QOP 011, book page numbers 235 or 428, and observe how I have implemented this quality objective.

5.6.2   REVIEW INPUT

Review (by assessing) current performance and improvement opportunities based on:

- *Results of audits*
- *Customer feedback*
- *Process performance and product conformance*
- *Status of preventive/corrective actions*
- *Follow-up actions from previous reviews*
- *Changes impacting the system*
- *Improvement opportunities*

COMMENTS
There is no equivalent requirement in ISO 9001/94 in clause 4.1.3.

The review input criteria listed under this clause relates to those quality objective procedures through which we control the realization processes to meet internal and external customer requirements. If we had implemented a process approach quality management system based on the scenario explained above under clause 5.6.1, we would be presenting to top management now the results of an effectively operating system. Customer feedback, process performance and product conformance, corrective/preventive actions, follow-up actions, change controls and others are all objective requirements in a correctly managed process-approach quality system. How to implement and fulfill these requirements should be defined in the applicable quality objective procedures. The internal audit results should provide top management not the problems it had found, but the improvements it had made in the processes, methods, and the products as a result of internal audits.

The review input in our management review agenda should include an additional requirement to that of the new standard, – fulfillment of management responsibility.

What to do:
Please follow the same action as explained under clause 5.6.1 above.

5.6.3   REVIEW OUTPUT

The outputs from management reviews shall include actions:
- *To improve effectiveness*
- *To improve customer related requirements*
- *Resource requirements*

COMMENTS
There is no equivalent requirement in ISO 9001/94 under clause 4.1.3.

The output results from management reviews are directly proportional to input results. The new standard requires that we conclude management reviews with action items relating to improving the quality management system and its processes, improving product related customer requirements, and determining resource needs to implement them. These requirements should be everyday achievements woven into the product realization processes and taken care of under process controls for the same reasons as explained above under clause 5.6.1.

What to do:
Please follow the same action as explained under clause 5.6.1 above.

6   RESOURCE MANAGEMENT

6.2.2   COMPETENCE, AWARENESS AND TRAINING

Evaluate the effectiveness of training provided (c). Maintain records (e). Ensure awareness of quality requirements

COMMENTS
These requirements were not in ISO 9001/94 under clause 4.18. (We must differentiate between general and specific training.)

Evaluation of the effectiveness of training is meant to test the understanding of quality requirements by certain employees engaged in specific work assignments. The competency requirement of these employees should include understanding the quality requirements impacting their work assignments to maintain product quality. The method of test is optional so long as there is a record of the test results on file, case by case, to prove that it was done. The awareness requirement is repeated again. This was already discussed under clause 5.5.2 above.

What to do:

Amend your current quality manual under "Training" in line with clause 4.18 of the '94 standard. A documented procedure is not a requirement. Create a form suitable to record the test results of the employee evaluated. Please see book page number 289 as to how I have inserted the "evaluation" requirement. See also the form on book page number 498 created to maintain record of it.

Ensure that employees are aware of the relevance and importance of their activities and how they contribute to the achievement of the quality objectives

Comments

Awareness of quality requirements in work assignments is meant to enact the requirements of quality objectives. Therefore, the procedures written for achieving quality objectives should include the process instructions in order that the employees understand the relevance and importance of the quality requirements pertaining to their work assignments. Then, when performance results are monitored, it would become evident whether the employee followed the process instructions or not in meeting the relevant quality requirements. The subject of employee awareness is highly dependent on enforcing internal communication, which we have already discussed under clause 5.5.3 above. Next, employee awareness to understand requirements was also the topic of discussion under 'Management Representative,' clause 5.5.2 above. Please go back and refresh your mind, now that the new standard keeps on reiterating the same requirement under this clause.

What to do:

Amend your current quality manual under "Training" in line with clause 4.18 of the '94 standard and insert the "awareness" requirement (see book page number 289). A documented procedure is not a requirement. You will be amazed as I tell you how seriously I believe in training process owners to understand process and product related quality requirements. I have devoted approximately 230 pages of flowcharts in this book for that cause.
(See 'Training Metrics' starting on book page number 42.)

6.4   WORK ENVIRONMENT

Determine and manage the work environment needed to achieve conformity of product

COMMENTS

The emphasis is on "conformity of product." Any Health and Safety rules are relevant only within the meaning of conformity of the product. This requirement is a rephrasing of "suitable working environment" under clause 4.9 (b) of ISO 9001/94.

This clause drives home the fact that the work environment should be a safe place to achieve product conformity. OSHA and EPA violations alone may not be a cause for noncompliance reporting by ISO third party auditors. But, if the violation prevents the achievement of product conformity, there is a valid reason for requiring corrective action. Safety related requirements that are part of meeting product conformity should be defined in the Job Travelers and/or Operation Sheets, product by product and process by process, as they are applicable to the work assignment.

For general information, all safety requirements should be posted on bulletin boards as well as on equipment. Employee awareness regarding safety rules should be top priority during training sessions. Evidence to this effect should be recorded and signed by all employees on a 'Signature Sheet.' This, of course, is not a requirement of ISO 9001/2000.

What to do:

Amend your current quality manual under the "Process Control" provisions in line with clause 4.9b of the '94 standard. Insert there how you will ensure compliance with this requirement. A documented procedure is not a requirement. Please see book page number 281 and clause 4.9b to observe how I have inserted this requirement. Also, turn to book page number 302, clause 3.03, item number 2, in order to observe how this requirement has been flown down to Process Engineering for implementation into the processing related documents. Then, turn to book page number 375, clause 3.03 and observe there how process monitoring for this objective is implemented. (Please note that the complete flow-down of requirements is a must in any effectively implemented process-approach quality management system.)

7   PRODUCT REALIZATION

7.1   PLANNING OF PRODUCT REALIZATION

19

Plan the processes for realization of product. Determine quality objectives for the product. Determine the need for documentation, verification, validation and the criteria for acceptability, including the resources specific to the product. Determine the records necessary to provide evidence of conformity of the processes and product

COMMENTS

This clause is the pinnacle of ISO 9001/ 2000 as to how to plan meeting customer requirements. It should be understood in relation to all other planning clauses given under 5.4.2 (System Planning), 7.3.1 (Design and Development Planning), and 8.1 (General).

As we review the clauses of ISO 9001/94 under 4.2.3 (Quality Planning) and 4.9 (Process Control), what we find here is a clear consolidation of the same planning requirements and more. This is good, for it will prevent incomplete quality planning that was possible under ISO 9001/94.

The control of product realization processes is the heart of ISO 9001/2000. While the general quality planning requirements were stated under clause 5.4.2, this clause will present us with specific planning requirements for product realization. Complying with these planning requirements, suddenly we are faced with the monumental task of integrating responsibilities and activities within relevant functions and levels of the organization to manage shop related interactions for product realization. Determining the quality objectives for the product, project or contract will bring together the totality of core departmental controls through the establishment of interactive quality objective procedures in order to maintain consistent operation. All in all, this will require the identification of the necessary work processes, process owners' responsibility and the appropriate documentation to prove work results. Most importantly, product acceptance will have to be proven by verification and monitoring records. Providing resource needs to meet the conformity of the product is no longer a marketing slogan, but rather, specific to product realization. The orderly interaction of product related work assignments and their fulfillment should demonstrate that the product realization processes are cohesively integrated and controlled.

Auditors will look for: 1) identification of the quality objective requirements in the quality manual and the reference to procedures or forms. 2) the quality objective procedures which implement the fulfillment processes of the quality objectives. 3) the measurable work results documented in the various forms.

What to do:

Add clarifying statements (enhancement) in your current quality manual under "Process Control" clause 4.9, and "Quality Planning" clause 4.2.3, regarding planning activities. Then list the quality objective procedures in each sub-clause, as applicable. Please see book page numbers 277, clause 4.2.3, and 281, clause 4.9, in order to observe how enhancement is made possible. Then, look up any of the referenced quality objective procedures to understand how planning is linked to everything we do in quality management.

Although the new standard doesn't require any documented procedures, it would be suicidal not to have them. Don't throw away any of your current quality plans, documented procedures or flowcharts. Update or rewrite them to fall in line with the flow-down and integration scheme of process approach. This whole book serves as a model for you. Take advantage of it.

7.2   CUSTOMER-RELATED PROCESSES

7.2.1   DETERMINATION OF REQUIREMENTS RELATED TO THE PRODUCT

Determine customer requirements:

- *Product requirements stated*
- *Product requirements not stated but necessary for intended or specified use*
- *Statutory and regulatory requirements relate to product*
- *Additional internal requirements*

(20)

COMMENTS

ISO 9001/94 did contain similar requirements under clause 4.3 (Contract Review) except for the last three items.

Item one is the general requirement that we have been accustomed to doing under clause 4.3.2 (Review) of the '94 standard. What was missing from this clause was item two, three, and four. Item two: what were we supposed to do in cases where the customer sent in a contract and said: "Make it to the drawing." This is a very common event in the manufacturing industry and many times the source of misunderstanding, leading to legal action between parties. The new standard recognizes this omission and stipulates that we must do something about *not stated* quality requirements. Items three and four are non-specific requirements and should be weighted in relation to the interpretation of the fluctuating product quality requirements in the awarded contracts. Meeting customer requirements is the principal objective of ISO 9001/2000. It should also be the same of any quality management system tailored to meet customer satisfaction, irrespective of ISO requirements. Both, the identification of quality requirements and the product realization planning start here, under the various phases of Contract Administration. These should form the basis for all quality objectives, and the establishment of the appropriate procedures in which we flow down requirements that give us the necessary information required in maintaining product conformity to the stated as well as to the not stated quality requirements of a contract. If we miss doing these things at this stage in the product's flow-cycle, the entire quality management system could become ineffective. We would be backtracking to solve problems from the end of the line only to discover that we could have prevented them, if we had paid serious attention to quality planning requirements at the beginning of the flow cycle. This is the RFQ phase in business transactions. This is where we put down the final cost figures for all the product realization steps that will determine later how efficiently we must do our work at every stage in the product realization cycle to stay within the established budget. This is where <u>over-effectiveness</u> in the imposition of quality requirements could be the straw that broke the camel's back. This is known as cost overdrive. This could be the single most important quality-planning requirement that the new standard said nothing about. Ensure that you carefully plan the implementation of all quality related requirements, because meeting the effectiveness criteria of ISO 9001/2000 could, alone, become the hidden factor in cost overdrive.

What to do:

Since items two, three and four are entirely new requirements, they should be enfolded in the "Contract Review" provisions under the RFQ and /or Purchase Order Review sections. They are part of clause 4.3 (Contract Review) of the '94 standard. The quality objective procedures written to cover these requirements should provide process instructions to those doing the RFQ and PO reviews as to what is expected of them in light of these new items. Please see book page number 278, clause 4.3.2a, in order to observe this addition. Then, go to book page number 302, clause 3.02, and observe how the entire quality requirements of a contract are flown down by the application of Form A-001. Pay attention to "Note 5" regarding the handling of *not stated* requirements.

7.2.3   CUSTOMER COMMUNICATION

(21) *Identify and implement measures for communicating with customers relating to:*

- *Product information*
- *Enquiries, contracts or order handling, including amendments*
- *Customer complaints and feedback*

COMMENTS

Item one and two are basic and routine exchange of information between contracting parties. The '94 standard in clause 4.3 implied this as such, but without identifying the internal and external communication specifics involved daily between the customer and the organization. At last, ISO 9001/2000 is making it clear that we should *determine and implement effective arrangements* to control this part of the business management also.

Item three is a completely new requirement, significantly impacting the service sector. The manufacturing sector has, to a degree, in-house methods of handling customer complaints and feedback information, albeit without written procedures in a lot of places. We must recognize here that a lot of manufacturers had procedures to handle customer complaints and feedback information long before ISO 9000 was adopted. Quality system procedures under Mil-Q-9858 took care of it all (see MIL-STD-1535).

The new requirement under this clause is to "identify and implement arrangements" as to how communication takes place between the customer and the organization. This boils down to a pretty complex integration effort as there are very few things in contract manufacturing that customer requirements don't impact, either directly or indirectly.

What to do:

This clause imposes a completely new requirement regarding communication with customers. Since this effort involves different functions and levels within the organization, every procedure that includes customer related activities should be changed to bring about directing interactions between the customer and the various members of the organization. First, we should identify the requirement in our ISO '94 quality manual under "Contract Review" (4.3). Then, flow down and integrate all responsibilities in other quality objective procedures. First, in "Document and Data Control" (4.5 – QOP 002); second, in "Process Control" (4.9 – QOP 006, Sec Three); third, in "Inspection and Testing" (4.10 – QOP 006); and fourth, in "Control of Nonconforming Product" (4.13 – QOP 009). We should create similar quality objective procedures that govern these activities in order to bring about a cohesive interaction between the members of the organization and the customer.

If you ever wondered how system integration plays into the hand of effective quality management, this will be the moment of truth for you as I will demonstrate in this book. We will be focusing here on contract related matters only. But, please keep in mind that every one of the other clauses listed above has its own quality objective procedure, clearly defining specific activities dealing with customer communication through interactive processes. So, if you wish to fully implement customer communication into your quality management system, the framework on how to do it is already implemented in this book. Just follow the linkages referenced from procedure to procedure and you will find everything you need.

Contract Review (4.3)
Please go to book page number 278 (Quality Manual) and read all the sub-clauses (4.3.1, 4.3.2, 4.3.3, and 4.3.4). Note in sub-clause 4.3.1 the identification of QOP 001. This document is the quality objective procedure for contract related activities, which defines all interactions for every member in the organization regarding task specific responsibilities, communication, flow-down requirements, and linkages to other procedures. It has four sections, each dealing with a separate topic, regarding customer purchase order activities. Go to book page 295 and read all sections of QOP 001. After, you may wish to go to the other QOPs from here to convince yourself of the importance of how customer requirements should be integrated in just about

anything we do in a manufacturing environment to bring about completing the communication cycle within a quality management system.

7.3   DESIGN AND DEVELOPMENT

7.3.1   DESIGN AND DEVELOPMENT PLANNING

Plan and control the review, verification and validation activities of each design and development stage. Manage effective communication and assignment of responsibilities

COMMENTS

This is a new requirement in addition to what we find in ISO 9001/94 under clause 4.4.2 Design and Development Planning, and in 4.4.3 Organizational and Technical Interfaces. Planning to control the review, verification, validation, communication, and assignment of responsibility impacts only those companies that are engaged in design activities.

What to do:

Amend your current quality manual to include *review, verification, validation, communication, and assignment of responsibilities* under "Design and Development Planning" in line with clause 4.4.2 of the '94 standard. Then, either amend or rewrite the applicable quality objective procedures in order to define how you are going to implement and control them.

7.3.7   CONTROL OF DESIGN AND DEVELOPMENT CHANGES

Verify and validate changes before implementation, as appropriate. Evaluate the effect of the changes on constituent parts and delivered products. Maintain records.

COMMENTS

ISO 9001/94, clause 4.4.9, did not include these requirements. The control of design and/or development changes now include verification and validation requirements before implementation, as appropriate. Inasmuch as this is now the requirement, it does not mean that every change will have to be subjected to it. The phrase "as appropriate" allows the design activity to decide when to enforce it. On the other hand, evaluation of the effect of changes on constituent parts and delivered products must be enforced. The Standard reworded this clause in ISO/FDIS 9001 and *as appropriate* is no longer inclusive to the last sentence of the paragraph as it was before. Record maintenance applies to all the activities involving changes.

What to do:

Amend your current quality manual to include the control of design and development changes regarding these new requirements. Emphasize the phrase: <u>as appropriate</u> for the verification and validation part of this clause. Add this revision under "Design Changes" in line with clause 4.4.9 of the '94 standard. Update your applicable quality objective procedures and state how you are going to implement and control them.

7.4   PURCHASING

7.4.1   PURCHASING PROCESS

 Establish re-evaluation requirements

COMMENTS

This requirement can no longer be presumed. It is explicitly stated. ISO 9001/94 did not clearly state this requirement under clause 4.6.2.

Periodic evaluation is a new requirement added to the others we had under clause 4.6.2 in the '94 standard. Most of us would call this a "paper change" as we are already doing periodic re-surveys of our suppliers, frequently by mail-in surveys. The intent of the new standard under this clause is to periodically determine if the supplier is still capable of meeting the organization's product quality requirements. Re-survey helps us to detect whether the supplier underwent organizational changes impacting his previously approved quality system. The result of re-survey may trigger additional action if it indicates deterioration in the supplier's quality system. We could easily detect this also by reviewing the supplier's previous inspection records on file, even before we wait for the re-survey results.

What to do:

If you don't already have the *re-evaluation* included in your quality system, add this requirement in your current quality manual under "Evaluation of Subcontractors" in line with clause 4.6.2 of the '94 standard. Revise also your quality objective procedures dealing with controlling subcontractors and state there how you are going to implement and control re-surveys. Please turn to book page number 279 (Quality Manual), clause 4.6.2a, to observe how I have inserted subject requirement. Then, look up the quality objective procedure, QOP 003 on book page 342, and read clause 3.02.1 or 3. 02.2 in order to observe the implementation process in fulfilling this requirement.

7.6   CONTROL OF MONITORING AND MEASURING DEVICES

Software used for measuring and monitoring of specified requirements shall be confirmed prior to use and periodically reconfirmed

COMMENTS

Confirmation is the new requirement. The rest of the basic calibration requirements have been carried over from ISO 9001/94, clause 4.11, to the new standard. This means that we are required to control the approval, release and application of software if we want to use it in measuring, monitoring and accepting products.

Remember, there is a vast difference whether you design and manufacture software or you're just programming software. In the first one, you should control confirmation through certification from the manufacturer. In the second one, you should either subcontract the confirmation requirement and obtain certification for it, or control it in-house through a quality objective procedure. Changes done to software through reprogramming should be controlled in the same manner as you would control document changes. To comply with this clause, you have to make sure that your calibration records can prove that confirmation was done prior to the use of the software for product acceptance. If you have software on line during transition to ISO 9001/2000, you can continue using it, but your calibration records must indicate that the software has been accepted through "continual use" verification, known as "grandfather-ing."

What to do:

Amend your current quality manual under "Control of Inspection, Measuring and Test Equipment" to include the control of software confirmation requirement in line with clause 4.11 of the '94 standard. Then, revise, rewrite or establish the appropriate procedures to show how software confirmation will be implemented and controlled.

In this book, I'm only covering the control of software programming change. Please turn to book page 284 (Quality Manual) and observe clause 4.11.1 as to how I have inserted this requirement. Then, go to book page 320 to the quality objective procedure QOP 002, Sec Two, clause 3.03.2.3 and observe there how I have implemented controlling software changes as part of the software confirmation requirement.

8   MEASUREMENT, ANALYSIS AND IMPROVEMENT

8.1   GENERAL

Plan and implement the measurement, monitoring, and analysis activities to demonstrate conformity of the product and the quality management system to continually improve effectiveness

COMMENTS

There is no corresponding requirement in ISO 9001/94. This is typical process control for all the defined quality objectives requiring documented work results to demonstrate compliance through measurement. Deficiencies found should start the implementation of continual improvement.

We have here the fourth occasion involving quality planning. Under this clause, we are faced with measuring performance results in work processes. One of the reasons for doing this is to find process related problems and implement improvements. The other is to prevent problems from migrating from process to process. And the third one is to achieve process performance satisfaction that would eventually lead to customer satisfaction at the end of the production line. This is no longer just measuring product dimensions to drawing requirements. This involves being able to measure what the quality objectives are required to achieve through process controls. Methodologies, including statistical techniques, are all part of effective process control. In determining process performance results, the word *measurement* also means monitoring the processes to evaluate whether we have achieved expected results. Continual improvement is meant to be implemented at all stages in the product realization processes where non-conformity to *customer* requirements has been identified. Keep always in mind that *customer* in the eyes of ISO means both internal (organizational members) and external (buyer) parties. The full meaning of clause 8.1 (General) will not be understood until we relate it to all the other clauses in section 8, especially clauses 8.2.3 (Monitoring and Measurement of Processes) and 8.2.4. (Monitoring and Measurement of Product).

What to do:

In order to be able to apply measurement techniques to evaluate work results in all the processes of the quality management system, we need to determine the construction method of our quality objective procedures. These procedures must provide the necessary process instructions to enable all process owners to understand what is expected from them, for they will not be able to monitor and record work results. They will not be able to measure them and adjust for errors. What we have to keep in mind here is that we are not only providing quality plans to inspection personnel, but also to other process owners in the organization (including subcontractors) who are expected to achieve measurable and acceptable work results in other continuing work processes.

The layout of these procedures should include the identification of process owners and process requirements. These procedures should provide instructions when to measure and record work results to prove conformity to requirements. And should provide linkage information needed to integrate process continuity requirements in other procedures and forms. Thus making the entire quality system one continuously looped, integrated system, product by product. (Which means that you should have enough quality objective procedures in place to satisfy both the minimum "not stated," clause 7.2.1b, as well as the maximum "specified," clause 7.2.1a, quality requirements of a contract.) The standardization of quality objective procedures should become a reality, consistently providing process instructions to those employees executing customer requirements.

This type of procedure construction method is basic to a process-approach quality management system presented in this book and falls in line with ISO 9001/2000 requirements to effectively manage the processes of a quality system. My research concludes that there is nothing like this available on the market today outside of this book. What is available, however, falls far short of quality system integration in the manufacturing sector, for other books merely provide partial repair schemes without global linkage to all the processes required in manufacturing quality systems.

What to do:

To include this new requirement in the quality manual, we should add the *measurement, analysis, and improvement* definitions under applicable clauses like "Process Control" (4.9); "Inspection and Testing" (4.10); "Control of Non-conforming Product" (4.13); and under "Internal Audit" (4.17). The quality procedures that govern the quality objectives under these titles should also meet the criteria under this new clause by the very nature of the included information, regarding measurement requirements.

Let's illustrate integration of these in this book. Please turn to book page number 281, clause 4.9c and 4.9d and observe the inserted information. Then, go to the referenced documents and observe there the implementation processes. After that, let's verify how the other quality objective procedures took care of the same requirement. Turn to book page 282, clause 4.10 and read the entire sub-clauses. Then, turn to book page 357 and review from QOP 006 the extensive coverage in fulfilling the "measurement" requirements. The sections of this quality objective procedure have been designed to take care of the requirements under this ISO clause. I think it is obvious by now how to handle the other clauses (4.13 and 4.17) mentioned above.

8.2   MONITORING AND MEASUREMENT

8.2.1   CUSTOMER SATISFACTION

Determine and monitor information on customer requirements as one of the measurements of performance of the quality management system. Determine the methodologies

COMMENTS

No corresponding clause exists in ISO 9001/94. This clause should be understood in line with clauses 7.2.1 (Determination of Requirements Related to the Product); 7.2.2 (Review of Requirements Related to the Product); and 7.2.3 (Customer Communication) of the ISO 9001/2000 requirements.

Measuring customer satisfaction in manufacturing is vastly different from that of the service industry. In manufacturing, customer satisfaction is ensured by the very nature of controlling the product realization processes to comply with customer requirements <u>imposed by contract</u>. Records serve as <u>objective</u> evidence to demonstrate compliance with contractual requirements. Before we ship a product, however, we should know first hand the status of customer satisfaction. If we don't, we don't have an effective quality management system. But there are exceptions.

Even under the best of intentions to satisfy our customers, something of a minor nature could surface that we missed and the customer provided the feedback information to us via telephone, circumventing documentation and the formal implementation of corrective action. The recurrence of the same problem has not been eliminated. To avoid this type of embarrassment, we should request feedback information on our own initiative from our customers regarding any product we delivered.

To implement this type of customer feedback, I recommend the use of a simple form. The application of such a form, however, should never be taken as a substitution for the established procedural system implemented to control across-the-board occurring non-conformities. But, it should be complementary to the established system.

Keep in mind, though, that we are required to include customer feedback results in our management reviews. (See "Review Input," clause 5.6.2.)

What to do:

Add a sub-clause in your quality manual under "Controlling Non-conforming Product" in line with clause 4.13 of the '94 standard. Title it as 'Evaluation of Customer Feedback.' It could also be combined with customer complaints and product returns. Then, the title would change to 'Customer Complaints, Product Returns and Customer Satisfaction.' I have opted to use the latter one. The entire undertaking of measuring customer satisfaction should be handled under one quality objective, – Controlling Non-conforming Products.

The procedure written to implement customer feedback should have, then, a separate section dealing with 'Customer Satisfaction Reporting'.

Let's check out how the implementation of this clause takes place under a process-approach quality management system, – more precisely, under the 'Control of Non-conforming Product' provisions of the ISO 9001/94 standard. Please turn to book page number 286, sub-clause 4.13.3, and observe how this requirement is handled. Then, go to the referenced quality objective procedure QOP 009, Sec Three, clause 3.04, on book page 419. Please read all the sub-clauses of clause 3.04 and observe there how the "feedback" reporting requirement is made simple in a manufacturing environment.

8.4   ANALYSIS OF DATA

(28)

Collect and analyze data to demonstrate the suitability and effectiveness of the quality management system.
Identify improvements. Provide information on:
a) Customer satisfaction
b) Product conformance to requirements
c) Characteristics (status) of processes, products and their trends for preventive action opportunities
d) Suppliers (subcontractors)

COMMENTS

There are no equivalent requirements of this clause under ISO 9001/94. These requirements should be understood in line with ISO 9001/2000, clauses 8.2.3 (Monitoring and Measurement of Processes); 8.2.4 (Monitoring and Measurement of Product); and 8.3 (Control of Non-conforming Product). This means a complete integration of the quality system processes to accomplish it.

> **Special Note:** Clause 8 is the best example of the
> PLAN-DO-CHECK-ACT cycle in ISO 9001/2000
> - Plan as per clause 8.1
> - Do as per clauses 8.2.3, 8.2.4 and 8.3
> - Check as per clause 8.4 (analysis)
> - Act as per clause 8.5 (corrective action)

In order to be able to analyze data regarding the four areas listed under this clause, we need to go back to clause 8.2.3 (Monitoring and Measurement of

Processes) and to clause 8.2.4 (Monitoring and Measurement of Product). To fulfill the requirements under these clauses, we are required to create the necessary forms on which the pertinent data could be recorded. This alone brings us back to the framework construction of the entire quality management system (see "Quality Policy," clause 5.3c). Being able to measure work results in all the activities involving the processes of the quality management system, we have to engineer the data recording forms and filing control requirements into the quality objective procedures written to control these processes as part of the quality system framework.

We can accomplish this by combining the various forms with their respective quality objective procedures, or keeping them controlled separately in a forms manual. In this book, I controlled the quality objective procedures separately and also the designed forms separately. The reason for this is very important. The year 2000 Standard requires a documented procedure to control records (forms), (see clause 4.2.4). Experience proves that when the records (forms) are combined with the quality objective procedures, even though they belong together, the handling and maintenance of them becomes a convoluted process, leading to loss of records and disorganized files.

To simplify document control all around in the quality system, I have kept the various document groups (the quality manual, the quality operating procedures and the forms) separated from each other. Then, I linked them back by referencing in accordance with requirements from the identified processes of the quality manual and the quality objective procedures. Now, I can manage the three separate groups of documents under their own files independently, but in an integrated fashion. This process also has problems such as flow-down control when making changes and continual back and forth navigation to find the right information. Nevertheless, these can be efficiently and effectively managed once the routines are established. But, not being able to find records is a more serious problem than the two drawbacks combined. The lack of records to prove work results is tantamount to lack of control. It was for this reason I created a separate filing control for the records of the quality management system presented in this book. Having this type of record filing control allows immediate gathering of the needed information to support the data analysis requirement of this ISO clause.

What we have to keep in mind regarding this clause on "Analysis of Data" is that compliance with it doesn't happen directly. A brief policy statement in the quality manual under the designated process clauses will be adequate to acknowledge it, for implementation of the four requirements will take place under specific quality objective procedures (See the last paragraph below).

What to do:
1) Amend your current quality manual under "Purchasing" (4.6, see book page 279) and enfold there the requirement for data recording and collec-

tion of records regarding subcontractors (suppliers). This will be a response to sub-clause 8.4d of the new standard.

2) Amend your quality manual under "Process Control" (4.9, see book page 281) and enfold there process control requirements. This will be a response to sub-clause 8.4c of the new standard.

3) Amend your quality manual under "Inspection and Testing" (4.10, see book page 282) and enfold there in general terms the analysis of data requirement for both process and product compliance. This will be a response to all sub-clauses 8.4a,b,c,d of the new standard.

4) Amend your quality manual under "Control of Non-conforming Product" (4.13, see book page 285) and enfold there in general terms the implementation of improvements through corrective action (clause 4.14 on book page 286) in the entire quality system. This will be a response to satisfy the intent of clause 8.4 of the new standard.

5) Amend your quality manual under "Internal Quality Audits" (4.17, see book page 289) and enfold there in general terms how the entire quality system is audited to ensure compliance with all the requirements of the documented quality system. This will be a response in combination with all sub-clauses 8.4a,b,c,d of the new standard.

6) Amend your quality manual under "Statistical Techniques" (4.20, see book page 289) and enfold there in general terms that statistical techniques are applicable as required to enforce the intent of clause 8.4 of the new standard.

To gain the experience necessary to understand how subject clause (8.4) is implemented for maintaining an effectively operated quality system, please go back to each of the above given titles and clauses (1 through 6) and examine the quality objective procedures referenced in the quality manual under the applicable paragraphs. Each quality objective procedure will also reference the applicable Form(s) that are used to record the relevant data required to support the "Analysis of Data" requirement of ISO 9001/2000.

8.5   IMPROVEMENT

8.5.1   CONTINUAL IMPROVEMENT

(29) ***Continually improve the effectiveness of the quality management system by the application of quality policy and objectives, audit results, analysis of data, corrective/preventive actions, and management review***

COMMENTS

There is no equivalent requirement that exists for this clause in ISO 9001/94. Under the '94 standard, continual improvement was implied under phrases like "…continuing suitability…" (clause 4.1.3 Management Review) and

"...take timely corrective action..." (clause 4.17 Internal Quality Audits). In the new standard, however, continual improvement occupies a central roll as one of the policy objectives of the quality management system (see sub-clause 5.3b). Continual improvement, therefore, is a pervasive undertaking equally applied at any phase during and after the product realization processes when non-conformities prevent the achievement of internal and external customer satisfaction. It is no longer just relegated to the decision-making results of management reviews, but also implemented as an in-process driven objective, when and where required. A documented procedure to implement it is not a requirement, but when clauses 8.5.2 (Corrective Action) and clause 8.5.3 (Preventive Action) are implemented, they are best done by combining all the improvement efforts through one cohesively tailored documented procedure. Please review QOP 009 on book pages 403-420 in order that you fully understand the many areas we need to worry about to have a system-wide improvement undertaking effectively managed.

We should not underestimate the 'global' implication of continual improvement. In this clause, continual improvement is intended to embrace the entire quality management system, because every one of us could make a mistake in any of our work processes that could be missed and easily transferred from process to process, eventually impacting just about anybody's work results, including product quality.

The immediate question that we must answer is: How are we going to implement an organization-wide quality improvement system. Are we going to set up a new department and thereby create a watchdog task force? Absolutely not. Since everybody is a contributor to continual improvement, everybody should be responsible for his/her own work through monitoring his/her own work performance output to meet the quality requirements stated in every applicable quality objective procedure.

Continual improvement should be an ongoing endeavor, set into motion when we detect something wrong in what we do. Process owners are the first to observe when something is wrong. They are the ones putting quality into the product. As a result of this understanding, the quality objective procedures must provide the necessary information, either directly or indirectly (by reference) as to how to handle continual improvement in every process they are performing.

Under the new standard, nonconformity takes on a special meaning which could evidence itself in any process people perform. When discovered, it should be the alarm sounded for continual improvement. What we have to be mindful of is the right moment to step in and correct the nonconformity, for if we don't, we could create bottleneck situations down-stream in the realization processes.

The reason I'm providing all this information for you is to emphasize that we no longer are just handling non-conforming materials (parts, etc.) as we basically did under ISO 9001/1994. We are required to control non-conformities of all types that impact the identified processes in quality objectives. Since the processes in the quality management system are interactive, we must also have interactive procedures to implement corrective actions where and when needed during process realizations. Every quality objective is made up of several continuing processes interlinked to other processes, upstream/downstream in the realization cycle, before it can be called completed. For this reason, continual improvement should also be interactive in all processes of the quality management system to cure what's wrong when and where problems happen.

What to do:

Add a statement in your quality manual under "Corrective and Preventive Action" (4.14.1) and describe there the general intent of maintaining continual improvement throughout the quality management system. Add also a similar statement under "Control of Non-conforming Product" (4.13). Then, create the controlling procedure(s) as to how to implement them. Since a documented procedure is required for controlling non-conforming product, combine the requirements here with other activities, causing continual improvement to result from corrective actions.

Since corrective action could impact all activities in a quality management system, I have also designed the implementation process for it to apply to the entire system. Let's prove it.

Please turn to book page number 285, sub-clauses 4.13.1 and 4.13.2, and observe the evolution of the continual improvement implementation process. Turn now to book page 286, sub-clauses 4.14.1, 4.14.2, and 4.14.3, and observe there the follow-up interactions, implementing the requirements stated in sub-clauses 4.13.1 and 4.13.2. Note that these are just policy declarations in the quality manual in response to the requirements of clause 8.5.1 of the new standard. To follow up on how implementation takes place, please review the quality objective procedures referenced under each sub-clause in the book, pages 285 and 286.

PART TWO

REDUCTION AND/OR CONTRACTION IN REQUIREMENTS

4.2   DOCUMENTATION REQUIREMENTS

ISO 9001/2000 requires "documented procedures" in only five areas:

- Document Control
- Control of Quality Records
- Internal Audits
- Control of Nonconformity
- Corrective and Preventive Action

Note: ISO 9001/94, clause 4.2.2, stated only that documented procedures be consistent with requirements…then, effectively implement them

These five documented procedures (known as SOPs or QOPs) are the minimum required under ISO 9001/2000. This reduction was made to accommodate the Service Industry. Controlling manufacturing operations will require a few more documented procedures to fulfill the many quality objective requirements associated with manufacturing.

Examples: Contract Review, Inspection and Test, Calibration, Control of Purchases, Communications and more…

ISO 9004/2000 is not a model to implement ISO 9001/2000. It can be used as a guide for improvement beyond those given in ISO 9001/2000.

5.5   RESPONSIBILITY, AUTHORITY AND COMMUNICATION

5.5.1   RESPONSIBILIY AND AUTHORITY

"organizational freedom" is no longer a requirement as it was under ISO 9001/94 in clause 4.1.2.1

Under the definition of this clause, the new standard implies the creation of an interactive process-approach quality management system. Organizational freedom has been removed in order to ensure a more effective cohesion between functions and their interrelations within the organization.

4.2.3   CONTROL OF DOCUMENTS

"Master List" (ISO 9001/94, clause 4.5.2) is no longer explicitly required in ISO 9001/2000. But, the indication of current revision levels of documents is still a requirement.

The requirement for a Master List has been removed. Remember, without it, we would not be able to show effective revision control of the many procedures stacked in the Procedures Manual. This is a very important navigational tool to identify the status at a glance regarding changes done to individual procedures. It is also a summary index, indicating what type of procedures we have in the quality system. (Look this up on book page 293.) I recommend that you keep using it.

7.4   PURCHASING

7.4.2   PURCHASING INFORMATION

"review and approve purchasing documents" (ISO 9001/94, clause 4.6.3) is no longer a requirement. This phrase has been replaced by "ensure the adequacy of specified requirements" in ISO 9001/2000, clause 7.4.2. Thus, the controlling authority may be circumvented.

Removing the review and approval requirement of purchasing information is an invitation to disaster in manufacturing. This is a very important quality function to ensure that purchasing information contains the accurate flow-down of all customer quality requirements. Stick to the established procedures as sector specific standards will impose this requirement. This removal was made to accommodate the Service Industry.

7.4.3   VERIFICATION OF PURCHASED PRODUCT

"ensure that incoming product…has been inspected or otherwise verified…" (ISO 9001/94, clause 4.10.2). This phrase has been replaced by "establish and implement the inspection or other activities necessary for ensuring that purchased product meets the specified purchase requirements" under ISO 9001/2000, clause 7.4.3.

Rephrasing the verification of purchased product does not remove the requirement. It allows doing it in any controlled fashion suitable for the purpose of the organization. Stay with the already established procedure if it works for you.

7.5   PRODUCTION AND SERVICE PROVISION

7.5.4   CUSTOMER PROPERTY

"establish and maintain <u>documented procedures</u> for the control of verification, storage, and maintenance of customer-supplied product…" (ISO 9001/94, clause 4.7). This phrase has been replaced in ISO 9001/2000, clause 7.5.4, by "identify, verify, protect and safeguard customer property…" This includes all customer-supplied products not just supplies to be incorporated into the product.

The requirement for documented procedures is removed. Addition of the word "identify" rules out partial listing of customer property that we were able to do before. Now we have to control all customer supplied property. The best place to do this identification, verification, and recording is at Receiving. (Check out form A-015 on book page 470)

7.5.5   PRESERVATION OF PRODUCT

"…establish and maintain <u>documented procedures</u> for handling, storage, packaging, preservation, and delivery of product." (ISO 9001/94, clause 4.15.1). The individual clause listing has been combined into one sentence under ISO 9001/2000, clause 7.5.5 (…preserve the conformity of product during internal processing and delivery…shall include identification, handling, packaging, storage and protection…preservation shall also apply to the constituent parts of a product.).

The individual clauses found in the '94 standard have been consolidated under one clause. This will not remove any of the previously established requirements. Stay with the already established system. It is perceived that the Sector Specific Standards will reinstate everything, plus more, of what we already did under the '94 standard.

8   MEASUREMENT, ANALYSIS AND IMPROVEMENT

8.2.2   INTERNAL AUDIT

The requirement under clause 4.17 in ISO 9001/1994 that "…personnel independent of those having direct responsibility for the activity being audited." has been replaced by "…ensure objectivity and impartiality of the results of the audit process" Also: No longer required that internal quality audits determine the effectiveness of the quality management system. But, there is a stark reality in place here. Follow-up action is a requirement to verify actions and report verification results. Furthermore, management is under the gun to act *without undue delay* to eliminate non-conformities and their causes.

In an integrated quality management system the word "independent" does not fit into the organizational structure. Anybody trained to do internal audits can do it as long as he/she has not performed what's being audited. The removal of determining "effectiveness" by the audit scope does not remove the requirement for ensuring the effective operation and control of the quality management system's processes, invoked in other clauses in ISO 9001/2000, (see clauses 4.1c, 4.2.1d, 5.6.1).

8.2.4   MONITORING AND MEASUREMENT OF PRODUCT

<u>Receiving, in-process, and final inspections</u> are no longer titled as such under ISO 9001/2000. These have been contracted into one paragraph without mentioning titles. "The organization shall measure and monitor the characteristics of the product to verify that product requirements have been met. This shall be carried out at appropriate stages of the product realization process…."

The removal of inspection titles in the new standard does not remove inspection requirements as we already practice them in manufacturing. This is another move by the writers of the new standard to wipe out the long standing labeling of the '94 standard as a manufacturing standard. Stick with the same inspection titles we already have whenever writing new procedures to fulfill customer and internal requirements.

8.3   CONTROL OF NONCONFORMING PRODUCT

<u>Review and disposition</u> of non-conforming product (ISO 9001/94, clause 4.13) have been removed as a requirement under ISO 9001/2000 clause 8.3. These activities are now shifted over to paragraph 8.5.2 Corrective Action.

Removal of the <u>review and disposition</u> criteria in controlling non-conforming products is a self-defeatist proposition on the part of the new standard. We all know that to implement continuous improvement when nonconformance has been identified, cannot be done without reviewing and dispositioning the nonconformance. In order to determine corrective/preventive actions and implement continual improvement in line with the Plan-Do-Check-Act provisions, this is a must to do. The new standard is playing with semantics here because it puts this requirement under clauses 8.5.2 and 8.5.3 – corrective/preventive action – while forgetting that review and disposition is done for implementing other actions as well as corrective/preventive actions. It is true that the Service Industry has only selected need for reviewing and dispositioning nonconforming materials. The Manufacturing Industry, on the other hand, cannot do without it, for there would be questionable materials all over the shop.

Do not respond to this shifting strategy as the review and disposition of non-conforming products occur prior to the imposition of corrective/preventive actions. Stay with the P-D-C-A rules in managing all non-conformities. It is certain that the Sector Specific Standards will have detailed requirements on how to handle non-conforming materials in the manufacturing industry.